OPEN-HOLE LOG ANALYSIS AND FORMATION EVALUATION

OPEN-HOLE LOG ANALYSIS AND FORMATION EVALUATION

Richard M. Bateman

PRESIDENT, VIZILOG, INC.

International Human Resources Development Corporation

BOSTON

Library of Congress Cataloging in Publication Data
Bateman, Richard M., 1940–
    Open-hole log analysis and formation evaluation.

    Bibliography: p.
    Includes index.
    1. Oil well logging. 2. Geology, Stratigraphic.
3. Petroleum—Geology. I. Title.
TN871.35.B44   1985   622′.18282   85-8302
ISBN 0-88746-060-7 (U.S.)
ISBN 90-277-2132-7 (D. Reidel)
Printed in the United States of America

CONTENTS

PREFACE

Formation evaluation is an extremely broad term and can encompass many different disciplines. Thus, any work that includes the phrase in its title runs a double risk. An attempt to cover all possible methods of data gathering and analysis would appear presumptuous and overly ambitious, and yet the omission of any of the methods would rightly leave the work incomplete. In its broadest sense, formation evaluation can include everything from macroscopic studies of an entire geologic basin down to microscopic studies of individual mineral grains. Central to both extremes are borehole geophysical well logs. In the practical sense, each formation evaluation method leans on its neighbor for support. Seismic interpretation can be refined in the light of logging data, log response can be calibrated by reference to core analysis, flow testing can refine log-based productivity estimates, etc.

It is this interaction between a spectrum of different formation evaluation methods that makes the whole process of evaluating subsurface formations a dynamic and viable science. Thus, although logs and log analysis are central to this work, it should never be thought that logs alone are sufficient to define entirely the properties of a formation.

The student of formation evaluation should therefore be familiar with geology, geophysics, geochemistry, petrophysics, reservoir engineering, drilling technology, computer science, economics, mathematics, and probability theory. Few occupations require such a wide overall familiarity with so many disciplines. Unfortunately, this has led the formation evaluator to be labeled as the jack of all trades and master of none. However, because of his awareness of the interaction between so many disciplines, he is in fact a better master of many trades than many of the jacks of individual ones. It is hoped that this work reflects these sentiments by stressing the fruitfulness of combining neighboring sciences to produce a whole that is greater than the sum of its parts.

This book is divided into sections that address a number of formation evaluation themes. Part I, *Methods of Gathering Formation Evaluation Data,* includes chapters 1 through 5. These chapters address methods of gathering data needed to evaluate a formation. Chapters 6 through 8 comprise Part II, *Methods of Analysis and Application of Results,* and discuss analysis methods in general and the end use of the results of analysis. Part III, *Open-Hole Logging Measurements,* chapters 9 through 19, addresses open-hole logging measurements. Each tool's principle of operation and contribution to the overall evaluation problem is covered. Part IV, *Analysis of Logs and Cores,* includes chapters 20 through 28. These chapters address more subtle details of log analysis and its integration with core analysis. Part V is *Formation Testing,* chapters 29 and 30, which covers formation testing. Part VI, *Integrated Formation Evaluation,* consists of chapter 31. This final chapter gives an overall "game plan" for the formation evaluator. An appendix to the volume summarizes important log analysis equations.

Some of the material appearing in this book has been taken from two previous and related works, *Log Quality Control* (IHRDC Press 1985) and *Cased-Hole Log Analysis and Reservoir Performance Monitoring* (IHRDC Press 1985). This reflects the interdependence of many disciplines in any treatment of formation evaluation, a theme of this particular work. In particular, the chapter on gamma ray logs is universally applicable in either open or cased hole and has been copied almost verbatim from the earlier work on cased hole.

ACKNOWLEDGMENTS

This work is based on many years of practical experience with the vicissitudes of evaluating subsurface formation properties. In the course of compiling materials, I have leaned heavily on friends, coworkers, other authors, oil companies, service companies, and professional societies. Where humanly possible, I have given credit where credit is due. In some cases, it has proved difficult to trace the origin of some figures, and I must thank those who unknowingly have assisted me with such material. I also owe a debt of gratitude to the many students who have attended my courses and provided me with valuable feedback on both the style and content of this work.

I also owe a debt of gratitude to Annette Joseph of IHRDC and Julie Hotchkiss of Custom Editorial Productions without whose tireless efforts publication of this work would have been impossible.

Richard M. Bateman
HOUSTON
JANUARY 1985

This book is dedicated to my daughters, Karen and Michelle, and to the memory of my parents without whose example as scientists and teachers this work would not have been possible.

PART I

METHODS OF GATHERING FORMATION EVALUATION DATA

CHAPTER ONE

FORMATION EVALUATION OVERVIEW

THE SCOPE OF FORMATION EVALUATION

Formation evaluation covers a very large range of measurement and analytic techniques. The purpose of this work is to expose the reader to all the available methods and give them a perspective by reference to borehole geophysics, or well logs. Although the emphasis will be on well logging techniques and log analysis methods, these are not the only tools available to the formation evaluator. Well logs are central only in the sense that they are universally recorded in practically all wellbores and are directly relatable to *all* the other parameters available from the associated sciences. For example, a geophysicist needs borehole measurements to determine a time-depth relationship and a petrophysicist needs core analysis to define log response properly; but a thin section or SEM (scanning electron microscope) photo of a rock sample is of no direct help to the interpretation of a seismic section, nor is a VSP (vertical seismic profile) of any help in determining relative permeability. However, *all* the measurements are pertinent to the complete task of defining a reservoir's limits, storage capacity, hydrocarbon content, produceability, and economic value.

To place the various disciplines in perspective, it is valuable to consider the overall problem of formation evaluation in terms of orders of magnitude. If one meter is taken as a unit of measurement, then each formation evaluation technique can be placed in order as shown in table 1.1.

Thus, formation evaluation techniques cover at least 12 orders of magnitude. The physical principles employed to make the basic measurements are equally far ranging. An enlightening way of viewing the vast spread is to consider the frequency employed by the measuring processes available, as illustrated in table 1.2. Again, a span of many orders of magnitude. Few other sciences require, or use, such a wide range of measurement techniques over such a wide range of physical dimensions.

FORMATION EVALUATION

The initial discovery of a reservoir lies squarely in the hands of the explorationist using seismic records, gravity, and magnetics. Formation evaluation presupposes that a reservoir has been located and is to be defined by drilling as few wells as possible. In those wells, enough data should be gathered to extrapolate reservoir parameters fieldwide to arrive at realistic figures for both the economic evalua-

3

TABLE 1.1 *Formation Evaluation Perspective*

Order of Magnitude (meters)	Formation Evaluation Technique	Purpose
10^6	Satellite imagery	
10^5	Basin geologic studies	Gross structure
10^4	Seismic, gravity, magnetics	
10^3	Borehole gravimeter, Ultra long spacing electric logs	Local structure
10^2	Drillstem tests	Productivity
10^1	Wireline formation tests	and reserves
10^0	Full-diameter cores	Local porosity,
10^{-1}	Sidewall cores, most conventional well logs, measurements while drilling	permeability, and lithology
10^{-2}	Micro-focused logs, coreplug analysis	
10^{-3}	Cuttings analysis (mud logging)	Local hydrocarbon content
10^{-4}	Core analysis	Rock properties
10^{-5}	X-ray minerology	Rock and clay typing
10^{-6}	Scanning electron microscope (SEM)	Micro pore structure

TABLE 1.2 *Range of Physical Principles Used in Formation Evaluation*

Measurement Frequency (hertz)	Measurement Type
10^9	Dielectric (EPT)
10^7	Dielectric (DCL)
10^3	Induction, sonic
10^2	Laterolog, seismic
10^1	Old electric logs
10^{-4}	Step-rate well testing
10^{-8} to 10^{-9}	Material balance

tion of the reservoir and the planning of the optimum recovery method. Formation evaluation offers a way of gathering the data needed for both economic analysis and production planning.

What then are the parameters the manager, the geologist, the geophysicist, and the reservoir and production engineers need? Which of them can be provided by seismic records, by coring, by mud logging, by testing, or by conventional logging?

The geophysicist needs to know the time-depth relationship in order to calibrate conventional seismic and vertical seismic profile (VSP) surveys. The geologist needs to know the stratigraphy of the formations, the structural and sedimentary features, and the mineralogy of the formations through which the well was drilled. The reservoir engineer needs to know the vertical and lateral extent of the reservoir and its porosity (type of porosity) and permeability, fluid content, and recoverability. The production engineer needs to know the rock properties; be aware of overpressure if it exists; and be able to assess sanding and associated problems and the need for secondary recovery efforts or pressure maintenance. Once the well is on production, the production engineer will also need to know the dynamic behavior of the well under production conditions and need to diagnose problems as the well ages. The engineer may also need to know formation injectivity and residual water saturations to plan waterflooding and be able to monitor the progress of the waterflood when it is operational.

The manager needs to know the vital inputs to an economic study, namely, the original hydrocarbon in place; recoverability; cost of development; and, based on those factors, the profitability of producing the reservoir. Log measurements, when properly calibrated, can give the majority of the parameters required. Specifically, logs can provide either a direct measurement or a good indication of:

Porosity, both primary and secondary (fractures and vugs)
Permeability
Water saturation and hydrocarbon movability
Hydrocarbon type (oil, gas, or condensate)
Lithology
Formation dip and structure
Sedimentary environment
Travel times of elastic waves in a formation

From these data, good estimates may be made of the reservoir size and the hydrocarbons in place.

Logging techniques in cased holes can provide much of the data needed to monitor primary production and also to gauge the applicability of waterflooding and monitor its progress when installed. In producing wells, logging can provide measurements of:

Flow rates
Fluid type
Pressure
Residual oil saturations

From these measurements, dynamic well behavior can be understood better, remedial work planned, and secondary or tertiary recovery proposals evaluated and monitored.

In summary, when properly applied logging can answer a great many questions from a wide spectrum of special-interest groups on topics ranging from basic geology to economics. Of equal importance, however, is the fact that logging by itself cannot answer all formation evaluation problems. Coring, core analysis, and formation testing are integral parts of any formation evaluation effort.

FORMATION EVALUATION METHODS

In practice the order in which formation evaluation methods are used tends to follow the orders-of-magnitude table, that is, from the macroscopic to the microscopic. Thus, a prospective structure will first be defined by seismic records, gravity, and/or magnetics. A wellbore drilled through such a structure may employ mud logging and/or measurements while drilling (MWD); and, in that wellbore, cores may be cut or sidewall samples taken. Once the well has reached some prescribed depth, logs will be run. An initial analysis of mud log shows, together with initial log analysis, may indicate zones that merit testing either by wireline formation testing or by drillstem testing. Should such tests prove the formation to be productive, more exhaustive analysis is made of all available data including core analysis. The whole process can be summarized by table 1.3.

The exploration phase involving seismic records, gravity, and magnetics will not be covered in this book. However, the use of logs to refine seismic interpretation will be covered, as will the vertical seismic profile. The drilling phase will be covered by discussion of mud logging, coring, and measurements while drilling. The testing phase will be covered by sections dealing with both drillstem tests and wireline formation tests. Core analysis will be covered to the extent that results of core analysis have a direct effect on log analysis.

TABLE 1.3 *Formation Evaluation Overview*

Phase	Activity	Formation Evaluation Methods
Exploration	Define structure	Seismic, gravity, magnetics
Drilling	Drill well	Mud logging, coring, measurements while drilling (MWD)
Logging	Log well	Open-hole logs
Primary evaluation	Log analysis and testing	Sidewall cores, vertical seismic profile (VSP), wireline formation testing, drillstem testing
Analysis	Core analysis	Laboratory studies
Feedback	Refinement of seismic model and log analysis	Log calibration via core analysis results, seismic calibration from log analysis results
Exploitation	Producing hydrocarbons	Material balance analysis
Secondary recovery	Water or gas injection and production logging	Production log analysis, flood efficiency analysis, microrock property analysis
Abandonment	Economic decisions	

Production logging merits a book of its own, and the reader is referred to a companion volume to this work by the same author, *Cased-Hole Log Analysis and Reservoir Performance Monitoring* (IHRDC Press 1985).

Economic analysis is outside the scope of this book, but the vital inputs to any economic analysis still must come from the formation evaluation methods discussed here. For those unfamiliar with some or all of the formation evaluation methods mentioned, the following brief summaries will serve as introduction. Those familiar with the topics under discussion may forge ahead in the text to the relevant sections where each is discussed in detail.

MUD LOGGING

Mud logging, more elegantly referred to as hydrocarbon mud logging, is a process whereby the circulating mud and cuttings in a well being drilled are continuously monitored by a variety of sensors. The combined analysis of all the measurements provides indications of the rock type and its fluid content. The sundry measurements are displayed on a log as curves or notations as a function of depth. Not all wells are logged in this manner. Development wells, for example, are usually drilled and logged by wireline logging tools only. Wildcat wells, however, are nearly always monitored by the mud-logging process. The great merits of mud logging include the availability on a semicontinuous basis of actual formation cuttings analysis (which in turn gives immediate indications of rock type and hydrocarbon presence) and the ability to predict drilling problems (such as overpressure) before they become unmanageable.

CORING

A number of methods are in use to cut cores in a wellbore. Conventional cores are cut using a special core bit whereby a long core barrel is retrieved and brought back to surface. The sample of the formation so recovered may undergo physical changes on its journey from the bottom of the well, where it is cut, to the surface, where it can be analyzed. More sophisticated coring mechanisms are now in use that conserve either the orientation, the pressure, or the original fluid saturations of the rock sample gathered. An awareness of these methods is essential to an understanding of core analysis results.

Other coring methods are available for cases where additional rock samples are required after the well has been drilled and before it has been cased. These methods require wireline tools that cut core plugs from the side of the well.

Many of the parameters needed to interpret open-hole wireline logs correctly can only be determined from accurate core analysis. This presupposes that cores have been cut. Thus, in the initial stages of a field development, coring plays an essential part.

MEASUREMENTS WHILE DRILLING

Increasingly today, formation properties are being measured at the time the formation is drilled by use of special drill collars that house measuring devices. These measurement-while-drilling (MWD) tools are particularly valuable in deviated offshore wells where wellbore path control is critical and where an immediate knowledge of the formation properties is vital for decision making on such matters as the choice of logging and casing points. Although not as complete as open-hole logs, the measurements obtained by MWD are rapidly becoming just as accurate and usable in log analysis procedures.

TESTING

Formation testing is the proof of the pudding. If the well flows hydrocarbons on a drillstem test, no amount of logging data or core analysis can deny that a productive zone has been found. However, a drillstem test (DST) not only provides proof that hydrocarbons exist in the formation and will flow but also supplies vital data regarding both the capacity of the reservoir and its ability to produce in the long term. Correct interpretation of pressure records from drillstem tests adds immensely to the overall formation evaluation task.

Wireline formation testers complement drillstem tests by their ability to sample many different horizons in the well and produce not only fluid samples but also detailed formation pressure data that are almost impossible to obtain from a DST alone.

OPEN-HOLE LOGGING

Open-hole logging provides the great meeting place of all of the other formation evaluation methods. Only through open-hole logging can a continuous record of measurement versus depth be made of so many formation properties. In particular, wireline logs can record formation electrical resistivity, bulk density, natural and induced radioactivity, hydrogen content, and elastic modulae. These raw measurements can then be interpreted to give a continuous measurement-versus-depth record of formation properties such as porosity, water saturation, and rock type. Almost without exception, every well drilled for hydrocarbons is logged with wireline instruments. Unfortunately, the logs so acquired are not always analyzed in detail or are incorrectly analyzed because of a lack of training on the part of the analyst or a lack of understanding of where wireline logs fit with relation to the other methods of formation evaluation.

All too often, logs are seen as an end in themselves and are considered in isolation. It is hoped that this book will assist the reader in taking the broader view of log analysis in the context of overall formation evaluation. Figure 1.1 attempts to illustrate the formation evaluation picture and the central role of open-hole logging and log analysis.

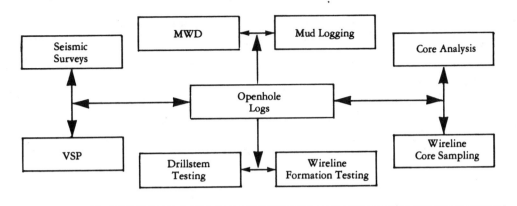

FIGURE 1.1 *Interrelationships between Formation Evaluation Methods.*

MODERN LOGGING TOOLS

The actual running of a log involves as much the tool on the end of the logging cable as the cable itself and the surface controlling and recording apparatus. So before discussing downhole tools, the common elements of all logs will be presented. Figure 1.2 illustrates the basic components of any logging system. A sensor, incorporated in a sonde together with its associated electronics, is suspended in the hole by a multiconductor cable. The sensor is separated from virgin formation by a portion of the mud column, by mudcake, and, more often than not, by an invaded zone in the surrounding formations. The signals from the sensor are conditioned by the electronics for transmissions up the cable to the control panel, which in turn conditions the signals for the recorder. As the cable is raised or lowered, it activates some depth-measuring device, a sheave wheel, for example, which in turn activates a recording device, either an optical camera (making a film) or tape deck (making a digital recording on magnetic tape). Finally, some form of reproduction takes place to provide a hard copy of the recorded data.

In general, well logging jargon distinguishes between a *logging survey*, a *logging tool*, and a *log* as well as a *curve*. There is frequently some confusion about these terms when logging matters are discussed.

A *logging survey* is provided by a logging-service company for a client. During the course of the survey, the logger may employ several different logging tools, and record several different logs on each of which are presented several different curves. The logging tools used, in turn, may be composed of multiple sensors. Figure 1.3 illustrates these terms and their interrelationship.

The open-hole logging tools in use today are:

Formation Fluid Content Indicators
   Induction
   Laterolog

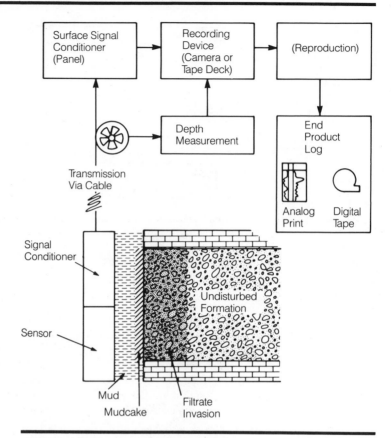

FIGURE 1.2 *Components of a Logging System.*

Microfocused (microresistivity)
Dielectric
Pulsed neutron
Inelastic gamma

Porosity-Lithology Indicators
Sonic (acoustic)
Density and lithologic density
Neutron
Natural gamma ray
Spectral gamma ray

Reservoir Geometry Indicators
Dipmeter
Borehole gravimeter
Ultra long spacing electric

Formation Productivity Indicators
Formation tester

Logging Survey

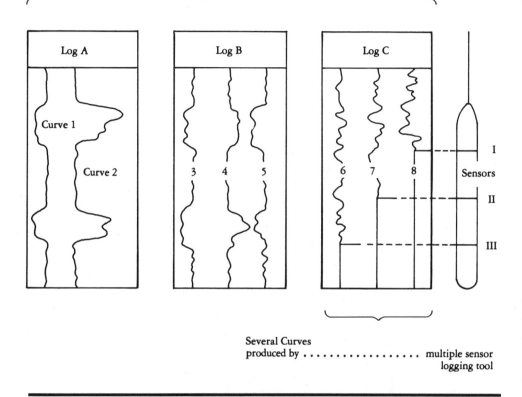

FIGURE 1.3 *Logging Terms.*

These basic devices will answer 90% of the questions about the formation. Omitted from the list are various types of old logs (such as electric logs) and some standard auxiliary devices which, while important, do not rate as separate tools since they always piggy-back along with one or another of the basic tools. Among those *auxiliary tools* are:

> Spontaneous potential (SP) log
> Caliper log

A discussion of each of the basic tools follows.

Induction Tools

Induction tools belong to the resistivity-tool family and measure formation resistivity. They work like mine detectors by inducing currents in the formation. Induction logs are called by a variety of names and initials such as:

Induction log
IES or ISF
DIL or DISF
6FF40

They may be run simultaneously with a spontaneous potential (SP) and/or gamma ray (GR) log and, in the case of the ISF, with a caliper log. The IES (induction electric survey) is used for the old standby of induction–short normal log combination; ISF refers to the newer induction-spherically focused log. DIL refers to the dual induction log, which will record two induction curves, a deep and a medium; the DIL also includes a Laterolog-8 (LL8) or a spherically focused log (SFL), which is a shallow recording device. The 6FF40 refers to the standard type of induction sonde used (six focused coils with a nominal spacing of 40 in.).

Curves recorded on induction logs are called by various names and initials such as:

ILd	induction log deep	
RILd	resistivity from a deep induction log	Deep and/or medium investigation of resistivity
ILm	induction log medium	
RILm	resistivity from a medium induction log	

RSN, SN	short normal resistivity or 16 in. SN	
LL8	laterolog-8	Shallow investigation or resistivity
RSFL, SFL	spherically focused resistivity	

SP	spontaneous potential	
GR	gamma ray	Auxiliary curves
CAL	caliper	

Laterolog Tools

The most important tool in the family of laterolog tools is a combination tool—the dual laterolog-MSFL (microspherically focused log). Older versions such as the Laterolog-3 or Laterolog-7 are not widely used anymore. The dual laterolog (DLL) tool is rather confusingly referred to sometimes as a DST, which stands for "dual simultaneous tool" in reference to the fact that it records both laterolog curves simultaneously. So if a dual laterolog is desired, ask for:

Dual laterolog
DLL
DST

This tool can be run with SP, GR, and caliper logs if required. The resistivity curves recorded are:

LLd	laterolog deep
RLLd	resistivity from a deep laterolog
LLs	laterolog shallow
RLLs	resistivity from a shallow laterolog
MSFL (or R_{xo})	microspherically focused log

Auxiliary curves are the same as for the induction tools.

Microresistivity Devices

There are four microresistivity devices one should be familiar with:

PL	proximity log (Prox, Proxy)
MLL	microlaterolog
MSFL	microspherically focused log
ML	microlog

The MSFL is usually run with the dual laterolog and not separately, but the proximity (PL) and microlaterolog (MLL) tools are run as separate surveys, usually together with a microlog (ML). The microlog is an old type of electric log that will probably never go out of style. Both the proximity log and the microlaterolog are run in combination with the microlog if desired. The services then are called:

P-ML	proximity–microlog
MLL-ML	microlaterolog–microlog

and the curves recorded are:

RPL	resistivity proximity log
RMLL	resistivity microlaterolog
RMSFL	resistivity microspherically focused log
$R_{2\,in.}$	micronormal
$R_{1\,in.\,\times\,1\,in.}$	microinverse
—	microcaliper

These curves comprise a microlog

Sonic Tools

The modern sonic log is known as a borehole compensated sonic or BHC log. It may be run with GR, SP, and caliper logs, or with an induction tool to make an ISF–sonic survey (induction and sonic all in one run). The curve recorded is:

Δt sonic travel time

Various other sonic-related parameters can also be recorded either simultaneously or on a separate run. Sonic amplitude logs are used for fracture detection. They may be recorded by various arrangements of gatings for the received wave trains. The borehole compensated sonic tool is also used for cement bond logs (CBL), in which case recorded curves are:

Δt a single-receiver travel time
—   amplitude
VDL   a variable display of wave trains

Density Tools

Compensated formation density (FDC) tools are also known as gamma-gamma tools in some parts of the globe (because their mode of operation is to send gamma rays to the formation and detect gamma rays coming back). They record two basic curves:

ρ_b bulk density curve
$\Delta\rho$ correction curve

Gamma ray and caliper logs are normally run at the same time. Additionally, an apparent-porosity curve can be generated and recorded, and, from that data, a formation-factor (F) curve can be generated and recorded as well. A density-derived F is referred to as F_d.

A modern version of the density tool is the litho-density tool (LDT). In addition to measuring bulk density, the LDT also measures the photoelectric factor P_e, which is a direct indicator of formation lithology and, as such, is a valuable adjunct to the basic density measurement.

Neutron Tools

There are several types of neutron tools. Today's standard is the compensated neutron log (CNL), which records the neutron porosity index ϕ_N. This index is normally recorded for a particular assumed lithology. Reading the porosity curve requires close attention to the porosity scale and the assumed matrix. Normally, a gamma ray will be recorded simultaneously. If the CNL is run on its own, no caliper is available; however, the standard presentation is a combination density–neutron log, in which the caliper from the density tool is available.

Another important group of neutron-based tools includes the pulsed neutron logs (NLL, TDT, and TMD), which measure complicated radiation parameters. The end result is a measurement that helps distinguish oil from salt water in the formation in cased holes. These logs may also be used in open hole. The curves actually appearing on the log are:

Σ the formation neutron capture cross section
τ the thermal neutron decay time
ϕ a porosity-type ratio curve

Dipmeter

There are several kinds of dipmeters: four-arm dipmeters, six-arm dipmeters, and eight-electrode types. The three-arm type is outdated and should not be used. High resolution dipmeters (HDT) record all the necessary curves for computing formation dip and hole drift and azimuth.

Wireline Formation Testers

Several types of wireline formation testers are available. These devices allow a limited sample of formation fluid to be sucked out of the formation and brought to the surface for analysis. These wireline formation testers also allow multiple formation pressure tests in one run into the hole and are proving to be a valuable addition to the formation evaluation arsenal.

Miscellaneous Logging Tools

Carbon/Oxygen Logging

Inelastic fast neutron scattering is used in carbon/oxygen logs in an attempt to measure directly the relative abundance of carbon and oxygen in a formation. This is a relatively new service and some controversy exists about its effectiveness. Carbon/oxygen logging is used in cased holes and is a natural candidate for those parts of the world where fresh formation waters preclude the use of a pulsed neutron logging survey.

Gamma Ray Spectral Logging

Gamma ray spectrometry measures the number and energy of naturally occurring gamma rays in the formation and distinguishes between elements and daughter products of three main radioactive families. It can distinguish uranium from thorium and potassium. Since these elements and/or their decay products are associated with certain distinct types of mineralogy, sedimentology, and formation waters, gamma ray spectral logging has obvious appeal.

Borehole Gravimeter

The borehole gravimeter (BHGM) measures perturbations in the gravitational acceleration constant caused by the proximity to the borehole of formation material that is lighter or heavier than normal. Thus, this tool can spot higher porosities, gas-bearing zones, and so on some distance from the borehole. Its use requires an exacting set of prerequisites regarding depth, temperature, time, and so on, and may not be available or applicable to all wells everywhere.

Dielectric Logging Tools

Dielectric logging tools measure the speed of electromagnetic waves sent along the wall of the formation. From the speed, the dielectric constant of the formation is deduced. Oil and water, having very different dielectric constants, can be distinguished. Other measurements, such as attenuation of the microwaves, can indicate porosity and resistivity related parameters. The tools' applications are in open holes where formation waters are fresh. They are limited by a very shallow depth of investigation.

Nuclear Magnetic Resonance

The nuclear magnetic resonance (NMR) tool has a long history and has been resurrected several times. It requires some doping of the mud column if it is to work and is a voracious consumer of rig time. It measures the precession rate of atomic nuclei after the removal of an intense magnetic field. The measured quantity is related to the free water content of the formation.

SUMMARY

In order to put all these tools, surveys, and curves in perspective, the appendix to this chapter sets out a summary of all the common logging tools, their measurements, and the uses to which they may be put.

APPENDIX 1A: OPEN-HOLE LOGGING TOOLS AND MEASUREMENTS AND THEIR USES

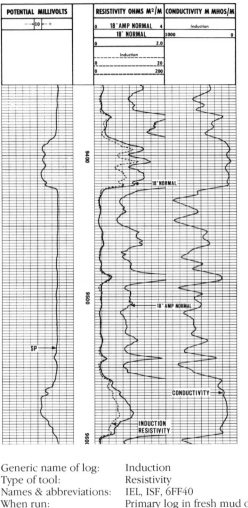

Generic name of log:	Induction
Type of tool:	Resistivity
Names & abbreviations:	IEL, ISF, 6FF40
When run:	Primary log in fresh mud or oil-based mud
Purpose:	Measures formation resistivity, R_t
Limitations:	Behaves badly in salt muds and/or large boreholes
Often combined with:	Sonic
Operating principle:	20-kHz coil induces current in formation
Curves recorded:	Deep induction
	Conductivity
	SFL or short normal
	SP
	GR

FIGURE 1A.1 *Induction Log. Courtesy WELEX, a Halliburton Company.*

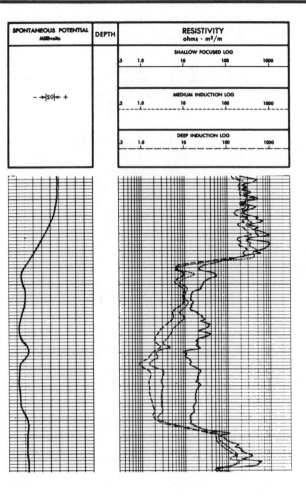

Generic name of log: Dual induction
Type of tool: Resistivity
Names & abbreviations: ISF, DISF
When run: Primary log in fresh mud
Purpose: Measures formation resistivity, R_t
Limitations: Behaves badly in salt muds and/or large
                                boreholes

Often combined with: Sonic
Operating principle: 20-kHz coil induces current in formation
Curves recorded: RID—deep induction
                              RIM—medium induction
                              RSFL—spherically focused
                              SP
                              GR

FIGURE 1A.2 *Dual Induction Focused Log. Courtesy Dresser Atlas, Dresser Industries, Inc.*

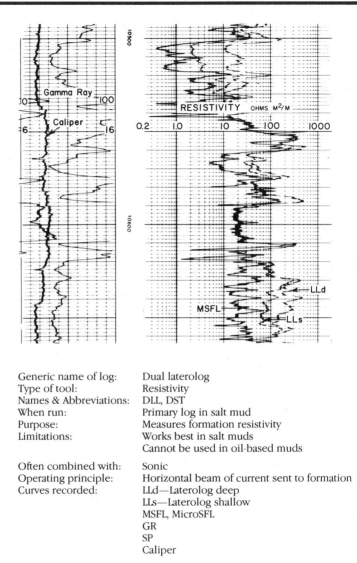

Generic name of log:	Dual laterolog
Type of tool:	Resistivity
Names & Abbreviations:	DLL, DST
When run:	Primary log in salt mud
Purpose:	Measures formation resistivity
Limitations:	Works best in salt muds
	Cannot be used in oil-based muds

Often combined with:	Sonic
Operating principle:	Horizontal beam of current sent to formation
Curves recorded:	LLd—Laterolog deep
	LLs—Laterolog shallow
	MSFL, MicroSFL
	GR
	SP
	Caliper

FIGURE 1A.3 *Dual Laterolog. Courtesy Schlumberger Well Services.*

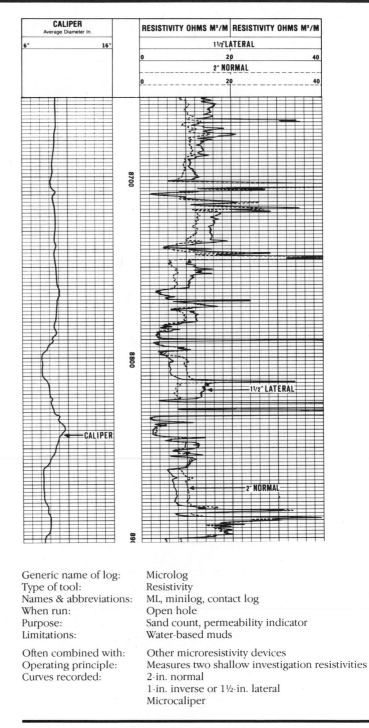

Generic name of log: Microlog
Type of tool: Resistivity
Names & abbreviations: ML, minilog, contact log
When run: Open hole
Purpose: Sand count, permeability indicator
Limitations: Water-based muds

Often combined with: Other microresistivity devices
Operating principle: Measures two shallow investigation resistivities
Curves recorded: 2-in. normal
                            1-in. inverse or 1½-in. lateral
                            Microcaliper

FIGURE 1A.4 *Microlog. Courtesy WELEX, a Halliburton Company.*

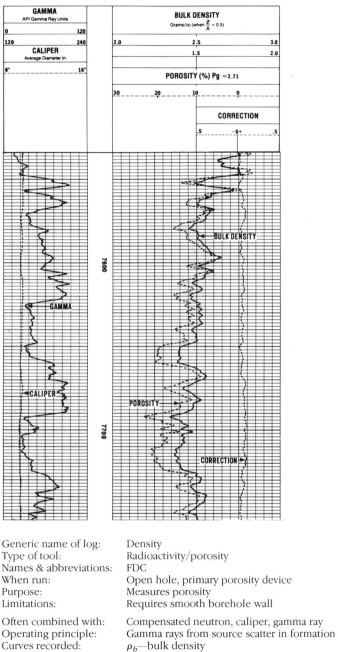

Generic name of log:	Density
Type of tool:	Radioactivity/porosity
Names & abbreviations:	FDC
When run:	Open hole, primary porosity device
Purpose:	Measures porosity
Limitations:	Requires smooth borehole wall
Often combined with:	Compensated neutron, caliper, gamma ray
Operating principle:	Gamma rays from source scatter in formation
Curves recorded:	ρ_b—bulk density
	$\Delta\rho$—correction
	ϕ_D—apparent porosity

FIGURE 1A.5 *Formation Density Log. Courtesy WELEX, a Halliburton Company.*

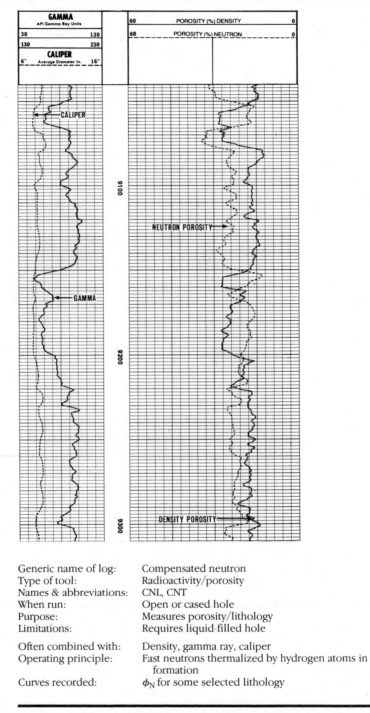

Generic name of log:	Compensated neutron
Type of tool:	Radioactivity/porosity
Names & abbreviations:	CNL, CNT
When run:	Open or cased hole
Purpose:	Measures porosity/lithology
Limitations:	Requires liquid-filled hole
Often combined with:	Density, gamma ray, caliper
Operating principle:	Fast neutrons thermalized by hydrogen atoms in formation
Curves recorded:	ϕ_N for some selected lithology

FIGURE 1A.6 *Compensated Neutron Log. Courtesy WELEX, a Halliburton Company.*

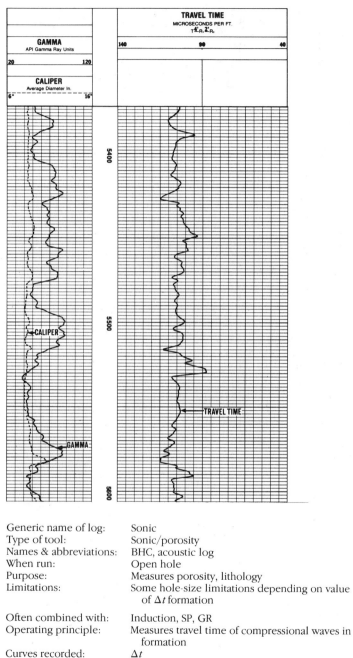

Generic name of log:	Sonic
Type of tool:	Sonic/porosity
Names & abbreviations:	BHC, acoustic log
When run:	Open hole
Purpose:	Measures porosity, lithology
Limitations:	Some hole-size limitations depending on value of Δt formation
Often combined with:	Induction, SP, GR
Operating principle:	Measures travel time of compressional waves in formation
Curves recorded:	Δt
	Caliper

FIGURE 1A.7 *Sonic Log. Courtesy WELEX, a Halliburton Company.*

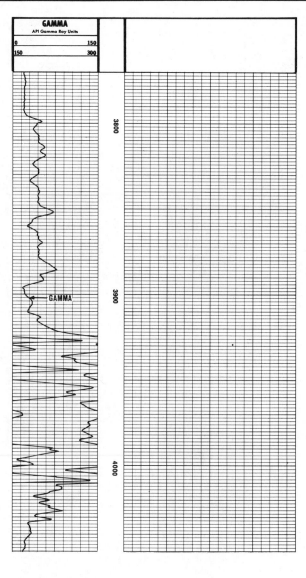

Generic name of log:	Gamma ray
Type of tool:	Radioactivity/lithology
Names & abbreviations:	GR
When run:	Open or cased hole
Purpose:	Sand/shale discriminator
Limitations:	None
Often combined with:	Any and all open- and cased-hole tools
Operating principle:	Scintillation detector measures natural formation gamma ray activity
Curves recorded:	GR

FIGURE 1A.8 *Gamma Ray Log. Courtesy WELEX, a Halliburton Company.*

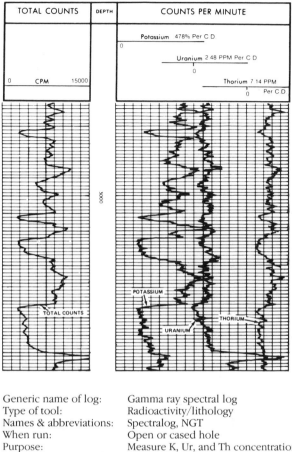

Generic name of log:	Gamma ray spectral log
Type of tool:	Radioactivity/lithology
Names & abbreviations:	Spectralog, NGT
When run:	Open or cased hole
Purpose:	Measure K, Ur, and Th concentrations in formation
Limitations:	Slow logging speed
Often combined with:	Neutron–density
Operating principle:	Gamma ray energy spectrum characterizes source of gamma rays
Curves recorded:	Total counts
	Uranium
	Potassium
	Thorium

FIGURE 1A.9 *Gamma Ray Spectral Log. Courtesy Dresser Atlas, Dresser Industries, Inc.*

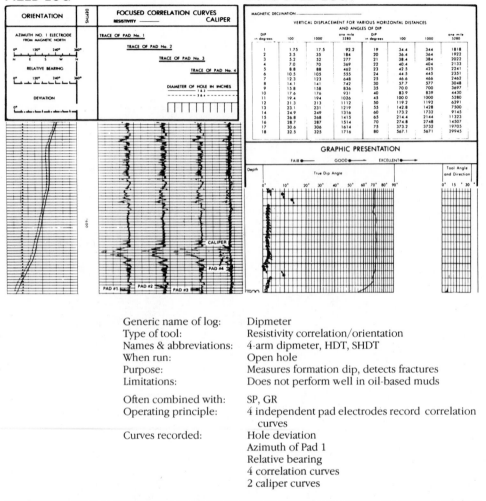

Generic name of log: Dipmeter
Type of tool: Resistivity correlation/orientation
Names & abbreviations: 4-arm dipmeter, HDT, SHDT
When run: Open hole
Purpose: Measures formation dip, detects fractures
Limitations: Does not perform well in oil-based muds

Often combined with: SP, GR
Operating principle: 4 independent pad electrodes record correlation
                            curves
Curves recorded: Hole deviation
                            Azimuth of Pad 1
                            Relative bearing
                            4 correlation curves
                            2 caliper curves

FIGURE 1A.10 *Dipmeter. Courtesy Dresser Atlas, Dresser Industries, Inc.*

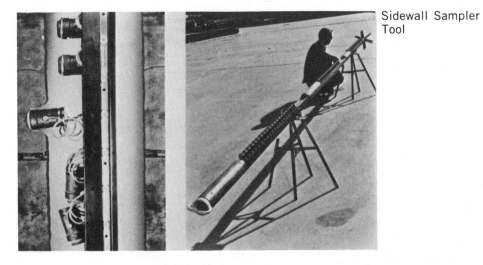

Sidewall Sampler Tool

Generic name of log:	Sidewall sampler
Type of tool:	Percussion core cutter
Names & abbreviations:	CST
When run:	Open hole
Purpose:	To retrieve samples of the formation
Limitations:	Limited number of cores (typically 30) per trip in hole
Often combined with:	SP for depth control
Operating principle:	Explosive charge propels hollow cylinder into formation
Curves recorded:	Auxiliary SP curve only, for depth control
	Recovery report

FIGURE 1A.11 *Sidewall Sampler Tool. Courtesy Schlumberger Well Services.*

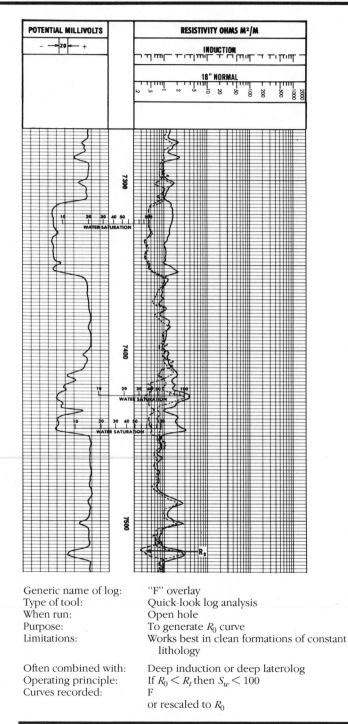

Generic name of log:	"F" overlay
Type of tool:	Quick-look log analysis
When run:	Open hole
Purpose:	To generate R_0 curve
Limitations:	Works best in clean formations of constant lithology
Often combined with:	Deep induction or deep laterolog
Operating principle:	If $R_0 < R_t$ then $S_w < 100$
Curves recorded:	F
	or rescaled to R_0

FIGURE 1A.12 *"F" Overlay. Courtesy WELEX, a Halliburton Company.*

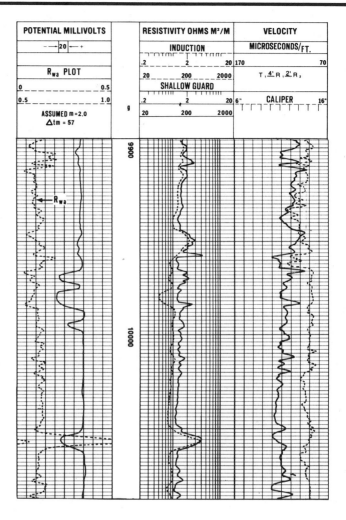

Generic name of log: R_{wa}
Type of tool: Quick-look log analysis
Names & abbreviations: R_{wa}, R_{waS}, R_{waD}
When run: Open hole
Purpose: Distinguish water- and oil-bearing rocks, find R_w
Limitations: Needs careful attention to detail in gas-bearing
                            formations and/or shales, depending on
                            porosity tool used

Often combined with: ISF, sonic
Operating principle: Apparent porosity and R_t combined to give
                            apparent water resistivity
Curves recorded: R_{wa}

FIGURE 1A.13 *R_{wa} Log. Courtesy WELEX, a Halliburton Company.*

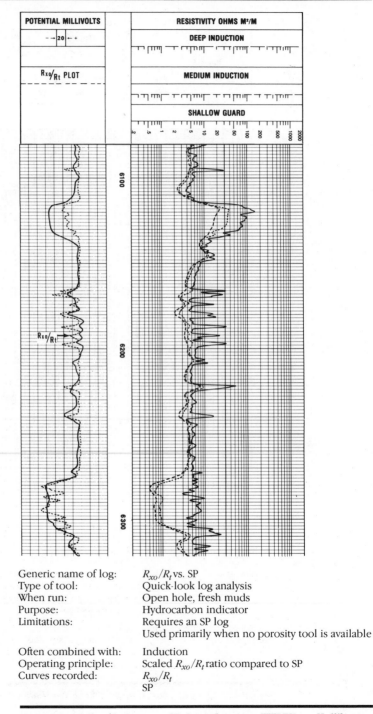

Generic name of log:	R_{xo}/R_t vs. SP
Type of tool:	Quick-look log analysis
When run:	Open hole, fresh muds
Purpose:	Hydrocarbon indicator
Limitations:	Requires an SP log
	Used primarily when no porosity tool is available
Often combined with:	Induction
Operating principle:	Scaled R_{xo}/R_t ratio compared to SP
Curves recorded:	R_{xo}/R_t
	SP

FIGURE 1A.14 *R_{xo}/R_t Log vs. SP Log. Courtesy WELEX, a Halliburton Company.*

BIBLIOGRAPHY

Dresser Atlas: "Log Review 1" (1974).
Dresser Atlas: "Wireline Service Catalog" (1984).
Schlumberger: "Engineered Open Hole Services" (undated).
Schlumberger: "Service Catalog," C-12001 (1977).
Welex: "Open Hole Services," G-6003 (1983).

CHAPTER TWO

MUD LOGGING

The old saying goes: "If you throw enough mud some of it is bound to stick." In the case of mud logging, the name stuck. Today, however, mud logging embraces far more than just monitoring mud returns during the drilling process. It would be more elegant and more informative to refer to it as "hydrocarbon well logging." Mud logging includes measurements of both the progress of the drilling operation itself and measurement of the contents and properties of the formations drilled. Some of the information gathered is vital to the drilling engineer but of little use to the formation evaluator. Other items are vitally important to the seeker of porous and permeable formations with recoverable hydrocarbons.

Not all wells are drilled with a mud logging unit in operation; and where mud logging equipment is present, all possible measurements are not necessarily recorded. The objective here is to mention the most commonly recorded parameters and review their use, interpretation, and integration into the overall formation evaluation effort.

As a well is drilled, the drill bit crushes the rock encountered and cuttings are circulated to the surface by the mud. If the rock being drilled is hydrocarbon-bearing, traces of oil and/or gas will reach the surface in both the mud and in the bit cuttings. These traces may be quantified, recorded, and interpreted by a number of measurement techniques the purpose of which is to provide indications of formation porosity, lithology, and hydrocarbon content (fig. 2.1). It is important to bear in mind when discussing mud logging that actual samples of the formation and its contents are physically available for inspection, usually within a relatively short time after drilling through them. No other form of formation evaluation (except coring) can offer such incontrovertible evidence—unless an uncontrolled blowout, offering proof positive, is included as a formation evaluation method, something most would prefer to avoid.

MEASUREMENTS RELATED TO THE DRILLING PROCESS

Many of the measurements appearing on a mud log are of direct interest only to the drilling engineer. Such measurements might include continuous monitoring of:

Mud properties (weight, viscosity, salinity, temperature, etc.)
Weight on bit, hook load, pump speed, etc.
Pit levels
The presence of H_2S

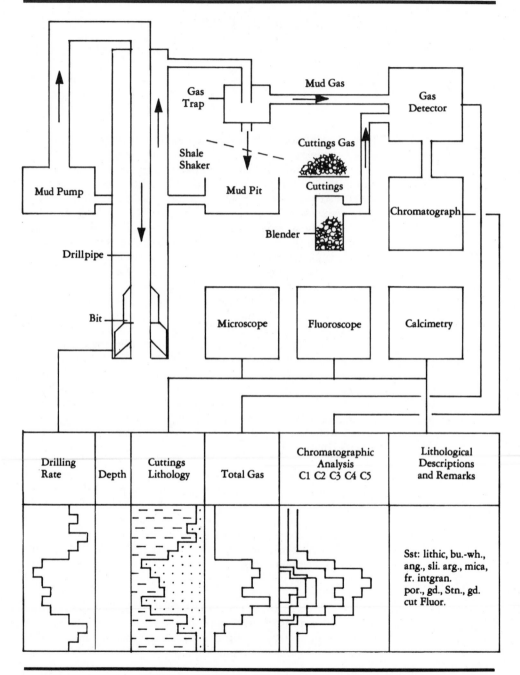

FIGURE 2.1 *Schematic of Mud Logging Data Gathering and Presentation.*

Many of these parameters may be combined in an interpretative fashion to give indirect indications of porosity and pore pressure. Through the use of the "d" exponent, for example, the driller may be forewarned of the approach of an overpressured zone and take appropriate action and increase the mud weight. While these measurements are of importance to those charged with the expedient and economic drilling of the well, the formation evaluator is more interested in records that give direct or indirect indications of hydrocarbons.

MEASUREMENTS RELEVANT TO THE FORMATION EVALUATION

The measurements of interest to the analyst are:

Rate of penetration
Mud gas detection and chromatographic gas analysis
Cutting gas detection and analysis
Cutting description and analysis for shows

With the exception of the rate of penetration, all these measurements are lagged (i.e., a time lag exists between the moment the bit liberates cuttings and fluids into the wellbore and the time these same fluids and bit cuttings actually reach the surface). Considerable care is therefore required to place mud and cuttings samples correctly at the proper depth in the well. Lag times in deep wells may reach several hours. Lag time may be calculated, provided that the pump capacity and pump strokes are known and the annular volume between the drillpipe and the formation and/or casing can be estimated. In washed-out holes this can be difficult to estimate. An alternative method is to drop a carbide pill into the drillstring at the surface. When it reaches the bit it is broken open and, reacting chemically with the water in the mud, produces acetylene gas, which is then circulated "bottoms up" to the surface and detected. Since the dimensions of the drillstring are accurately known, the time for the downward journey of the pill can be calculated exactly.

If properly lagged, all measurements should appear on the mud log on depth with each other. When correlating a mud log to a wireline log, the best curve to use is the rate of penetration (ROP). This usually shows good correlation with a gamma ray or SP curve.

The mud logger's sources of data include:

Gas extracted from the mud at the gas trap
Cuttings collected at the shale shaker
Gas liberated from cuttings in the blender

The types of measurement possible include:

Total gas concentration, from either the mud or the cuttings
Chromatographic analysis of gas
Visual inspection of cuttings by normal and ultraviolet light
Calcimetry for carbonate analysis

MEASUREMENT TECHNIQUES

Gas Concentration

The measurement of gas concentration in the mud is made using either a hot-wire detector, a flame-ionization detector, or a Katharometer. All three types of detectors produce a "total gas" reading. The modern norm (see SPWLA 1983) is to report total gas in universal volumetric units (i.e., in terms of volume/volume) and standardized to "equivalent methane in air," or EMA. Thus, the units on the mud log may be quoted as percent, ppk, or ppm.

Hot-Wire Gas Detector (Catalytic Combustion Detector, CCD)

The principle of the hot-wire detector is simple. A wire resistor is heated to a temperature that assures combustion of any hydrocarbons flowed past it in a mixture with air. The combustion of hydrocarbon gas, when present, heats the wire and thus raises its electrical resistance, which in turn is monitored by a bridge circuit. Such a device can be accurately calibrated. Regulators assure a constant flow of air and mud or cuttings gas, and the output can be continuously recorded as total gas on the mud log.

Flame-Ionization Detector (FID)

A high-temperature hydrogen flame is maintained in a chamber through which mud or cuttings gas is circulated. When hydrocarbons are encountered, the heat of the hydrogen flame is sufficient to ionize the hydrocarbons. A pair of electrodes in the chamber senses the presence of the ions by the small current that can flow through the ionized gas.

Katharometer (Thermal Conductivity Detector, TCD)

This device responds to the thermal conductivity of the mixture of mud or cuttings gas and the carrier gas (air), as measured by a heated platinum wire. It is less sensitive than either the FID or the CCD.

Detector Calibration

Calibration of any of these total-gas detectors should be made in terms of equivalent methane in air (EMA). The response of the device will then be a function of both the gas concentration and its composition. For example, if a mixture of 25% methane and 75% air passes through the detector then the response will be:

$$25 \times 1 = 25\% \text{ EMA}$$

If, on the other hand, a mixture containing gases with more carbon atoms (higher C numbers) passes through the detector, the percentage of each gas is weighted by its C number. For example:

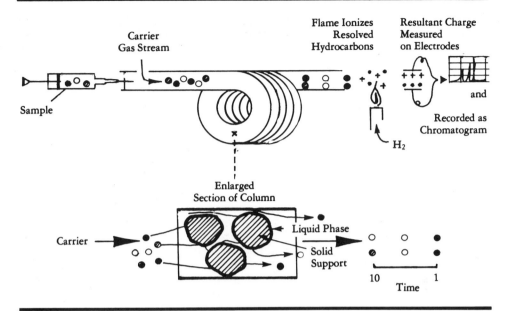

FIGURE 2.2 *Gas Chromatography. Courtesy Continental Laboratories.*

C Number	Gas	Formula	% Present	EMA
C 1	Methane	CH_4	8	$8 \times 1 = 8$
C 2	Ethane	C_2H_6	6	$6 \times 2 = 12$
C 3	Propane	C_3H_8	4	$4 \times 3 = 12$
i C 4	Iso-butane	C_4H_{10}	4	$6 \times 4 = 24$
n C 4	Butane	C_4H_{10}	2	
C 5	Pentane	C_5H_{12}	1	$5 \times 1 = 5$
Totals			25%	61% EMA

Thus, the same total gas reading could be obtained with only half as much ethane, a third as much propane, etc.

It is also common to correct the raw readings for a number of variables related to the drilling process. These include:

Mud-pump output
Penetration rate
Volume of mud/unit length of hole
Bit size

These are combined to obtain a "dilution ratio" that is multiplied by the raw total gas reading to give a normalized reading that is independent of the mechanical details of the drilling process.

Gas Chromatography

To determine the composition of the gas, a chromatograph is used. Gas is passed through a column packed with an inert powdered solid (fig. 2.2). Different components of the gas mixture take different times to travel the length of the column and on exiting may be detected by one or another of the gas detectors already discussed. Output from the chromatograph is recorded on the log in ppm or ppk of each component.

Visual Inspection

Cuttings are visually inspected on a regular basis in order to determine:

Approximate porosity, lithology, and texture
The presence of hydrocarbons
Chemical makeup

Since it takes about 15 minutes to process each sample, the frequency (by depth) of the reported results will be a function of the drilling rate. That is, if the drilling rate is 100 ft/hr the best that a mud logger can do to keep up is to report samples every 25 ft (4 per hour). Thus, in hard rock with a drilling rate of 8 ft/hr, samples can be handled every 2 ft.

WRITTEN DESCRIPTION. A written description follows a set of rules so that a fixed order is used thus:

Rock type—underlined and followed by classification
Color
Texture—including grain size, roundness, and sorting
Cement and/or matrix materials
Fossils and accessories
Sedimentary structures
Porosity and oil shows

For example:

Ls: ool, Grst, brn, med-crs, arg, Brach-Bry, glauc, gd intpar, por, gd Stn, gd cut Fluor

can be translated to mean:

limestone, oolitic, grainstone, brown, medium coarse, argillaceous, brachiopod bryozoa, glauconitic, good interparticle porosity, good stain, good cut fluorescence.

Appendix 2B to this chapter lists all the abbreviations commonly used to describe rocks in this geological telegraphese.

HYDROCARBON DETERMINATION. If hydrocarbons are still present in the cuttings after their journey up through the mud column they may be detected by a number of techniques. These include:

Smelling the sample
Visual inspection under ultraviolet light
Subjecting the sample to chemical reagents

A trained nose can distinguish an "oil odor" from a "condensate odor." In ordinary light, the cuttings may have a visible oil stain or oil may be seen bleeding from the rock fragments. Under ultraviolet light, these hydrocarbons will fluoresce (the API gravity of the oil will influence the color seen).

API Gravity	Color of Fluorescence
less than 15	brown
15 to 25	orange
25 to 35	cream or yellow
35 to 45	white
45 or over	blue-white to violet

If no fluorescence is seen on the untreated sample then a chemical reagent is added. This is usually a solvent (chlorothene or acetone, for example) that brings out any hydrocarbon still in the sample. Such solvent-extracted hydrocarbon is referred to as *cut*. Other chemicals used include hydrochloric acid, which will cause bubbling on contact with carbonates—a reaction that forms the basis for calcimetry.

CALCIMETRY. A weighed sample of cuttings (usually 1 g) is placed in contact with a standard volume of HCl (a few cc) in a closed chamber. If either limestone ($CaCO_3$) or dolomite ($CaMg\ CO_3$) is present, the acid reacts with the rock to form carbon dioxide gas (CO_2). The calcimeter measures both the volume of CO_2 liberated and the *rate* at which it is liberated. Since the reaction with dolomite proceeds more slowly than the reaction with limestone, it is possible to distinguish the relative concentrations of these carbonates in a sample.

EVALUATION OF SHOWS
When reading a mud log it is of use to be able to estimate formation productivity in a semiquantitative manner. A number of authors have suggested ways to combine the readings from chromatographic analysis in order to fingerprint the true formation content. One method (see Haworth et al. 1984) suggests using three ratios to type the formation. These are:

Gas wetness ratio (GWR)	=	C2 + / C1 +
Light-to-heavy-ratio (LHR)	=	(C1 + C2) / (C3 + C4 + C5)
Oil character qualifier (OCQ)	=	(C4 + C5) / C3

The diagnostics reported are as shown in table 2.1. The LHR correlates well with the API gravity—that is, a high LHR corresponds to low density (high API) hydrocarbons. The OCQ ratio is of use as a qualifier when excessive methane is present.

TABLE 2.1 *Hydrocarbon Potential*

Hydrocarbon	GWR (%)	LHR (%)	OCQ Ratio
Light dry gas	< 0.5%	100% +	very low
Medium density gas	0.5–17.5%	< 100%	< 0.5
Light oil gas	5–10%	17.5%	> 0.5
Medium gravity oil	17.5%–40%	< 10%	> 1.0
Residual oil	> 40%	5–10%	> 2.0
Coal bed	15–20%	< 100%	very low

Source: After Haworth et al. 1984.

Another technique used is to plot straight C-number ratios on a special logarithmic grid as shown in figure 2.3. The type of production is predicted according to the area of the graph on which the points fall.

SUMMARY
Hydrocarbon well logging can supply real-time data to both the formation evaluator and the drilling engineer and allows a hands-on investigation of both the rock and its pore fluids, subject to the vagaries of the process whereby the samples arrive in the analyst's hands. When the measurements are made using calibrated equipment, the analyst may hope to know, before running wireline logs, the location of reservoir rocks in the column drilled and to have a semiquantitative idea of their porosity and hydrocarbon content. Apart from these considerations, use of a mud logging unit will help in making decisions on coring and testing depths.

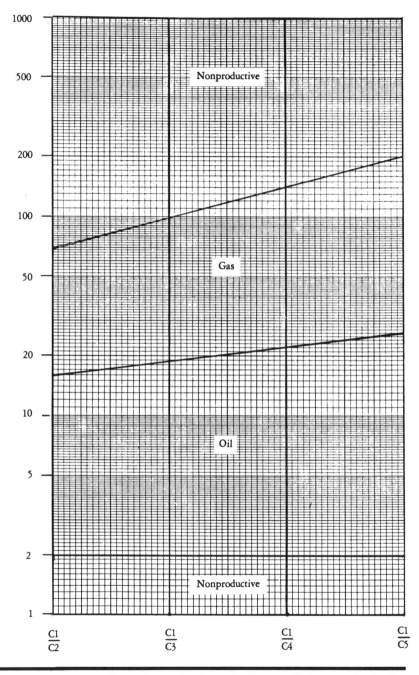

FIGURE 2.3 *Formation Production Potential from Carbon-Number Ratios.*

APPENDIX 2A: SYMBOLS USED TO DESCRIBE LITHOLOGY

Black and White Lithologic Symbols

To be used for Stratigraphic Columnar and Cross Sections

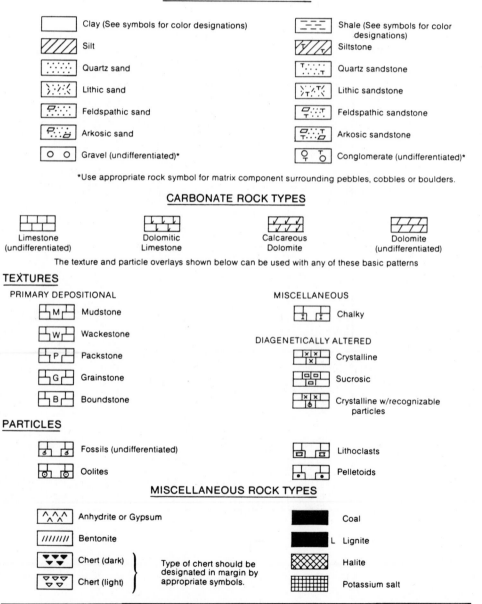

SILICICLASTIC ROCK TYPES

Clay (See symbols for color designations)

Silt

Quartz sand

Lithic sand

Feldspathic sand

Arkosic sand

Gravel (undifferentiated)*

Shale (See symbols for color designations)

Siltstone

Quartz sandstone

Lithic sandstone

Feldspathic sandstone

Arkosic sandstone

Conglomerate (undifferentiated)*

*Use appropriate rock symbol for matrix component surrounding pebbles, cobbles or boulders.

CARBONATE ROCK TYPES

Limestone (undifferentiated)

Dolomitic Limestone

Calcareous Dolomite

Dolomite (undifferentiated)

The texture and particle overlays shown below can be used with any of these basic patterns

TEXTURES

PRIMARY DEPOSITIONAL

M   Mudstone

W   Wackestone

P   Packstone

G   Grainstone

B   Boundstone

MISCELLANEOUS

Chalky

DIAGENETICALLY ALTERED

Crystalline

Sucrosic

Crystalline w/recognizable particles

PARTICLES

Fossils (undifferentiated)

Oolites

Lithoclasts

Pelletoids

MISCELLANEOUS ROCK TYPES

Anhydrite or Gypsum

Bentonite

Chert (dark)

Chert (light)

Type of chert should be designated in margin by appropriate symbols.

Coal

L   Lignite

Halite

Potassium salt

Reprinted by permission of the SPWLA from "Hydrocarbon Well Logging Recommended Practice," 2nd ed., Dec. 1983.

IGNEOUS AND METAMORPHIC ROCKS

`x x x / x x`	Extrusive	`/ / /`	Metamorphic
`+ + + / + +`	Intrusive	`‖ ‖ ‖ / ‖ ‖`	Pyroclastic

Specific rock types can be designated by using symbols and/or inserting first two letters of rock name in center of lithic column.

SECONDARY COMPONENTS

MODIFYING COMPONENTS

Symbol	Name	Symbol	Name	Symbol	Name
∧ ∧	Anhydritic	▽ ▽	Cherty	: : : :	Very sandy
— —	Argillaceous	⊥ ⊥	Dolomitic	: : :	Sandy
⊥ ⊥	Calcareous	⌣ ⌣	Micaceous	: :	Slightly sandy
‖ ‖ ‖	Carbonaceous	✳ ✳	Salt hoppers	/ /	Silty

CEMENTS

Symbol	Name	Symbol	Name	Symbol	Name	Symbol	Name	Symbol	Name
∧	Anhydrite	⊥	Calcite	▽	Chert	⊥	Dolomite	π	Quartz

STREAKS AND LENSES

Symbol	Name	Symbol	Name	Symbol	Name	Symbol	Name	Symbol	Name
= _	Argillaceous	⊥ ⊥	Calcareous	⊥ ⊥	Dolomitic	· · · ...	Sandy	// //	Silty

INTERBEDS AND INTERCALATIONS

Use same symbols as for streaks and lenses enclosed within above symbols.

CONCRETIONS AND NODULES

Symbol	Name	Symbol	Name	Symbol	Name
⊕ ⊕	anhy. Anhydritic	⊕ ⊕	calc. Calcareous	⊕ ⊕	ch Cherty or Siliceous
⊕ ⊕	fe. Ironstone or other ferruginous material	⊕ ⊕	phos. Phosphatic	⊕ ⊕	sid. Sideritic

MISCELLANEOUS SYMBOLS

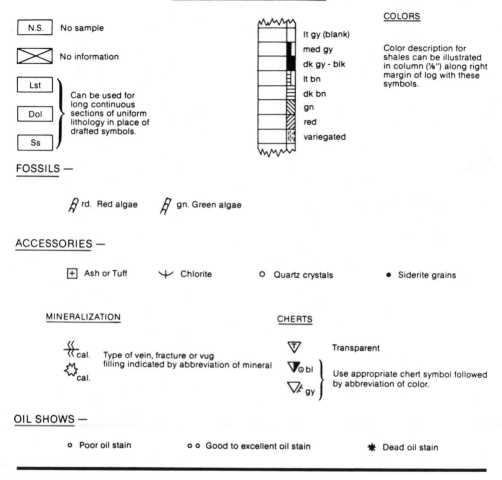

COLORS

Color description for shales can be illustrated in column (⅛") along right margin of log with these symbols.

N.S.	No sample
⊠	No information
Lst	
Dol	Can be used for long continuous sections of uniform lithology in place of drafted symbols.
Ss	

lt gy (blank)
med gy
dk gy - blk
lt bn
dk bn
gn
red
variegated

FOSSILS —

rd. Red algae   gn. Green algae

ACCESSORIES —

⊞ Ash or Tuff   ⅄ Chlorite   ○ Quartz crystals   ● Siderite grains

MINERALIZATION

cal. Type of vein, fracture or vug filling indicated by abbreviation of mineral
cal.

CHERTS

▽ Transparent

▽○ bl
▽ gy   Use appropriate chert symbol followed by abbreviation of color.

OIL SHOWS —

○ Poor oil stain   ○ ○ Good to excellent oil stain   ✴ Dead oil stain

APPENDIX 2B: ABBREVIATIONS USED TO DESCRIBE LITHOLOGY

above	ab	bioturbated	bioturb
absent	abs	birdseye	Bdeye
abundant	abd	bitumen (-inous)	Bit, bit
acicular	acic	black (-ish)	blk, blksh
after	aft	blade (-ed)	Bld, bld
agglomerate	Aglm	blocky	blky
aggregate	Agg	blue (-ish)	bl, blsh
algae, algal	Alg, alg	bored (-ing)	Bor, bor
allochem	Allo	botryoid (-al)	Bot, bot
altered	alt	bottom	Btm
alternating	altg	boudinage	boudg
amber	amb	boulder	Bld
ammonite	Amm	boundstone	Bdst
amorphous	amor	brachiopod	Brach
amount	amt	brackish	brak
amphipora	*Amph*	branching	brhg
and	&	break (-en)	Brk, brk
angular	ang	breccia (-ted)	Brec, brec
anhedral	ahd	brighter	brt
anhydrite (-ic)	Anhy, anhy	brittle	brit
anthracite	Anthr	brown	brn
aphanitic	aph	bryozoan	Bry
apparent	apr	bubble	Bubl
appears	ap	buff	bu
approximate	apprx	bulbous	bulb
aragonite	Arag	burrow (-ed)	Bur, bur
arenaceous	aren		
argillaceous	arg	calcarenite	Clcar
argillite	argl	calcareous	calc
arkose (-ic)	Ark, ark	calcilutite	Clcit
as above	a. a.	calcirudite	Clcrd
asphalt (-ic)	Asph, asph	calcisiltite	Clsit
assemblage	Assem	calcisilitite	Clsit
associated	assoc	*calcisphaera*	*Casph*
at	@	calcisphere	Clcsp
authigenic	authg	calcite (-ic)	Calc, calctc
average	Av, av	caliche	cche
		carbonaceous	carb
band (-ed)	Bnd, bnd	carbonate	crbnt
barite (-ic)	bar	carbonized	cb
basalt (-ic)	Bas, bas	cavern (-ous)	Cav, cav
basement	Bsmt	caving	Cvg
become (-ing)	bcm	cement (-ed, -ing)	Cmt, cmt
bed (-ed)	Bd, bd	center (-ed)	cntr
bedding	Bdg	cephalopod	Ceph
belemnites	*Belm*	*chaetetes*	*Chaet*
bentonite (-ic)	Bent, bent	chalcedony (-ic)	Chal, chal
bioclastic	biocl	chalk (-y)	Chk, chky
bioherm (-al)	Bioh, bioh	charophyte	Char
biomicrite	Biomi	chert (-y)	Cht, cht
biosparite	Biosp	chitin (-ous)	Chit, chit
biostrom (-al)	Biost, biost	chitinozoa	Chtz
biotite	Biot	chlorite (-ic)	Chlor, chlor

Reprinted by permission of the SPWLA from "Hydrocarbon Well Logging Recommended Practice," 2nd ed., Dec. 1983.

chocolate	choc	devitrified	devit
circulate (-ion)	circ, Circ	diabase	Db
clastic	clas	diagenesis (-etic)	Diagn, diagn
clay (-ey)	Cl, cl	diameter	Dia
claystone	Clst	disseminated	dissem
clean	cln	distillate	Dist
clear	clr	ditto	" or do
cleavage	Clvg	dolomite (-ic)	Dol, dol
cluster	Clus	dolostone	dolst
coal	C	dominant (-ly)	dom
coarse	crs	drillstem test	DST
coated (-ing)	cotd, cotg, Cotg	drilling	drig
coated grains	cotd gn	drusy	dru
cobble	Cbl		
colonial	coln	earthy	ea
color	Col, col	east	E
common	com	echinoid	Ech
compact	cpct	elevation	Elev
compare	cf	elongate	elong
concentric	cncn	embedded	embd
conchoidal	conch	*endothrya*	*Endo*
concretion (-ary)	Conc, conc	equant	eqnt
conglomerate (-ic)	Cgl, cgl	equivalent	Equiv
conodont	Cono	euhedral	euhd
conquina (-oid)	coqid	*euryamphipora*	*Euryamph*
considerable	cons	euxinic	eux
consolidated	consol	evaporite (-itic)	Evap, evap
conspicuous	conspic	excellent	ex
contact	Ctc	exposed	exp
contamination (-ed)	Contam, contam	extraclast (-ic)	Exclas, exclas
content	Cont	extremely	extr
contorted	cntrt	extrusive	exv
coral, coralline	Cor, corin		
core	c	facet (-ed)	Fac, fac
covered	cov	faint	fnt
cream	crm	fair	fr
crenulated	cren	fault (-ed)	Flt, flt
crinkled	crnk	fauna	Fau
crinoid (-al)	Crin, crinal	*favosites*	*Fvst*
cross	x	feet	Ft
cross-bedded	x-bd	feldspar (-athic)	Fspr, fspr
cross-laminated	x-lam	fenestra (-al)	Fen, fen
cross-stratified	x-strat	ferro-magnesian	Fe-mag
crumpled	crpld	ferruginous	ferr
crypridopsis	Cyp	fibrous	fibr
cryptocrystalline	crpxin	fill (-ed)	fld
crystal (-line)	Xl, xin	fine (-ly)	f, fnly
cube, cubic	Cub, cub	fissile	fis
cuttings	Ctgs	flaggy	flg
cypridopsis	*Cyp*	flake, flaky	Flk, flk
dark (-er)	dk, dkr	flat	fl
dead	dd	flesh	fls
debris	Deb	floating	fltg
decrease (-ing)	Decr, decr	flora	Flo
dendrite (-ic)	dend	fluorescence (-ent)	Fluor, fluor
dense	dns	foliated	fol
depauperate	depau	foot	Ft
description	Descr	foraminifer	Foram
desiccation	dess	foraminiferal	foram
detrital	detr	formation	Fm

fossil (-iferous)	Foss, foss	inch	in
fracture (-d)	Frac, frac	inclusion (-ded)	Incl, incl
fragment (-al)	Frag, frag	increasing	incr
framework	frmwk	indistinct	indst
frequent	freq	indurated	ind
fresh	frs	*inoceramus*	*inoc*
friable	fri	insoluble	insl
fringe (-ing)	Frg, frg	interbedded	intbd
frosted	fros	intercalated	intercal
frosted quartz	F.Q.G.	intercrystalline	intxin
fucoid (-al)	Fuc, fuc	interfragmental	intfrag
fusulinid	Fus	intergranular	intgran
		intergrown	intgn
gabbro	Gab	interlaminated	intrlam
galeolaria	*Gal*	interparticle	intpar
gas	G	interpretation	intpt
gastropod	Gast	intersticies	inst, intstl
generally	gen	interval	intvl
geopetal	gept	intraclast (-ic)	Intclas, intclas
gilsonite	Gil	intraparticle	intrapar
girvanella	*Girv*	intrusive	Intr, intr
glass (-y)	Glas, glas	invertebrate	invtb
glauconite (-itic)	Glauc, glauc	iridescent	irid
globigerina (-inal)	*Glob, glob*	ironstone	Fe-st
gloss (-y)	Glos, glos	irregular (-ly)	irr
gneiss (-ic)	Gns, gns	isopachous	iso
good	gd	*ivanovia*	*ivan*
grading	grad		
grain (-s, -ed)	Gr, gr	jasper	jasp
grainstone	Grst	joint (-ed, -ing)	Jt, jt
granite	Grt		
granite wash	G.W.	kaolin (-itic)	Kao, kao
granule (-ar)	Gran, gran		
grapestone	Grapst	lacustrine	lac
graptolite	Grap	lamina (-itions, -ated)	Lam, lam
gravel	Grv	large	lge
gray, grey (-ish)	gry, grysh	laterite (-itic)	Lat, lat
graywacke	Gwke	lavender	lav
greasy	gsy	layer	Lyr
green (-ish)	gn, gnsh	leached	lchd
grit (-ty)	Gt, gt	lens, lenticular	Len, lent
gypsum (-iferous)	Gyp, gyp	lentil (-cular)	len
		light	lt
hackly	hkl	lignite (-itic)	Lig, lig
halite (-iferous)	Hal, hal	limestone	Ls
hard	hd	limonite (-itic)	Lim, lim
heavy	hvy	limy	lmy
hematite (-ic)	Hem, hem	lithic	lit
heterogeneous	*hetr*	lithographic	lithgr
heterostegina	Het	lithology (-ic)	Lith, lith
hexagonal	hex	little	Ltl
high (-ly)	hi	littoral	litt
homogeneous	hom	local	loc
horizontal	hor	long	lg
hornblende	hornbd	loose	lse
hydrocarbon	Hydc	lower	l
igneous rock	Ig, ig	lumpy	lmpy
imbedded	imbd	luster	Lstr
impression	imp	lustre	Lstr
in part	I.P.	lutite	Lut

macrofossil	Macrofos	ooid (-al)	Oo, oo
magnetite, magnetic	Mag, mag	oolicast (-ic)	Ooc, ooc
manganese	Mn, mn	oolite (-itic)	Ool, ool
marble	Mbl	oomold (-ic)	Oomol, oomol
marine	marn	opaque	op
marl (-y)	Mrl, mrl	orange (-ish)	or, orsh
marlstone	Mrlst	*orbitolina*	*Orbit*
maroon	mar	organic	org
massive	mass	orthoclase	Orth
material	Mat	orthoquartzite	O-Qtz
matrix	Mtrx	ostracod	Ostr
maximum	max	overgrowth	ovgth
medium	m or med	oxidized	ox
member	Mbr	oyster	Oyst
meniscus	men		
metamorphic (-osed)	meta, metaph	packstone	Pkst
metamorphic rock	Meta	paper (-y)	Pap, pap
metasomatic	msm	*paraparchites*	*Para*
mica (-aceous)	Mic, mic	part (-ly)	Pt, pt
micrite (-ic)	Micr, micr	particle	Par, par
micro	mic	parting	Ptg
micro-oolite	Microol	parts per million	PPM
microcrystalline	microxin	patch (-y)	Pch, pch
microfossil	microfos	pearly	prly
micrograined	micgr	pebble (-ly)	Pbl, pbl
micropore (-osity)	micropor	pelecypod	Pelec
microspar	Microspr	pellet (-al)	Pel, pel
microstylolite	Microstyl	pelletoid (-al)	Peld, peld
middle	Mid	pendular (-ous)	Pend, pend
millolid	Millid	*pentamerus*	*Pent*
milky	mky	permeability (-able)	Perm, k, perm
mineral (-ized)	Min, min	petroleum	Pet, pet
minor	mnr	phlogopite	Phlog
minute	mnut	phosphate (-atic)	Phos, phos
moderate	mod	phreatic	phr
mold (-ic)	Mol, mol	phyllite, (-itic)	Phyl, phyl
mollusc	Moll	pin-point (porosity)	p.p.
mosaic	mos	pink	pk
mottled	mott	pinkish	pkish
mud (-dy)	md, mdy	pisold (-al)	Piso, piso
mudstone	Mdst	pisolite (-itic)	Pisol, pisol
muscovite	Musc	pitted	pit
		plagioclase	Plag
nacreous	nac	plant	Plt
no sample	n.s.	plastic	plas
no show	n/s	platy	plty
no visible porosity	n.v.p.	polish, polished	Pol, pol
nodules (-ar)	Nod, nod	pollen	Poln
north	N	polygonal	poly
novaculite	Novac	poor (-ly)	p
numerous	num	porcelaneous	porcel
		porosity, porous	Por, por
occasional	occ	porous (-sity)	por
ochre	och	porphyry (-itic)	prphy
odor	od	possible (-ly)	poss
oil	O	predominant (-ly)	pred
oil source rock	OSR	preserved	pres
olive	olv	primary	prim
olivine	olvn	prism (-atic)	pris
oncolite (-oidal)	Onc, onc	probable (-ly)	prob

production	Prod	shelter porosity	Shlt por
prominent	prom	show	shw
pseudo-oolite (-ic)	Psool, psool	siderite (-itic)	Sid, sid
pseudo-	ps	sidewall core	S.W.C.
pumice stone	Pst	silica (-iceous)	Sil, sil
purple	purp	silky	slky
pyrite (-ized, -itic)	Pyr, pyr	silt (-y)	Slt, slty
pyrobitumen	Pybit	siltstone	Sltst
pyroclastic	pyrcl	similar	sim
pyroxene	pyrxn	size	sz
		skeletal	skel
quartz (-ose)	Qtz, qtz	slabby	slb
quartzite (-ic)	Qtzt, qtzt	slate (-y)	Sl, sl
		slickenside (-d)	Slick, slick
radial (-ating)	Rad, rad	slight (-ly)	sil, silly
radiaxial	Radax	small	sml
range	rng	smooth	sm
rare	r	soft	sft
recemented	recem	*solenpora*	*Solen*
recovery (-ered)	Rec, rec	solitary	sol
recrystallized	rexizd	solution, soluble	Sol, sol
red (-ish)	rd, rdsh	somewhat	smwt
reef (-oid)	Rf, rf	sorted (-ing)	srt, srtg
remains	Rem	south	S
renalcis	*Ren*	spar (-ry)	Spr, spr
replaced (-ment)	rep, Repl	sparse (-ly)	sps, spsly
residue (-ual)	Res, res	speck (-led)	Spk, spkld
resinous	rsns	*sphaerocodium*	*Sphaer*
rhomb (-ic)	Rhb, rhb	sphalerite	Sphal
ripple	Rpl	spherule (-itic)	Spher, spher
rounded, frosted, pitted	r.f.p.	spicule	Spic, spic
rock	Rk	splintery	Splin
round (-ed)	rnd, rndd	sponge	Spg
rubble (-bly)	Rbl, rbl	spore	Spo
rudistid	Rud	spotted (-y)	sptd, spty
rugose	rug	*stachyode*	*Stach*
		stain (-ed, -ing)	Stn, stn
saccharoidal	sacc	stalactitic	stal
salt (-y)	Sa, sa	strata (-ified)	Strat, strat
salt and pepper	s&p	streak (-ed)	Strk, strk
salt cast (-ic)	sa-c	streaming	stmg
salt water	S.W.	striae (-ted)	Strl, strl
same as above	a.a.	stringer	strgr
sample	Spl	stromatolite	Stromlt, stromlt
sand (-y)	Sd, sdy	stromatoporoid	Strom
sandstone	Sst	structure	Str
saturation (-ated)	Sat, sat	*styliolina*	*Stylio*
scales	sc	stylolite (-itic)	Styl, styl
scaphopod	Scaph	sub	sb
scarce	scs	subangular	sbang
scattered	scat	sublithic	sblit
schist (-ose)	Sch, sch	subrounded	sbrndd
scolecodont	Scol	sucrosic	suc
secondary	sec	sugary	sug
sediment (-ary)	Sed, sed	sulfur (-ous)	Su, su
selenite	Sel	superficial oolite	Spfool, spfool
septate	sept	surface	Surf
shadow	shad	syntaxial	syn
shale (-ly)	Sh, sh	*syringopora*	*Syring*
shell	Shl		

tabular (-ate)	tab	variegated	vgt
tan	tn	varved	vrvd
tasmanites	*Tas*	vein (-ing, -ed)	Vn, vn
tension	tns	veinlet	Vnlet
tentaculites	*Tent*	vermillon	verm
terriginous	ter	vertebrate	vrtb
texture (-d)	Tex, tex	vertical	vert
thamnopora	*Tham*	very	v
thick	thk	very poor sample	V.P.S.
thin	thn	vesicular	ves
thin section	T.S.	violet	vl
thin-bedded	t.b.	visible	vis
throughout	thru	vitreous (-ified)	vit
tight	ti	volatile	volat
top	Tp	volcanic rock,	Volc, volc
tough	tgh	vug (-gy)	Vug, vug
trace	Tr		
translucent	trnsl	wackestone	Wkst
transparent	trnsp	washed residue	W.R.
trilobite	Tril	water	Wtr
tripoli (-itic)	Trip, trip	wavy	wvy
tube (-ular)	Tub, tub	waxy	wxy
tuff (-aceous)	Tf, tf	weak	wk
type (-ical)	Typ, typ	weathered	wthd
		well	Wl, wl
unconformity	Unconf	west	W
unconsolidated	uncons	white	wh
underclay	Uc	with	w/
underlying	undly	without	w/o
unidentfiable	unident	wood	Wd
uniform	uni		
upper	u	yellow (-ish)	yel, yelsh
vadose	Vad, vad	zeolite	zeo
variation (-able)	Var, var	zircon	Zr
varicolored	varic	zone	Zn

APPENDIX 2C: ABBREVIATIONS USED BY ENGINEERS

absolute open flow	AOP	heavy oil	HO
barrel of oil	BO	initial air blow	IAB
barrels of oil per day	BOPD	initial production	IP
barrels of oil per hour	BOPH		
barrels of water	BW	kelly bushing	KB
barrels of water per day	BWPD		
barrels of water per hour	BWPH	legal subdivision	LSD
bottom hole flow pressure	BHFP	location	loc
bottom hole pressure	BHP		
bottom hole shut in	BHSIP	million cubic feet of gas	MCFG
pressure		mud cut oil	MCO
bottom hole temperature	BHT	mud cut water	MCW
brackish	brk	mud resistivity,	RM
		OHM-METER	
casing	csg		
choke	ck	new bit	NB
circulate (-ed) (-tion)	circ	new core bit	NCB
circulated out	CO	no returns	NR
completed (tion)	comp		
connection gas	CG	oil and gas	O&G
cored	crd	oil and salt water	O&SW
		oil cut	OC
decreasing	decr	oil cut mud	OCM
depth correction	DC	oil flecked mud	OFM
derrick floor	DF	oil to surface	OTS
development	(D)	old total depth	OTD
directional survey	DS	old well drilled deeper	OWDD
distillate	dist	old well plugged back	OWPB
drillstem test	DST	old well worked over	OWWO
driller	drlr	open	op
dry and abandoned	D&A		
		packer	pkr
estimated	est	per day	PD
		per hour	PH
faint air blow	FTAB	perforated	perf
fair air blow	FAB	plugged back	PB
filter cake	CK	pounds per square inch	psi
filtrate, API, cc's	F	pump pressure	PP
flowed (ing)	fl/	pump strokes	SPM
flowing pressure	FP		
flowline temperature	F/T	recovered	rec
		rotary speed	RPM
gas and oil cut mud	G&OCM	rotary table	RT
gas cut mud	GCM		
gas cut water	GCW	salinity — PPMCL	CL
gas to surface	GTS	salt water	SW
gas-to-oil ratio	GOR	show of oil	SO
gauged	ga	show of oil and gas	SO&G
good air blow	GAB	show of oil and water	SO&W
good initial puff	GIP	shut in	SI
gravity	gty	shut in pressure	SIP
ground	GR	slight gas cut mud	SGCM
ground level	GL	slight gas cut water	SGCW

Reprinted by permission of the SPWLA from "Hydrocarbon Well Logging Recommended Practice," 2nd ed., Dec. 1983.

slight oil cut mud	SOCM	valve open	V.op
slight oil cut water	SOCW	viscosity, API, sec.	V
slight show of oil	SSO		
squeezed	sqz		
strong air blow	SAB		
suction temperature	S/T	water	wtr
swabbed	swbd	water cushion	wtr cush
		water cut mud	WCM
testing	tstg	weak air blow	WAB
thousand cubic feet of gas	MCFG	weak initial puff	WIP
too small to measure	TSTM	weight of mud	W
total depth	T.D.	weight on bit	WOB
trip gas	TG	wildcat	(W)

BIBLIOGRAPHY

AAPG: "Sample Examination Manual," R. G. Swanson, a part of the Methods in Exploration Series, American Association of Petroleum Geologists, Tulsa (May 1981).

API: "Recommended Practice—Standard Hydrocarbon Mud Log Form," API RP 34, American Petroleum Institute, Production Department, 300 Corrigan Tower Building, Dallas, TX 75201 (November 1958).

API: "Conversion of Operational and Process Measurement Units to the Metric (SI) System," API 2564, 1st ed., American Petroleum Institute, Publications Section, 1801 K Street N.W., Washington, DC, 20006 (March 1974).

API: "Recommended Practice and Standard Form for Electrical Logs," API RP 31, 3rd ed., American Petroleum Institute, Production Department, 300 Corrigan Tower Building, Dallas, TX 75201 (reissued November 1976).

Association of Desk and Derrick Clubs (compile): "D&D Standard Oil Abbreviator," 2nd ed., Penn Well Publishing Company, Tulsa (1973).

Haworth, H.L., Gurvis, R.L., and Sellens, M.P.: "Reservoir Characterization by Mathematical Treatment of Hydrocarbon Gas Show," SPWLA 25th Annual Logging Symposium, New Orleans, June 10–13, 1984.

Helender, D. P.: "Drilling-Mud Logging Becoming More Important in Formation Evaluation." *Oil and Gas Equipment,* four part series (February, March, April, May 1969).

Mercer, R.F.: "Liberated, Produced, Recycled or Contaminated?" SPWLA 15th Annual Logging Symposium June 2–5, 1974.

Pixler, B.O.: "Formations Evaluation by Analysis of Hydrocarbon Ratios," *J. Pet. Tech.* (June 1969) 665–670.

SPWLA: "Hydrocarbon Well Logging Recommended Practice," 2nd ed., Society of Professional Well Log Analysts, Houston (December 1983).

CHAPTER THREE

MEASUREMENTS WHILE DRILLING

Measurements while drilling play an increasingly important role in modern drilling practices. They allow an operator almost immediate feedback regarding both the geometry of the hole being drilled and the characteristics of the formations penetrated. Without MWD this kind of information is available only from conventional sources such as deviation surveys and logs that, a priori, must be run after the drilling has taken place. MWD is of particular benefit in that it can be applied when drilling directional wells and/or when overpressured formations are of concern.

By having the kind of information MWD can supply more or less in real time, the driller may take appropriate action such as changing the weight on bit, increasing the mud weight, or pulling out of the hole for a conventional logging run once the objective formation has been reached.

MWD MEASUREMENTS

Many different MWD measuring systems are in commercial use today; they all have the following common characteristics:

A downhole sensor sub
A power source
A telemetry system
Surface equipment

The downhole sensor subs may contain instrumentation capable of measuring some of the following parameters:

Torque
Weight on bit
Borehole pressure
Borehole temperature
Tool-face angle
Natural formation gamma ray activity
Formation acoustic travel time
Formation resistivity
Hole deviation from vertical
Hole azimuth with respect to geographic coordinates

MWD TELEMETRY

The power source for activation of the sensors and the telemetry system can be one of three types:

Reprinted, with permission, from chapter 56 of the forthcoming *Petroleum Engineering Handbook,* © Society of Petroleum Engineers.

A surface power source
A downhole turbine
Downhole batteries

In the case of a surface power source, it is necessary to make electrical connections between the surface and the downhole sensors; this, in turn, requires either special drillpipe or an electric cable.

In the case of a downhole turbine, the circulating mud itself drives an electric generator located in the MWD drill collar. This in turn leads to an increase in the hydraulic horsepower required of the mud pumps in order to maintain circulation.

In the case of batteries, no special cabling or additional mud pumping is required, but the MWD system is limited by the life of the batteries used. Once the batteries are discharged, no further measurements can be made and the MWD sub must be retrieved and redressed with fresh batteries.

The telemetry system most commonly used is that of coded mud-pressure pulses. The output from a specific sensor is converted from analog to digital form and encoded as a series of pressure pulses that are detected and decoded at the surface. The pressure pulses may be in the form of overpressure or underpressure anomalies introduced respectively by either a relief valve "shorting" the mud circulation or a check valve "choking" it. Coded mud-pressure pulses are not the only means available for telemetry, however. Other methods, either in use or under experimentation, include: electromagnetic e-mode (electric current) or h-mode (magnetic field); acoustic telemetry through drillpipe and/or tubing in straight hole, or through the earth by seismic waves; hardwire systems; systems with self-energizing repeaters; and hybrid systems that combine various transmissions methods.

The surface equipment consists of a decoder of the mud pulses (or other parameter depending on the telemetry system in use) together with signal-processing hardware and software which together produce the output that the drilling engineer wishes to use. Output may be in the form of a visual display (either on the rig floor or at a remote site) or as a hard-copy listing, or log, of the parameters recorded. Data may also be recorded on magnetic tape for future use.

In most systems the transmission of data to the surface is selective. For example, a measurement of hole deviation and azimuth may require that the drilling process be temporarily suspended and the drillstring held motionless for a short period. Readings are then accumulated in a buffer and transmitted to the surface only when mud circulation recommences.

Figure 3.1 illustrates an MWD downhole assembly with its mud-pulse transmitter, turbine generator, and sensor sub. Figure 3.2 shows a data-transmission schematic for MWD. Typically, each measurement or "word" is transmitted as part of a data frame, which, in turn, consists of a synchronization word and 15 measurement

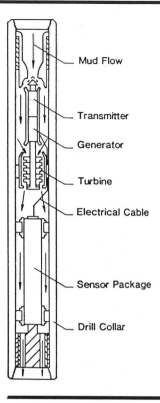

FIGURE 3.1 *Typical MWD Downhole Assembly. Reprinted by permission of the SPE-AIME from Grosso et al. 1983, fig. 1, p. 900. © 1983 SPE-AIME.*

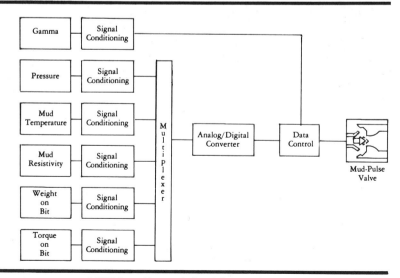

FIGURE 3.2 *MWD Data Transmission Schematic. Reprinted by permission of the SPE-AIME from Grosso et al. 1983, fig. 2, p. 901. © 1983 SPE-AIME.*

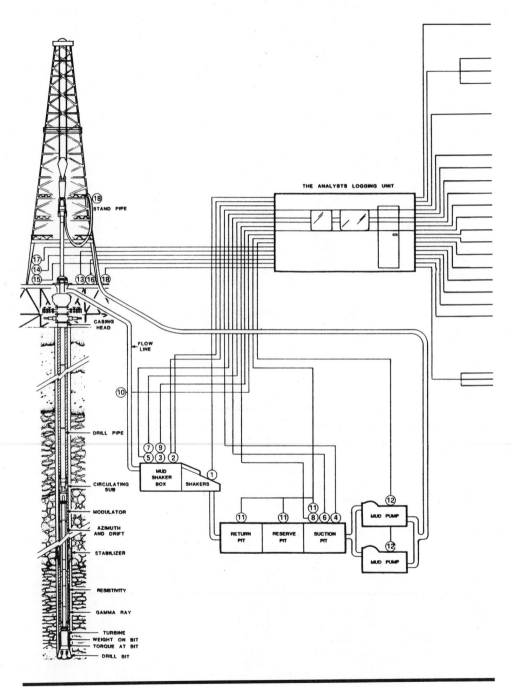

FIGURE 3.3 *MWD Schematic. Courtesy Anadrill/Schlumberger.*

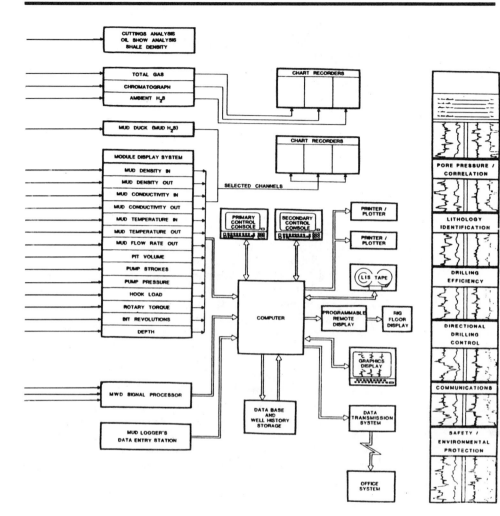

Sensor Identification

1. Sample Catching	7. Conductivity Probe	13. Pump Pressure Sensor
2. Gas Trap	8. Temperature Probe	14. Hook Load Sensor
3. Mud Duck Probe	9. Temperature Probe	15. Torque Sensor
4. Density Probe	10. Flow Detector	16. Bit Revolutions Sensor
5. Density Probe	11. Pit Level Sensors	17. Depth Encoder
6. Conductivity Probe	12. Pump Stroke Counters	18. MWD Pressure Sensors

words. Some measurements are transmitted more than once in each frame. Current telemetry systems are capable of transmitting a complete frame in a matter of one or two minutes. The actual sampling rate in terms of measurements per unit of depth is inversely proportional to the rate of penetration (unit of depth/per unit of time). Figure 3.3 shows a complete MWD logging system and integrates surface and downhole sensors, the telemetry system, the surface hardware and software for computer processing of the data, and the final product in the form of a log.

LOGS PROVIDED BY MWD

Figure 3.4 shows an MWD log on which is displayed gamma ray, short normal resistivity, annular temperature, downhole weight on bit, surface weight on bit, and computed directional data (drift and azimuth). Figure 3.5 shows a comparison between an MWD-generated computed directional survey and a multishot run in the same well. Figure 3.6 illustrates a directional survey listing corresponding to the plane view shown in figure 3.5. Figure 3.7 shows a comparison between an MWD gamma ray log and a wireline gamma ray log. Obviously, the MWD log is more than adequate for formation top picking and basic lithology differentiation.

THE FUTURE OF MWD

It seems safe to assume that in the future the use of MWD for formation evaluation will become the norm. Technological advances will allow more wireline-type logging measurements to be made by MWD subs. Telemetry will improve to the point where a full suite of logs will be available effectively in real time. This in turn will allow almost instant evaluation of a formation within minutes of drilling. Such advances are to be encouraged since any improvement in one formation evaluation technique helps all the others.

BIBLIOGRAPHY

The Analysts: "Measurements While Drilling (M.W.D.) Technical Specifications," Schlumberger (no date).

Grosso, Donald S., Raynal, Jean C., and Radar, Dennis: "Report on MWD Experimental Downhole Sensors," *J. Pet. Tech.* (May 1983).

Kamp, Anthony W.: "Downhole Telemetry from the User's Point of View," *J. Pet. Tech.* (October 1983).

Roberts, A., Newton, R., and Stone, F.: "MWD Field Use and Results in the Gulf of Mexico," SPE paper 11226, presented at the 57th Annual Technical Conference and Exhibition, New Orleans, Sept. 26-19, 1982.

Tanguy, D. R., and Zoeller, W. A.: "Applications of Measurements While Drilling," SPE paper 10324, presented at the 56th Annual Technical Conference and Exhibition, San Antonio, Oct. 5-7, 1981.

Zoeller, W. A.: "Pore Pressure Detection from the MWD Gamma Ray," SPE paper 12166, presented at the 58th Annual Technical Conference and Exhibition, San Francisco, Oct. 5-8, 1983.

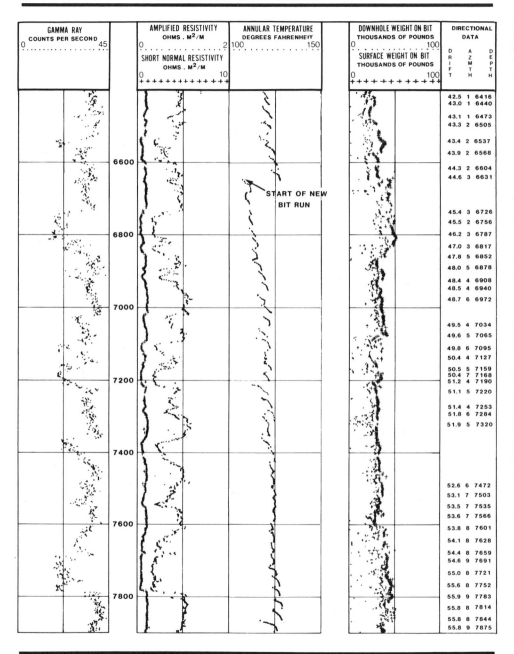

FIGURE 3.4 *An MWD Log. Reprinted by permission of the SPE-AIME from Tanguy and Zoeller 1981, fig. 2.* © *1981 SPE-AIME.*

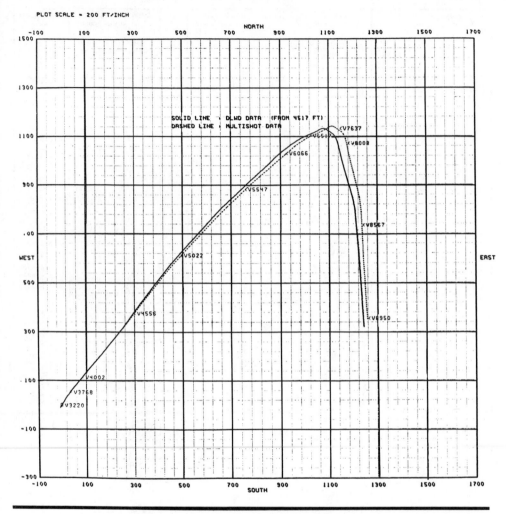

FIGURE 3.5 *Comparison of an MWD Directional Survey with a Multishot Directional Run. Courtesy EXLOG.*

	READINGS				ANALYSIS (CONFIDENCE LEVEL = 99.0%)							
SVY #	DEPTH	COURSE LENGTH	INCL. ANGLE	AZIMUTH ANGLE	DOGLEG /100	VERTICAL DEPTH	POSITION NORTH	POSITION EAST	VERT	MAJOR	MINOR	AXIS
88	5425.0	30.0	37.80	41.80	2.34	5180.4	696	559	1.3	10.0	9.5	343
89	5457.5	32.5	37.67	43.50	3.22	5206.1	710	573	1.3	10.0	9.5	344
90	5478.0	20.5	36.53	43.50	5.52*	5222.5	719	581	1.3	10.0	9.6	344
91	5509.0	31.0	35.73	44.00	2.75	5247.5	732	594	1.3	10.0	9.6	345
92	5540.0	31.0	35.72	41.80	4.14	5272.7	746	606	1.3	10.0	9.6	346
93	5571.0	31.0	35.08	42.40	2.33	5298.0	759	618	1.3	10.1	9.6	347
94	5589.0	18.0	35.08	42.40	0.00	5312.7	767	625	1.3	10.1	9.6	347
95	5605.0	16.0	34.35	42.80	4.79	5325.8	773	631	1.3	10.1	9.6	347
96	5631.0	26.0	33.90	43.30	2.03	5347.4	784	641	1.3	10.1	9.6	348
97	5695.0	64.0	32.17	46.30	3.72	5401.0	809	666	1.4	10.1	9.6	351
98	5735.0	40.0	31.28	47.40	2.64	5435.0	823	681	1.4	10.1	9.7	352
99	5756.0	21.0	30.97	46.60	2.47	5453.0	831	689	1.4	10.1	9.7	353
100	5821.0	65.0	28.87	45.10	3.43	5509.3	853	712	1.4	10.1	9.7	356
101	5852.0	31.0	27.88	47.20	4.52	5536.6	863	723	1.4	10.1	9.7	357
102	5949.0	97.0	26.52	45.40	1.64	5622.9	894	755	1.4	10.2	9.8	3
103	6032.7	83.7	24.65	46.10	2.26	5698.4	919	781	1.5	10.2	9.8	6
104	6090.0	57.3	23.63	48.90	2.67	5750.7	935	798	1.5	10.2	9.8	8
105	6117.0	27.0	22.80	47.00	4.15	5775.5	942	806	1.5	10.2	9.8	8
106	6150.2	33.2	22.45	46.10	1.48	5806.1	951	815	1.5	10.2	9.8	9
107	6216.2	66.0	21.27	46.80	1.83	5867.4	968	833	1.5	10.2	9.8	10
108	6241.8	25.6	20.93	46.00	1.71	5891.2	974	840	1.5	10.2	9.8	10
109	6272.0	30.2	19.68	48.20	4.85	5919.6	981	848	1.5	10.3	9.8	11
110	6302.0	30.0	19.78	44.80	3.84	5947.8	988	855	1.5	10.3	9.8	11
111	6337.0	35.0	19.30	46.10	1.85	5980.8	997	863	1.5	10.3	9.8	11
112	6402.0	65.0	18.28	41.80	2.64	6042.3	1012	878	1.5	10.3	9.8	12
113	6455.8	53.8	17.17	49.50	4.82	6093.6	1023	889	1.5	10.3	9.8	13
114	6553.3	97.5	15.87	50.10	1.34	6187.0	1041	911	1.5	10.3	9.8	14
115	6600.3	47.0	14.92	59.50	5.67*	6232.3	1048	921	1.5	10.3	9.8	14
116	6678.3	78.0	14.62	47.20	4.03	6307.8	1060	937	1.5	10.3	9.8	15
117	6708.3	30.0	14.78	53.50	5.35*	6336.8	1065	942	1.5	10.3	9.8	15
118	6740.3	32.0	14.32	55.90	2.38	6367.8	1069	949	1.5	10.3	9.9	15
119	6771.3	31.0	13.85	59.50	3.20	6397.8	1073	955	1.5	10.3	9.9	15
120	6838.3	67.0	13.18	57.90	1.14	6463.0	1082	969	1.5	10.4	9.9	16
121	6893.9	55.6	12.63	57.60	.99	6517.2	1088	979	1.5	10.4	9.9	16
122	6926.7	32.8	12.87	63.30	3.90	6549.2	1092	986	1.5	10.4	9.9	16
123	6989.4	62.7	12.55	67.80	1.66	6610.3	1097	998	1.6	10.4	9.9	16
124	7020.0	30.6	12.58	66.30	1.06	6640.2	1100	1004	1.6	10.4	9.9	16
125	7054.0	34.0	12.60	65.70	.35	6673.4	1103	1011	1.6	10.4	9.9	16
126	7084.4	30.4	12.40	67.10	1.18	6703.1	1106	1017	1.6	10.4	9.9	16
127	7114.2	29.8	11.50	66.90	3.02	6732.2	1108	1023	1.6	10.4	9.9	16
128	7153.9	39.7	10.90	66.90	1.51	6771.1	1111	1030	1.6	10.4	9.9	16
129	7184.8	30.9	10.00	68.60	3.08	6801.5	1113	1035	1.6	10.4	9.9	16
130	7240.1	55.3	7.90	68.90	3.80	6856.2	1116	1043	1.6	10.4	9.9	16
131	7272.7	32.6	7.40	63.00	2.85	6888.5	1118	1047	1.6	10.4	9.9	16
132	7285.8	13.1	7.00	66.90	4.81	6901.5	1119	1048	1.6	10.4	9.9	16
133	7316.9	31.1	7.40	58.70	3.54	6932.3	1121	1052	1.6	10.4	9.9	16
134	7346.0	29.1	5.90	58.40	5.15*	6961.2	1122	1055	1.6	10.4	9.9	16

FIGURE 3.6 *MWD Data Listing for Directional Survey. Courtesy EXLOG.*

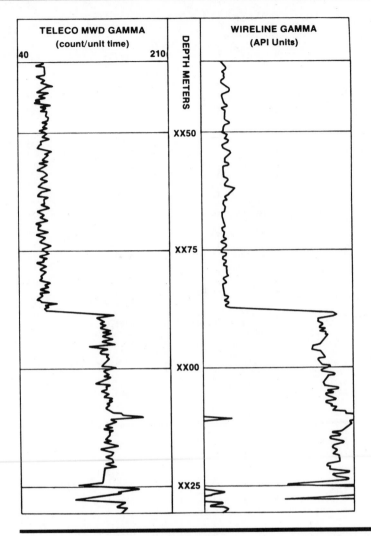

FIGURE 3.7 *Comparison between MWD and Wireline Gamma Ray Logs. Courtesy Teleco Oilfield Services Inc.*

CHAPTER FOUR

CORING

OBJECTIVES

The objectives of coring are to bring a sample of the formation and its pore fluids to the surface in an unaltered state, to preserve the sample, and then transport it to a laboratory for analysis. These objectives are hard to meet since the very act of cutting a core will, to some extent, alter both the properties of the rock itself and the saturation of the fluids in its pores.

A number of techniques exist for minimizing damage to formation samples. These will be discussed in this chapter. Other techniques, aimed at restoring the original state of the formation sample when it was at reservoir conditions, may be brought into play at the time the core is analyzed. These will be discussed in chapter 26, which deals with core analysis.

WIRELINE CORING

Two methods of retrieving formation samples with wireline tools are currently in use: the conventional sidewall core gun and a relatively new device, the core plugger.

Sidewall Cores

Figure 4.1 illustrates a sidewall core gun. The body of the gun carries a number of hollow steel bullets that can be fired selectively into the formation by means of explosive charges. Once lodged in the formation, the bullet can be retrieved by means of flexible steel wires attached to the bullet. Raising the gun in the borehole usually provides sufficient tension on the wires to dislodge the bullet. Note that the gun is equipped with an SP electrode. This allows the tool to be placed at the correct depth in the well prior to sampling by correlation of a short section of SP log with other open-hole logs already run.

These guns come in a variety of shapes and sizes. On average, they are capable of retrieving 60 samples in one trip in the hole. The diameter of the core barrel may be anywhere between ¾ and 1⅛ in. The length of the core retrieved is a function of many variables. The strength of the explosive charge used, the type of core barrel selected, and the hardness of the formation determine the length of the recovered sample; it may be as long as 2 in. or as short as nothing at all. Occasionally, the retainer wires used to retrieve the core barrel will break and the bullet will be lost in the hole. Figure 4.2 is a closeup view of a core gun.

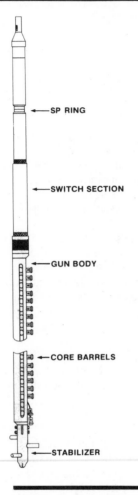

FIGURE 4.1 *Conventional Sidewall Core Gun. Courtesy WELEX, a Halliburton Company.*

When all the samples have been collected, the gun is raised to the surface and each core plug is stored in a glass jar marked with the well name and the depth from which it was cut. Subsequently, these cores may be analyzed for porosity, permeability, and hydrocarbon content. There are obvious limitations to the amount of data that can be obtained from sidewall cores. In the first place, the sample is taken from a part of the formation that has been flushed with mud filtrate. Second, the act of explosively firing the core barrel into the formation may induce local fracturing. Last, the trip up the hole to the surface involves a considerable amount of flushing through the mud column. Despite these drawbacks, sidewall cores are still good quick-look indicators of formation properties. It is normal practice to inspect

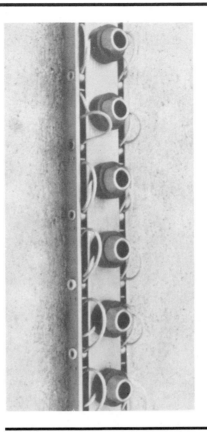

FIGURE 4.2 *Closeup View of Core Gun. Courtesy Dresser Atlas, Dresser Industries, Inc.*

these cores at the well site for hydrocarbon odor, fluorescence, stain, and cut, when a mud logging unit or geologist's dog house is available.

Core Plugger

The core plugger uses a motorized circular bit to bore physically into the wall of the formation in order to retrieve its samples. It is presently capable of cutting up to 12 core samples in one run in the hole. With this tool, core size is $^{15}/_{16}$ in. in diameter and 1¾ in. long. Each core takes about 5 minutes to cut. This device works better than the conventional sidewall core gun in consolidated formations and causes no physical damage to the sample.

CONVENTIONAL CORING

When the decision is reached to cut a conventional core (at a drilling break, a planned depth reached, etc.), the drillpipe is removed from the hole and dressed with a hollow core bit and a hollow barrel equipped with a nonrotating inner barrel as shown in figure 4.3.

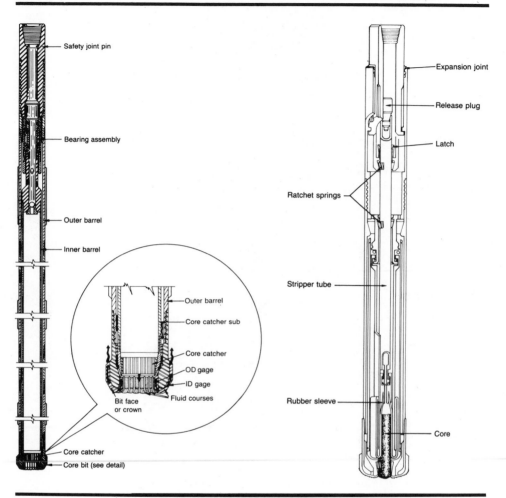

FIGURE 4.3 *Conventional Coring Devices. Reprinted by permission of* World Oil *from Park, April, 1985.*

Cores can be from 10 to 60 ft long and have a diameter of 1⅞ to 4½ in. If unconsolidated formations are to be cored, a rubber sleeve is used to hold the friable material more securely.

For some applications a special *pressure core* can be cut (see fig. 4.4). The core barrel is designed so that after the core is cut it is maintained at original reservoir pressure until it arrives at the laboratory for analysis. Normally, pressure cores are frozen at the wellsite for transportation. For other applications, *oriented cores* may be cut. The original orientation of the core relative to north is maintained by use of a special key that cuts a groove in the core. This kind of core is valuable in fractured reservoirs, where fracture orientation is of interest. Both bed dip and azimuth may be deduced from such a core.

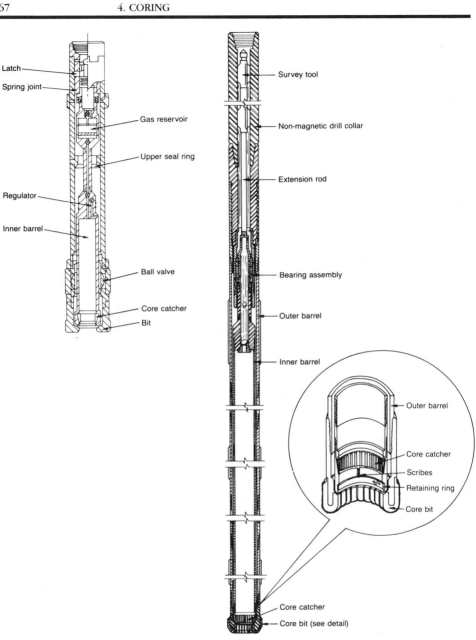

FIGURE 4.4 *Pressure Core Barrel. Reprinted by permission of* World Oil *from Park, April, 1985.*

TABLE 4.1 *Effects of Coring Fluids on Reservoir Fluid Saturation*

| Coring Fluid | Filtrate | Effect on Core Saturations | |
		Water	Hydrocarbon
Water based	Water	Increased	Decreased
Oil based	Oil	No change	Replaced
Inverted oil emulsion	Oil	No change	Replaced
Oil emulsion	Water	Increased	Decreased
Gas	Gas	No change	Replaced
Air	Uncertain	Uncertain	Decreased

TABLE 4.2 *Coring Fluids Required to Meet the Core Objectives of Analysis.*

Coring Fluid	Measurements Suited
Water based	ϕ, k, lithology, k_{rg}/k_{ro}, cap pressure
Oil based	S_w, relative permeability, water/oil cap pressure
Inverted oil emulsion	S_w
Gas	ϕ, k, lithology, S_w, wettability, relative permeability, cap pressure
Water	Wettability, k_{rw}/k_{ro}
Nonoxidized crude	S_w, wettability, relative permeability cap pressure

CORING FLUIDS

The coring fluid used will either increase or decrease the initial saturations of gas, oil, and water in the formation. Water-based muds, for example, produce a water filtrate that invades the core, displacing hydrocarbons. Oil-based muds having an oil filtrate may replace reservoir oil by the flushing mechanism but do not substantially alter the original oil saturation. Table 4.1 summarizes the effects of various coring fluids. The fact that fluid saturations are changed by the coring fluid is not always of consequence—for example, if the core is being cut only for porosity and permeability estimates the change of water saturation is not important. Table 4.2 lists the most suitable coring fluid for a given objective.

Saturation changes in cores can take place at three different times: at the time the core is cut, as it travels in the core barrel from the reservoir to the surface, and at the surface during transportation and storage. It is worthwhile to review the causes and approximate magnitudes of these saturation changes. Table 4.3 documents ten different cases covering a range of coring fluids and initial reservoir saturations. Once the core is retrieved at surface, precautions must be taken to ensure that no further saturation changes occur.

TABLE 4.3 *Changes in Saturation as a Function of Reservoir Content and Coring Fluid*

Case	Formation Content	Coring Fluid	Initial Saturations		Change Due to	Saturations in Core Barrel	Change Due to	Saturation at Surface	
1	Oil/water, oil productive	Water based, badly flushed	O	70	Flushing	30	Shrinkage	12	O
			G	0	Expansion	0	Expansion	40	G
			W	30	Expulsion	70	Expulsion	48	W
2	Oil/water, oil productive	Water based, unflushed	O	70	—	70	Shrinkage	20	O
			G	0	—	0	Expansion	50	G
			W	30	—	30	—	30	W
3	Oil/water, oil productive	Oil based, badly flushed	O	70	Flushing	70	Shrinkage	40	O
			G	0	—	0	Expansion	30	G
			W	30	—	30	—	30	W
4	Gas/water, gas productive	Water based, badly flushed	O	0	—	0	Condensation	1	O
			G	70	Flushing	30	Expansion	49	G
			W	30	Expulsion	70	Expulsion	50	W
5	Gas/water, gas productive	Water based, unflushed	O	0	—	0	Condensation	2	O
			G	70	—	70	—	68	G
			W	30	—	30	—	30	W
6	Gas/water, gas productive	Oil based, badly flushed	O	0	Invasion	50	Expulsion	40	O
			G	70	Flushing	20	Expansion	30	G
			W	30	—	30	—	30	W
7	Water, water productive	Water based	O	0	—	0	—	0	O
			G	0	—	0	Expansion	10	G
			W	100	Flushing	100	Expulsion	90	W
8	Water, water productive	Oil based, badly flushed	O	0	Invasion	40	Shrinkage	35	O
			G	0	—	0	Expansion	10	G
			W	100	Flushing	60	Expulsion	55	W
9	Oil/water, depleted reservoir	Water based, badly flushed	O	30	—	30	Shrinkage	12	O
			G	0	—	0	Expansion	40	G
			W	70	Flushing	70	Expulsion	48	W
10	Oil/water, depleted reservoir	Water based, badly flushed	O	55	Flushing	25	Shrinkage	22	O
			G	15	Flushing	5	Expansion	10	G
			W	30	Invasion	70	Expulsion	68	W

SPECIAL CORE HANDLING

On surface, the core is exposed to the atmosphere, and evaporation of water and light hydrocarbons can occur. Thus, it is normal practice to store the core immediately in a protective environment. Storage methods include:

Freezing the core with dry ice
Wrapping cores in plastic bags
Wrapping cores in plastic wrap and/or aluminum foil and sealing them with paraffin wax
Submergence under deaerate water
Submergence under nonoxidizing crude

Rubber sleeve cores are normally left in their sleeves although they may be cut into shorter lengths for ease of handling.

SUMMARY

Cores can be cut in a variety of ways. Great care should be taken to plan the coring fluid and in the transportation and storage of cores in order to preserve the fluid saturation of interest. The coring fluid used and/or the initial reservoir conditions may make the saturation of oil, water, or gas in the core arriving at the lab considerably higher, or lower, than it was in the formation. With poetic license from Ovid (or was it Livy?) *cave corum!*

BIBLIOGRAPHY

Dresser Atlas: "Services Catalog," 1984.
Hyland, C. R.: "Pressure Coring—An Oilfield Tool," SPE paper 12093, presented at the 58th Annual Technical Conference and Exhibition, San Francisco, Oct. 5–8, 1983.
Park, A.: "Coring," Part 2, *World Oil* (April 1985) **200**, 83–92.
Welex: "Open Hole Services," G-06003 (1983).

CHAPTER FIVE

WIRELINE LOGGING OPERATIONS

RIGGING UP TO RUN A LOG

Figure 5.1 shows a typical setup for a logging job. A logging truck is anchored about 100 to 200 ft from the well. Two sheaves are mounted in the derrick with one suspended from the crown block and the other chained down near the rotary table. The logging cable from the truck winch is then passed over the sheaves, attached to the logging-tool string and lowered into the hole. A more detailed diagram of this hookup is shown in figure 5.2.

Two mechanical details of this method of rigging up are worth noting. Between the top sheave and the elevators there is a tension device (fig. 5.3) that measures the strain on the logging cable and displays it in the logging truck. The tension on the elevators is twice the tension on the cable. The elevators should be securely locked and the traveling block braked and chained.

The tiedown chain for the lower sheave is also of great importance. When it breaks or comes untied, the cable quite probably will break and the sheave can be catapulted several hundred feet away (fig. 5.4).

QUESTION # 5.1

If the tension in the logging cable is 5000 lb, estimate the tension in the tiedown chain if it forms an angle of 45° with each of the two cable directions on either side of the sheave.

LOGGING TRUCKS

Logging service companies offer a variety of logging units. Each type of unit has the following components:

Logging cable
Winch to raise and lower the cable in the well
Self-contained 120-V AC generator
Set of surface control panels
Set of downhole tools (sondes and cartridges)
Recording mechanism (tape and/or film)

Figure 5.5 shows a cutaway of a typical logging truck. Land units are mounted on a specially adapted truck chassis reinforced to bear the load of a full winch of cable (up to 30,000 ft long). The instrument and recorder cabs are usually cramped, noisy, too hot or too cold,

FIGURE 5.1 *Setup for a Logging Job. Courtesy Schlumberger Well Services.*

and/or filled with ammonia fumes from an ozalid copier. Although not designed to do so, this keeps the merely curious onlooker away. Do not be deterred. If you have good reason to be in the logging unit, stay there and do what you must.

Offshore units are mounted on skids and bolted (or welded) to the deck of the drilling barge, vessel, or platform. Some logging units can be disassembled into many small components and flown into remote jungles suspended under helicopters. Nevertheless, all logging units are basically the same.

Good mechanical maintenance is required to avoid problems during logging operations. An engine that stops during logging operations will leave the logging tool dangling in the well. If it stays there too long without moving, it may become stuck. In that case, the traveling block from which the top sheave is suspended can be raised

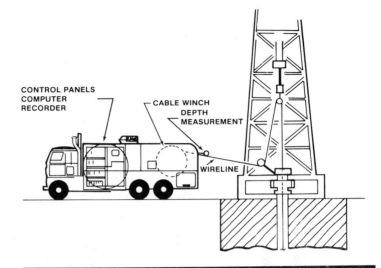

FIGURE 5.2 *Details of Wireline Logging Setup. Courtesy Gearhart Industries, Inc.*

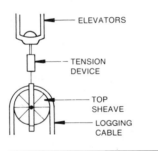

FIGURE 5.3 *Top Sheave and Elevator Arrangement for a Logging Job.*

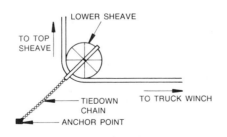

FIGURE 5.4 *Lower Sheave Tiedown Arrangement.*

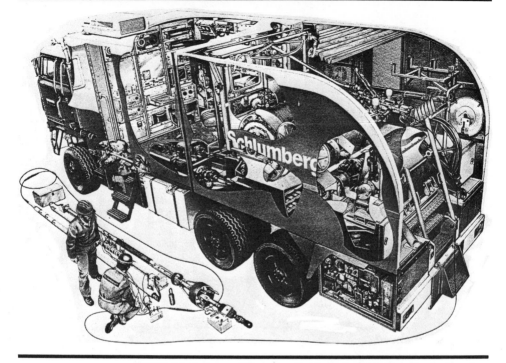

FIGURE 5.5 *Cutaway View of a Computerized Logging Truck. Courtesy Schlumberger Well Services.*

and lowered so that the tool is kept in motion. Another item that can cause grief is the 120-V generator. If it fails, all logging panels and tools go dead. Rig power may then be substituted.

LOGGING CABLES

Modern logging cables are of two types: monoconductor and multiconductor. Monoconductor cables are used for completion services—such as shooting perforating guns, setting wireline packers and plugs—and for production logging surveys—such as flowmeter and temperature logs in producing wells. Multiconductor cables are used by most logging-service companies for recording open-hole surveys.

The monoconductor cables are approximately ½ or ¼ in. in diameter—the smaller cable is used for jobs where high wellhead pressure is encountered. The multiconductor cables contain six or seven individual insulated conductors in the core. The outer sheath is composed of two counterwound layers of steel wire. Such a cable has a breaking strength of 14,000 to 18,000 lb and weighs 300 to 400 lb/per 1000 ft. It is quite elastic and has a stretch coefficient of around 10^{-6} ft/ft/lb.

It is worth noting that fiber optics is now being field tested as an alternative to conventional electric cables.

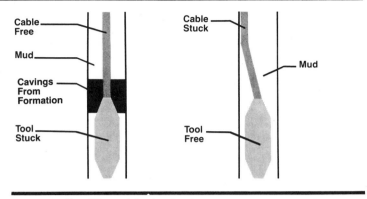

FIGURE 5.6 *Two Ways of Getting Stuck.*

The Head and the Weakpoint

The cable ends at the logging *head.* The head anchors the cable and attaches to the logging tool by means of a threaded ring. Thus, the head provides the electrical connection between the individual cable conductors and the various pins in the top of the tool and the mechanical connection. Built into the head is a *weakpoint,* which is a short length of aircraft cable designed to break at some given tension. The standard breaking point is 6000 lb, but deephole weakpoints break at 3500 lb. The weakpoint is necessary to provide a means to free the cable from the tool in case the tool has become irrevocably stuck in the wellbore. To explain fully the function of the weakpoint, a discussion of the various ways the tool can become stuck is in order.

Getting Stuck

There are two ways of getting stuck. Either the tool will stick and the cable in the hole above the tool will remain free, or the tool will remain free and the cable will become *key seated* further up the hole above the tool. Figure 5.6 illustrates the difference. Once the tool or cable is stuck, the first thing to do is to determine whether it is the tool or the cable that is stuck. The standard procedure is to put normal logging tension on the cable and let it sit for a few minutes while the following data are gathered:

a. the present depth of the tool
b. the surface tension that was on the cable just before it got stuck
c. the cable type, size, etc.
d. the cable-head weakpoint rating

Once these data have been gathered, make certain that the tiedown chain on the lower sheave is secure. Then mark the cable (using chalk or friction tape) at the rotary table. Securely position a T-bar clamp around the cable just above the rotary table. (If, in the subsequent tug of war, the cable breaks at the top sheave, this clamp will hold the cable at the surface and prevent it from snaking down the hole on top

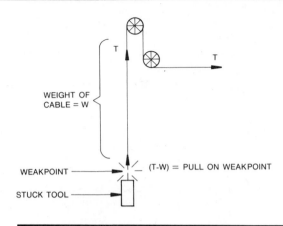

FIGURE 5.7 *Breaking the Cable at the Weakpoint.*

of the tool.) Next apply 1000 lb over the tension already on the cable and measure the distance the cable mark has moved. This will be the stretch produced in the elastic cable as a result of 1000 lb extra tension. Knowing this distance, the length of free cable can be estimated from a stretch chart or from knowledge of the stretch coefficient. If the length of free cable so determined proves to be the present logging depth, then the tool is evidently stuck and the cable is free. On the other hand, if the length of free cable is less than the present logging depth, the cable itself must be stuck higher up the hole. If the cable proves to be stuck, it is counterproductive to apply any further tension since this will merely compound the problem by aggravating the differential pressure sticking of the cable.

If it is the tool that is stuck, pulling on the cable will have one of three results. The tool will pop free, the weakpoint will break (leaving the tool in the hole but saving the cable), or the cable will break at the point of maximum tension at the top sheave. Of the three, the first is to be preferred. Of the other two, the breaking of the weakpoint is preferred. But which will occur first? Will the cable part at the surface before the weakpoint breaks? Figure 5.7 will help to explain the tensions involved.

If the cable weakpoint is 6000 lb, and 10,000 ft of cable weighing 3500 lb is in the hole, it will be necessary to apply 9500 lb (6000 + 3500) at the surface in order to apply 6000 lb at the weakpoint. At greater depths, the surface tension required to break the weakpoint will be even higher and may exceed the breaking strength (14,000 to 18,000 lb) of the cable itself. Thus, for deep holes a weakpoint of only 3500 lb is preferred.

Differential pressure sticking of the cable is caused when the cable cuts through the mudcake. One side of the cable is exposed to formation pressure while the other side is exposed to the hydrostatic mud column. When the hydrostatic pressure is significantly higher than the formation pressure, the cable is forced against the formation, and the resulting friction stops any further cable movement (fig. 5.8).

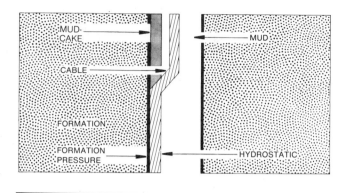

FIGURE 5.8 *Differential Pressure Sticking of the Cable.*

An example will illustrate this phenomenon: Mud weight is 10 lb/gal. At 5000 ft, the hydrostatic pressure is 2600 psi, but formation pressure is 2250 psi; there is thus a differential pressure of 350 psi. If the diameter of the cable is ½ in., and only 10 ft of the cable is stuck, the force on the cable will be 21,000 lb (350 psi × 120 in. × ½ in.). If the coefficient of friction is 0.1, the pull required to overcome the friction will be 2100 lb.

Fishing Alternatives

There are several alternatives available for recovering a stuck tool and/or cable:

1. Leave the cable attached to the tool and run a side-door overshot.
2. Use the cut-and-thread technique.
3. Break the weakpoint; recover the cable; and fish for the logging tool with the drillpipe, or push it to the bottom of the hole and mill it up.

Figure 5.9 illustrates these three methods. The side-door overshot is not recommended at depths greater than 3000 ft. Historically, the cut-and-thread technique is the surest way to recover a stuck logging tool.

LOGGING TOOLS

Logging tools are cylindrical tubes containing sensors and associated electronics. These tubes can be attached to the logging cable at the logging head. Although there are large variations in size and shape, a typical logging tool is 3⅝ in. in diameter and 10 to 30 ft long. They are built to withstand pressures up to 20,000 psi and temperatures of 300 to 400°F. The internal sensors and electronics are ruggedly built to withstand physical abuse. Modern tools are modularized to allow combination tool strings. By appropriate mixing and matching, various logging sensors can be connected together. This technique has obvious limitations; for example, very long tools are difficult to handle and the conductors in the cable have limited information-transmitting power.

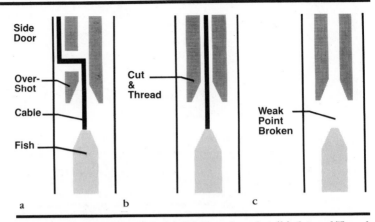

FIGURE 5.9 *Fishing Alternatives.* (a) *Side-Door Overshot.* (b) *Cut-and-Thread.* (c) *Fishing with Drillpipe.*

Because the multiple sensors of logging tools are located at different points along the axis of the tool, their respective measurements have to be memorized and placed on a common depth reference. Thus, the signal from the sensor highest on the tool must be "remembered" until the signal from the lowest sensor arrives (see figs. 5.10 and 5.11). In figure 5.11 the reference point for the survey is the sensor producing the curve marked A. Higher up on the tool are two more sensors, B and C. If recorded unmemorized, the formations appear off depth (e.g., Sand 2 appears at different depths for each of the sensors). Note that Sand 1 can only be logged by Sensor A. It is important, therefore, to make certain that all curves recorded simultaneously are on-depth on the log.

Another depth problem arises when several surveys are recorded on different trips into the hole. Unless care is taken, these surveys may not be on-depth with each other. The only method of assuring good depth control is to insist on a repeat section that passes a good marker bed. Each subsequent log should be placed on-depth using this repeat section as a depth reference before the main logging run is made.

THE BOREHOLE ENVIRONMENT

As the drill bit penetrates a permeable formation, an invasion process begins. Since the pressure in the mud column exceeds formation pressure, fluid from the mud will move into the formation (provided it is porous and permeable) and deposit a mudcake on the borehole wall. Figure 5.12 illustrates the process and the names used in logging literature for the various zones that surround the borehole.

It is important to distinguish between the resistivity of the fluid within the pore space and the resistivity of the rock/fluid system itself. The terms used in table 5.1 should be well known to everyone involved in the evaluation of well logs. The flushed (invaded) zone is

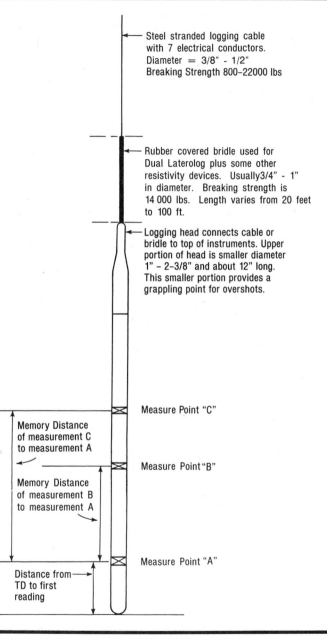

FIGURE 5.10 *Measure Points and Memorization Distances.*

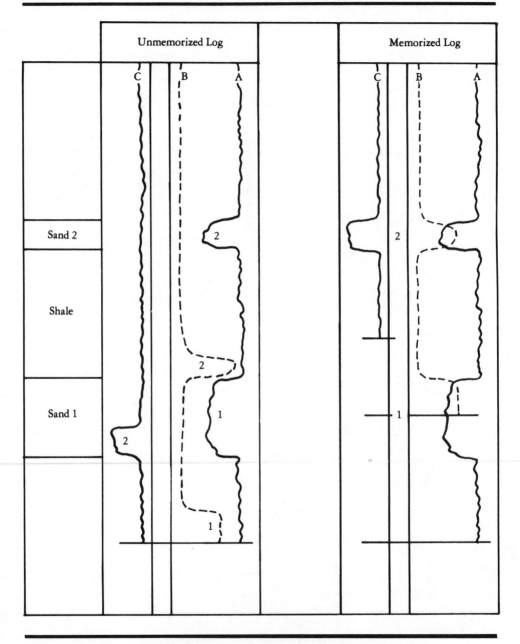

FIGURE 5.11 *Memorized and Unmemorized Logs.*

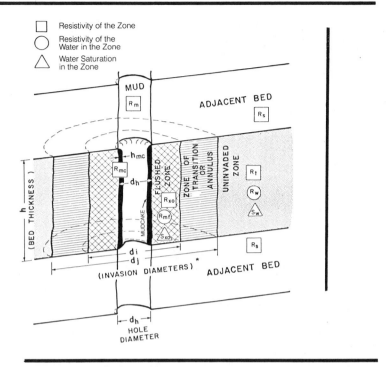

FIGURE 5.12 *Symbols Used in Log Interpretation. Courtesy Schlumberger Well Services.*

TABLE 5.1 *Nomenclature for Zones in and around the Borehole*

Name of Zone	Dimension	Fluid Content	Fluid/Water Resistivity	Rock Resistivity
Bed	h	—	—	—
Adjacent bed	—	—	—	R_s
Mud	d_h	Mud	R_m	—
Mudcake	h_{mc}	—	—	R_{mc}
Flushed or invaded	d_i	Mud filtrate + residual oil	R_{mf}	R_{xo}
Transition	d_j	Mixed + hydrocarbons	—	—
Uninvaded (undisturbed)		Connate Water + hydrocarbons	R_w	R_t

Source: Reprinted by permission of IHRDC Press from Bateman 1985.

important because it affects the readings of some logging tools and because it forms a reservoir of mud filtrate that will be recovered on a drillstem test before formation fluids are recovered.

QUESTION #5.2
A 30% porous sandstone bed is 100 ft thick. It is 100% saturated with mud filtrate ($S_{xo} = 100\%$) to an invasion diameter of 40 in. The hole diameter is 8½ in. Estimate the volume of filtrate that will be recovered from this sandstone before connate water or oil begins to be recovered.

QUESTION #5.3
How many linear feet of 3½ in. drillpipe will the volume calculated in question 5.2 occupy?
(Note: for this size drillpipe there are 27.1 linear feet/cubic foot.)

The invasion process may change the resistivity profile around the borehole. Whether resistivity increases or decreases with distance from the borehole wall depends on the type of mud used (oil-based or water-based) and the relative values of R_{mf} and R_w. A schematic illustration of what may be expected for a number of cases is shown in figure 5.13.

CHOOSING A LOGGING SUITE
Choice of a logging suite will be influenced by:

Type of well—wildcat or development well
Hole conditions—depth, deviation from vertical, hole size, mud type
Formation-fluid content—fresh or salt connate water
Economics—rig time, logging dollars, etc.

Each tool is designed for a specific set of conditions. Outside these limitations, the tool fails to provide the required measurements and its use is thus discouraged.

Depth, Pressure, and Temperature
The majority of logging tools are rated at 20,000 psi and 350°F. This is adequate for most holes to be logged. For higher temperatures, special tools are available from the logging-service companies.

Hole Size
Six inches is the standard minimum hole size for correct and safe operation of normal logging tools. A limited number of slim-line, small-diameter tools are available for smaller holes.
Maximum hole size is hard to define. Most pad-contact tools (compensated neutron logs, compensated formation density logs,

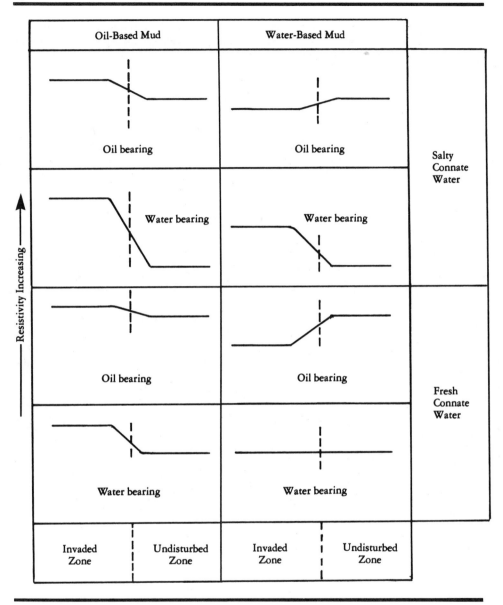

FIGURE 5.13 *Radial Resistivity Profile as a Function of Mud Type and Formation Content.*

microfocused logs, dipmeters, etc.) have spring-loaded, hydraulically operated arms that push the relevant sensor against the borehole wall. The arms will open to about 20 in., although this limit varies a little from tool to tool. If holes are deviated, good pad contact may still be obtained since the tool will "lean" on the low side of the hole; although this cannot be guaranteed. Running a pad-contact tool in a hole with a diameter of more than 20 in. is risky because the pad may not be able to make contact with the wall of the wellbore.

Resistivity devices such as induction logs and laterologs suffer in a progressive fashion as the borehole gets bigger. Theoretically, there is no fixed limit to the hole size, but practically there is a limit because borehole corrections to the raw data become so large that nothing useful can be determined from the logs. Logging of large-diameter surface holes may thus cause a problem and require logging in a specially drilled medium-sized hole, which is subsequently underreamed to the desired gauge.

Hole Deviation

In today's offshore environment, the deviated hole is the norm rather than the exception. The greater the angle of deviation of the hole from vertical, the greater the difficulties of physically getting a logging tool to the bottom of the hole. In general, hole deviation above 40° from vertical causes problems. A number of techniques have been tried to get logging tools safely to bottom; these include:

Keeping the open-hole section as short as possible
Removal of centralizers and standoff pads
Use of a hole finder—a rubber snout on the bottom of the logging-
    tool string
Use of logging tools specially adapted to be run to the bottom of the
    hole on drillpipe.

In difficult situations, the hole may have to be logged through open-ended drillpipe with a slim logging tool physically pumped down by mud circulation. Holes with deviations as high as 65° have been logged with this technique.

Types of Logs to Be Run

Logging combinations (chosen according to the type of mud, lithology, and information needed) generally consist of one resistivity device and one porosity device. Where hydrocarbon reservoirs are difficult to evaluate, two or more porosity devices are needed to differentiate oil from gas and rock types and to provide more accurate porosity data. Additional tools are also needed when the reservoir engineer, the completion engineer, and the geophysicist want additional information for evaluation and completion of the well. Use of computers in formation evaluation has increased utilization of the measurements recorded with such comprehensive logging programs.

A list of the programs recommended for most logging situations is provided in Appendix 5A. Mud resistivity, formation water resistivity, hole conditions, and formation types dictate the type of devices needed. The extent of the logging program is also a function of the information obtained on previous wells. Appendix 5B supplies cross references for relating the tool nomenclature of various service companies.

INFLUENCE OF THE MUD PROGRAM

Mud type influences the choice of a logging tool, especially the choice of a resistivity tool.

Air-drilled holes, which have nonconductive fluid in them, must be logged with an induction device. Likewise, holes drilled with oil can only be logged with an induction log. Where conductive fluids are in the borehole for logging operations, the choice between induction and laterolog devices is controlled by the salinities of the mud and the connate water in the formation. In general, fresh muds and salty formation waters favor the induction log and salty muds favor the laterolog.

All formations should be protected from excessive fluid losses so that porosity and saturation can be adequately determined. Bit-cutting sample recovery should be efficient enough to help in interpretation of lithology so that proper constants for log evaluation formula can be established. Thus, the mud program should be designed for both the drilling and the logging operations.

It is possible for a logging program to succeed or fail solely because of the way the mud program is designed. For example, filtrate from a high-water-loss mud can invade a formation so deeply as to mask the measurement of true resistivity, reduce the amplitude of the spontaneous potential curve, obscure the detection of the residual hydrocarbons, and result in water recovery on a drillstem test from zones that would otherwise produce oil. Invasion of oil from oil-based or oil-emulsion muds can affect resistivity readings and erroneously indicate oil in water-bearing formations or reduce the formation porosity values calculated from microresistivity devices.

The practice of mudding up just before reaching the objective zone can adversely affect interpretation when mud filtrate invades the formation beyond the radius of investigation of the resistivity device. Also, friable formations drilled with natural high-water-loss muds are usually badly washed out and can prevent the logging tools from going down the hole because they hang up on ledges and/or bridges. Borehole contact devices cannot obtain effective contact with the side of the borehole in highly rugose holes and will give erroneous measurements.

The decision to drill through shallow formations with natural high-water-loss muds is normally based on the erroneous assumption that the shallow formations are of no interest. Nevertheless, logs through the shallow formations are invariably consulted later to find

zones for recompletion, to determine prospects for new hydrocarbon-bearing zones in the area, to locate and evaluate high
pressure zones, and for general correlation work.

CHOOSING WHEN TO LOG

Logs should be run just prior to the running and setting of a casing
string. Once casing is set, the choices for logging are severely limited.
Logs should also be run if hole conditions suggest that a section of
hole could be lost (caving, washouts, etc., which would negate
running a logging tool), if cuttings indicate that an unexpected
formation has been encountered, or if one is otherwise lost
structurally.

However, one's enthusiasm for running logs should be tempered
somewhat by the economic and practical realities of service-company
price lists and fee structures. Each time a logging truck is called, some
kind of setup charge is assessed to cover costs of mobilization; a
depth charge is assessed per foot of hole from surface to total depth;
and a survey charge is assessed over the actual interval logged. The
full cost of a logging operation is, more than anything else, a function
of the depth of the well. To log a 100-ft section at 10,000 ft is an expensive proposition while a 4000-ft survey at 5000 ft total depth is
probably less expensive.

COMPUTERIZED LOGGING UNITS

Systems Available

Major service companies now offer logging services from computer-
based logging units. The available systems are:

CSU	Cyber Service Unit (Schlumberger)
DDL	Direct Digital Logging (Gearhart)
CLS	Computer Logging Service (Dresser)
DLS	Digital Logging Service (Welex)

The advantages of using these computer-based units are many, and
their use is to be encouraged.

Features of Computer-Based Logging Systems

In contrast to conventional logging systems, computer-based systems
offer the following features:

All logs are directly recorded on digital magnetic tape.
Logs can be recorded while logging up or logging down, with all
curves mutually on depth.
Calibrations are performed under programmed control and can be
performed more quickly and accurately than with conventional
systems.

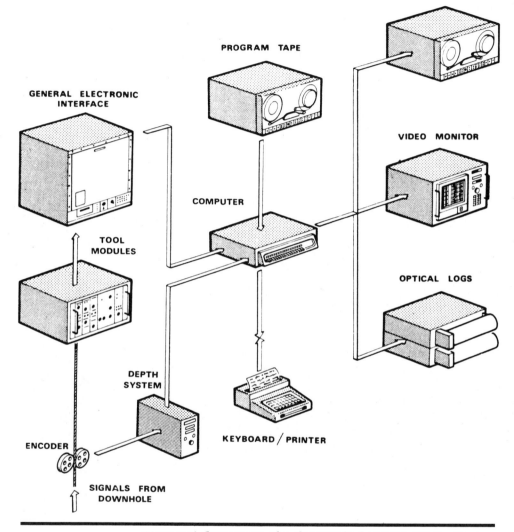

FIGURE 5.14 *Components of a Computerized Logging System. Courtesy Schlumberger Well Services.*

Logs can be played back from the data tapes on many different scales (both depth and response scales).

Wellsite computation of raw data is commonly available. These computations range from completion aids (hole-volume integration for cement volumes) to dipmeter computations and complete log analysis.

Figure 5.14 is a schematic of a computer-based logging system. The logging engineer interacts with the system using a keyboard. At a command, the computer will load programs to perform such functions as calibration, logging, computation, and playback.

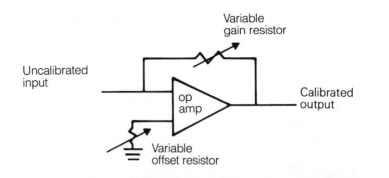

FIGURE 5.15 *Conventional Calibration Method.*

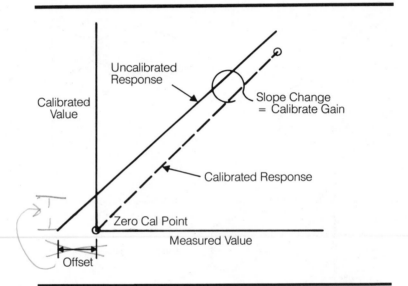

FIGURE 5.16 *Principle of Two-Point Calibration. Courtesy Schlumberger Well Services.*

Calibrations

Conventional logging systems require human operation of both sensitivity and zero offset control. The variable offset resistor (fig. 5.15) is adjusted when the logging sensor is at the low end of its range of measurement (e.g., the caliper tool in a 6-in. ring), and the variable gain resistor is adjusted when the sensor is at the high end (e.g., the caliper tool in a 12-in. ring).

Computer systems eliminate the need for human intervention—other than to place the tool to be calibrated in the correct environment (putting the 6-in. ring over the caliper arms, for example). The computer system accepts the raw, uncalibrated readings of the tool and computes a calibration equation to transform the raw data into calibrated data (see fig. 5.16). For example, a caliper may read 7.2 in. in an 8-in. ring and 13.6 in. in a 16-in. ring. The computer solves the equation relating calibrated readings to raw readings:

calibrated caliper $= A + B \times$ raw caliper,

where A is the required offset and B is the required gain. Thus, for the above readings,

$8 = A + B \times 7.2$, and

$16 = A + B \times 13.6$.

Simultaneous solution of these two equations gives

$A = -1$ and $B = 1.25$.

Thus, the caliper value recorded on the log will be $(1.25) \times$ (raw caliper) $- 1$, but the transformation will be carried out by software rather than hardware.

A typical log heading as generated by a computer system is shown in figure 5.17. Some typical computerized calibration records are shown in figure 5.18. Note that these calibration tails do not record galvanometer positions directly. Instead the actual value of some parameter is recorded as a number expressed in engineering units. Computerized calibrations fall into three categories:

1. shop calibrations
2. before survey calibration
3. after survey calibration

The important things to check include the agreement of all three calibration values within the specified tolerances as listed in figure 5.19. Note that these tolerance values refer to Schlumberger logs. Other service companies will use different numbers for their tools. Booklets explaining calibration techniques of a specific service company can be obtained from logging company sales personnel.

By way of example, consider a density log calibration. The calibration tail shown in figure 5.18 shows that on September 23, 1983, a shop calibration was performed for a set of density equipment identified by the serial numbers PGS: 310, PDH: 394, SFT: 354. Using the block, the far count rate (FFDC) was measured as 518 cps and was calibrated to 337 cps; the near count rate (NFDC) was measured as 914 cps and calibrated to 528 cps. The wellsite calibration jig was then placed on the tool and again the count rates were measured and calibrated at 337 cps and 528 cps, respectively.

At the wellsite, before the survey, these same count rates were calibrated to 335 and 526 cps, respectively. The question, therefore, is: "How much difference can be tolerated between shop calibrations and wellsite calibrations?" For the near count rates, there is a discrepancy of 2 cps (528–526), and for the far count rates there is also a

INDUCTION/LITHODENSITY/NEUTRON/GR

CSU

COMPANY:
WELL:
FIELD:
COUNTY:
STATE:
MATION:
LOCATION:

SEC: TWP: RGE:

OTHER SERVICES-
DIL/SLT
LDT/CNT/GR
EPT/PCD
HDT
RFT
CORES

PERMANENT DATUM: G.L.
ELEV. OF PERM. DATUM: 315.5 F
LOG MEASURED FROM: RKB
21.5 F ABOVE PERM. DATUM
DRLG. MEASURED FROM: RKB

ELEVATIONS-
KB: 337.0 F
DF: 336.0 F
GL: 315.5 F

PROGRAM
TAPE NO:
26.21F
SERVICE
ORDER NO:
427719

DATE: 9 SEP 84
RUN NO: 1

DEPTH-DRILLER: 6200.0 F
DEPTH-LOGGER: 6208.0 F
BTM. LOG INTERVAL: 6205.0 F
TOP LOG INTERVAL: 2364.0 F

CASING-DRILLER: 2360 F
CASING-LOGGER: 2364 F
CASING: 10.75"
WEIGHT: 40.50 LB/F
BIT SIZE: 9.875"
DEPTH: 6200 F
TYPE FLUID IN HOLE: CALEX
DENSITY: 10.5 LB/G
VISCOSITY: 38.0 S
PH: 12.5
FLUID LOSS: 5.2 C3
SOURCE OF SAMPLE: FLOWLINE
RM: .937 OHMM AT 75.0 DEGF
RMF: .634 OHMM AT 75.0 DEGF
RMC: 1.406 OHMM AT 75.0 DEGF
SOURCE RMF/RMC: MEAS. /CALC.
RM AT BHT: .449 OHMM AT 164. DEGF
RMF AT BHT: .304 OHMM AT 164. DEGF
RMC AT BHT: .674 OHMM AT 164. DEGF

TIME CIRC. STOPPED: 0030 9 SEP.
TIME LOGGER ON BTM.: 0500 9 SEP.

MAX. REC. TEMP: 164.0 DEGF

LOGGING UNIT NO:
LOGGING UNIT LOC:
RECORDED BY:
WITNESSED BY:

REMARKS:

MUD REPORT CHLORIDES = 1600 PPM CL-.
LOC. CONT.: 800' FNL & 1000' FWL OF SECTION 42

LOGS RUN SEPARATE AND MERGED ELECTRONICALLY.

EQUIPMENT NUMBERS-

DIS 555	DIC 1097	DRS 31	NSC 31
GSR 7125	PDH 2944	CNC 1311	CNB 3524
SGC 511	TCC 23	TCM 32	

ALL INTERPRETATIONS ARE OPINIONS BASED ON INFERENCES FROM ELECTRICAL OR OTHER MEASUREMENTS AND WE CANNOT, AND DO NOT GUARANTEE THE ACCURACY OR CORRECTNESS OF ANY INTERPRETATIONS, AND WE SHALL NOT, EXCEPT IN THE CASE OF GROSS OR WILLFUL NEGLIGENCE ON OUR PART, BE LIABLE OR RESPONSIBLE FOR ANY LOSS, COSTS, DAMAGES OR EXPENSES INCURRED OR SUSTAINED BY ANYONE RESULTING FROM ANY INTERPRETATION MADE BY ANY OF OUR OFFICERS, AGENTS OR EMPLOYEES. THESE INTERPRETATIONS ARE ALSO SUBJECT TO OUR GENERAL TERMS AND CONDITIONS AS SET OUT IN OUR CURRENT PRICE SCHEDULE.

FILE 103 09-SEP-84 13:03
DATA ACQUIRED 09-SEP-84 06:32

FIGURE 5.17 *Computerized Log Heading.*

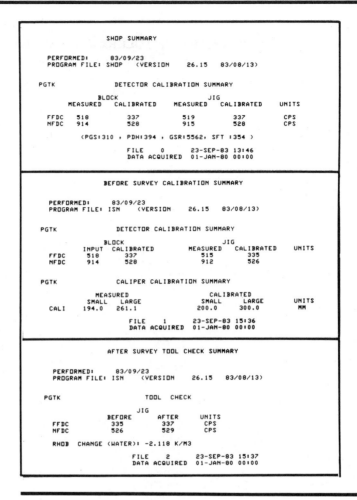

FIGURE 5.18 *Computer Unit Calibrations for a Density Log. Courtesy Schlumberger Well Services.*

discrepancy of 2 cps. What is the tolerance for the density tool? Referring to the tolerance table of figure 5.19, we see that the near count rates are allowed a variation of ±22 cps and the far count rates a variation of ±14 cps. The wellsite calibration in this case can, therefore, be considered good.

QUESTION #5.4
Read the difference in count rates between the before and after survey calibrations from figure 5.18 for both the near and far count rates.

a. Are they within allowable tolerances?
b. What effect will the drift have in terms of changes in the logged parameter, bulk density (ρ_B)?

SUMMARY OF TOLERANCES BETWEEN BEFORE AND AFTER SUMMARIES

Log	Curve	Tolerance Zero	Tolerance Plus	Units
ISF	ILD	±2	±20	MHO/M
	SFL	±2	±20	MHO/M
DIT	ILM	±2	±20	MHO/M
	ILD	±2	±20	MHO/M
IES	ILD	±2	±20	MHO/M
	16″ N	±0.08	± 1.05	ΩM
DLL	LLS		± 0.7	ΩM
	LLD		± 0.7	ΩM
MSFL	CMSF	± 2	± 20	MMHO/M
	I1	± 2	± 20	MMHO/M
Proximity	PROX	±1	±15	MMHO/M
MLL	MLL	±1	±15	MMHO/M
GR	GR		±15	CPS
CNL	NRAT		± 0.04	RATIO
Dual Porosity	NRAT		± 0.04	RATIO
Neutron	NECN			CPS
	FECN			CPS
SNP	SNP		±13	CPS
FDC	FFDC		±14	CPS
	NFDC		±22	CPS
LDT	LL		± 1	CPS
	LU		± 1	CPS
	LITH		± 0.3	CPS
	SS1		± 0.5	CPS
	SS2		± 0.5	CPS
NGS	W1NG		± 6	%
	W2NG		± 6	%
	W3NG		± 6	%
	W4NG		± 20	%
	W5NG		± 20	%

FIGURE 5.19 *Calibration Tolerances (Schlumberger Logs). Courtesy Schlumberger Well Services.*

For further details on log calibrations please refer to *Log Quality Control* (Bateman 1985).

LOG QUALITY CONTROL

The Need for Log Quality Control

The need for log quality control has been documented in various studies, such as the one by Neinast and Knox of Sun Oil Company (1973). Table 5.2 summarizes the results of their studies. As these study results show, poor log quality control can result in a large per-

TABLE 5.2 *Results of Log Quality Study by Sun Oil*

	Schlumberger	Dresser Atlas	Welex
Big Wells Field, Texas			
Induction errors 10 mmho	46%	81%	67%
Density errors 0.01 g/cc	46%	61%	70%
Upper Valley Field, Utah			
Sonic	23%	100%	—
Density } logs in error	25%	78%	—
Neutron	33%	56%	—
Verden Field, Oklahoma			
Density errors 0.01 g/cc	36%	43%	58%
0.02 g/cc	15%	26%	33%

Source: Reprinted by permission of the SPWLA from Neinast and Knox 1973.

centage of logs being in error. Frequently, the errors made in recording the logs render them useless as formation evaluation tools.

It is the function of the well operator's representative to ensure that the logs obtained are of the best quality. Service-company personnel expect a representative to be available and in the logging unit during the logging operation. Logging operations should be discussed with the logging engineer before and after the job. The most critical time during a logging operation is when the tool is within 1000 ft of the bottom of the well. Do not distract the logging engineer during this time. Even the best engineers may make mistakes when they are distracted. Give the engineer a chance to perform the operation with minimum interruption.

After each log is complete, discuss it with the engineer as thoroughly as possible. Ask for an explanation of any abnormal curve responses, equipment failure, or hole problems, and enter the information in the "remarks" column of the log heading. This may be done after the first print is made but before any more prints are made. If there is any question about validity, request a rerun of the log before the crew rigs down. Generally 200 ft of repeat in relatively smooth hole should be enough to verify the log. Everyone is reluctant to go back in the hole after rigging down, but once pipe is set, it will be impossible to get another resistivity survey of any type.

No matter how competent and conscientious an observer may be, there are ways in which bad logs can defy detection at the wellsite. To this degree log quality also depends on the competence and integrity of the logging company's engineer. Perhaps the most important objective is to develop relationships of mutual trust with the logging company personnel with whom you work. Further details of log quality control procedures are available in Bateman (1985).

Practical Log Quality Checks

Calibrations are the only completely objective verifications of log quality available. Learn what they mean and how to use them. The depth-related log measurements will include one or more repeat sections, usually of about 200 ft. These records are valuable, though not conclusive, indications of correct tool operation and should be examined carefully on every log run.

Acceptance Standards for Logs

Although a large percentage of all logs might contain erroneous data of some nature, it would be unfair to intimate that all such logs are worthless. Often a log can be corrected visually or mathematically, but sometimes the log must be rerun before valid conclusions can be made. If so, the cost of rerunning the log might outweigh the importance of the error. When the log is not rerun, the error should be noted on the heading in the "remarks" column and also noted on the log opposite any zone of interest.

For the most serious errors, it is a mistake to think a bad log is better than none; a bad log may influence important decisions. Consequently, it is imperative that the log be rerun. The problem, of course, is to determine whether to accept or reject the questionable log. One reliable method for determination of a bad log is to ask the question: "Is the interpretation accurate?" If in doubt, rerun the log. Also ask: "Can everyone who will use this log see the error and/or be able to perform an accurate interpretation?" If in doubt, rerun the log.

A definite criterion for acceptance or rejection of a log is difficult to establish; each situation will be somewhat different. Good judgment should outweigh written instructions when deciding whether to accept or rerun a log. The following guidelines should assist in making such decisions.

OVERALL TECHNICAL QUALITY. Many things may adversely affect the technical quality of logging data. The most obvious is equipment malfunction. Other possible causes of poor data are: rugose boreholes, sticking tools, logging engineer's errors, tool rotation, excess logging speed, deviated wells, poor centralization or eccentralization, and formation alteration.

Many times an anomaly over a logged interval will suggest the possibility of a malfunction. This should be resolved by repeating the log in that section. After all, it may be significant. It is interesting to recall that the SP was originally an anomaly that interfered with measurement of formation resistivity.

REPEATABILITY. Properly functioning resistivity tools, run under conditions that are within their capability, will nearly always repeat very well. As a functional check of the equipment, a repeat section of 200 ft or more is routinely run, and should be required except in unusual circumstances. Aside from equipment failures, factors that could cause poor repeats include:

Washed out holes, particularly those of extremely noncircular cross section

Variable tool centering, particularly in large holes with fairly high mud conductivities

TABLE 5.3 *Tool Response in Common Benchmarks*

| Benchmark | Expected Value for Various Tools | | | |
	Sonic (μs/ft)	Density (g/cc)	Compensated Neutron Limestone %	Photoelectric Factor (P_e)
Salt	67.0	2.04	−2.0	4.65
Anhydrite	50.0	2.98	−1.0	5.05
Sulfur	122.0	2.02	−3.0	5.43
Casing	56.0	—	—	—

The presence of metallic fish in the borehole

Comparing an up run with a down run (which may appear quite different with some types of equipment

The repeatability of a log run may be affected by time-related phenomena too, such as varying invasion profiles. Invading filtrate can penetrate deeper, migrate vertically, accumulate as annuli, or dissipate altogether with the passage of time. The log response, particularly of the shallow reading devices, may continue to change for many days after the well is logged. Though unusual, such changes can be very troublesome; but from the viewpoint of log quality, they are usually recognizable. The changes occur only in the invaded sections, not the shales or other impervious rocks.

OFFSET LOGS. If the well is in a developing field, or in a consistent geological block, available offset logs are likely to be useful. This is especially true in an unfamiliar area.

ABSOLUTE LOG VALUES. Comparison of log readings with known absolute values is seldom possible, but when it can be done, this positive crosscheck should be used. Formations that consist of pure, zero-porosity minerals such as halite, anhydrite, or limestone can be used to check log readings. These natural benchmarks are listed, for several of the more common tools, in table 5.3.

Casing can sometimes be used as a check. All caliper tools should read the same in casing. The diameter indicated is usually slightly greater than that of new casing owing to drillpipe wear. The two diameters measured by a four-arm caliper should be equal. The sonic should read about 56 μs/ft in unbonded casing.

DEPTH MEASUREMENTS. Measurement of depth is perhaps the logging company's most basic function, but one that tends to get lost among the more glamorous parameters. Absolute depth control is provided either by a calibrated sheave or by magnetic marks placed on the logging cable every 100 ft. In either case, the operational procedure for obtaining accurate depth control is rather rigorous and if followed properly will almost always result in accurate depth measurements. This is one of the places where it is advisable to be on

TABLE 5.4 *Recommended Logging Speeds*

Tool	Ft/min	Ft/hr	Remarks
Resistivity log	100	6000	
Resistivity + GR log	60	3600	GR for correlation
Radiation log	30	1800	
Sonic log	60	3600	Slower if the sonic is noisy
Dipmeter	60	3600	
Microresistivity	40	2400	

terms of mutual trust with the logging engineer. It may be possible to detect evidence of inaccurate depth measurements, but absolute verification is very difficult. Compare logger's TD (total depth) and casing depth with the depths reported by the driller, watch for excessive tie-in corrections with previous log runs, and check the apparent depths of known markers.

Relative depth control means ensuring that all measurements are on-depth with each other. All curves that are recorded on the same trip in the hole should be on-depth with each other within ±6 in. In addition, each subsequent log should match the base log within 2 ft in straight holes and 4 ft in highly deviated wells (greater than 30°).

LOGGING SPEEDS. The logging speed in feet per minute is indicated by gaps or ticks along the edge of the film track. Acceptable logging speeds depend on the type of log, the type of unit (computer or conventional), the intended use of the data, and the type of formation being logged. Normal routine logging speeds are given in table 5.4.

APPENDIX 5A: *Recommended General Logging Programs*

Condition	Data Desired	Recommended Services	Remarks
Fresh mud $\dfrac{R_{mf}}{R_w} > 2$	Correlation and lithology in sand/shale	Dual Induction–SFL–GR/SP; Induction Electrical; Dual Laterolog–GR/SP (low porosity and/or high resistivities); Sidewall Cores	The dual induction and dual laterolog devices are superior to single induction and single laterolog in all cases Sidewall cores give lithology in sand/shale
	Porosity; water saturation; lithology in carbonates and evaporites; hydrocarbon type	Density and/or Neutron and/or Sonic and GR; Formation Tester	For shaly sands or simple mixed lithologies, density–neutron or neutron–sonic. For complex mixed lithologies use all three. For hydrocarbon type as above and/or Formation Tester
	Producible hydrocarbons and permeability indicators	Proximity-Microlog; Microlaterolog or Micro-SFL; Dual Resistivity device; Sidewall Cores; Formation Tester	Proximity log used for thick mudcake; Microlaterolog for thin mudcake; Micro-SFL available with Dual Laterolog. Sidewall cores give permeability estimate in shaly sands
	Hydrocarbon indication at the wellsite	Wellsite Computer Products, including: apparent formation water resistivity, R_{wa}; formation factor/resistivity overlay; and R_{xo}/R_t vs. SP overlay	Available with computer units only. Others available with both units
	Formation dip magnitude and directional	Dipmeter	Wellbore directional information is also available
Salt mud $\dfrac{R_{mf}}{R_w} < 2$	Correlation and lithology in sand/shale	Dual Laterolog–Micro-SFL–GR	In some areas of low porosity and/or mixed lithology the neutron or density–neutron is usable as a correlation log
	Porosity, water saturation, and lithology in carbonates and evaporites; hydrocarbon type	Density and/or Neutron and/or Sonic and GR; Formation Tester	See remarks under fresh mud
	Producible hydrocarbons and permeability indicators	Proximity–Microlog; Microlaterolog-Microlog; Dual Laterolog–Micro-SFL; Sidewall Core; Formation Tester	In very salty muds or when R_{xo} is less than R_{mc}, the microlog results may be unsatisfactory. See remarks under fresh mud

APPENDIX 5A: (Continued)

Condition	Data Desired	Recommended Services	Remarks
Salt mud (continued)	Hydrocarbon indication at the wellsite	Wellsite Computer Products; R_{wa}, R_o Overlay	See remarks under fresh mud. Apparent formation water resistivity may not be satisfactory in low and/or mixed lithologies
	Formation dip magnitude and directional	Dipmeter	Wellbore directional information is also available
Oil-base mud	Correlation and lithology in sand/shale	Dual Induction–GR; Induction–GR; Sidewall Cores	High temperature (350 to 400°F) will restrict the use of the dual induction and sidewall cores
	Porosity, water saturation, lithology in carbonates and evaporites; hydrocarbon type	Density and/or Neutron and/or Sonic; Formation Tester	See remarks under fresh mud. High temperature (350°F) and small holes (6 in.) restrict the use of the formation tester
	Producible hydrocarbons and permeability indicators	Dual Induction; Sidewall Cores; Formation Tester	Sidewall cores give permeability estimates in shaly sands
	Hydrocarbon indication at the wellsite	Wellsite Computer Products; R_{wa}, R_o Overlay	See remarks under fresh mud
	Formation dip magnitude and directional	Dipmeter	See remarks under fresh mud
Air- or Gas-filled hole	Resistivity, correlation, and lithology in sand/shale	Dual Induction–GR; Induction–GR	
	Porosity, water saturation, gas saturation, estimated lithology in carbonates and evaporites	Density and/or Neutron	Dual spacing neutron CNL cannot be used; must use gamma ray-neutron (GRN) or preferably sidewall neutron porosity (SNP)
	Permeability indicator	Temperature Log; Noise Log	Gas entry
	Hydrocarbon indication at the wellsite	Wellsite Computer Products; R_o Overlay, R_{wa}	See remarks under fresh mud
Fresh or unknown formation water	Resistivity; correlation and estimated lithology in sand/shale	See fresh mud and salt mud	See fresh mud and salt mud

	Objective	Tools	Remarks
	Porosity water saturation; estimated lithology in carbonates and evaporites; hydrocarbon type	Density and/or Neutron and/or Sonic and GR; Electromagnetic Propagation Tool; Inelastic Neutron Scattering and Capture Gamma Ray Spectroscopy; Formation Tester Sidewall Cores; Formation Tester	Under conditions of no invasion the electromagnetic propagation tool will yield water saturation directly. See remarks under fresh mud
	Producible hydrocarbons and permeability indicators	Wellsite Computer Products	
	Hydrocarbon indication at the wellsite		See remarks under fresh mud. May use electromagnetic propagation tool
Cased hole	Fluid type and lithology	Pulsed Neutron Log; Inelastic Neutron Scattering and Capture Gamma Ray Spectroscopy; GR and Natural Gamma Ray Spectroscopy	
	Porosity and hydrocarbon type	Pulsed Neutron Log; Gamma Ray–Neutron; Density Formation Tester; Inelastic Neutron Scattering and Capture Gamma Ray Spectroscopy	Dual spacing neutron (CNL) cannot be used if hole is gas filled. Under favorable conditions the density may be used for porosity
	Permeability	Formation Tester	

APPENDIX 5B: *Service Company Nomenclature*

Schlumberger	Gearhart	Dresser Atlas	Welex
Electrical Log (ES)	Electric Log	Electrolog	Electric Log
Induction Electric Log	Induction Electrical Log	Induction Electrolog	Induction Electric Log
Induction Spherically Focused Log			
Dual Induction Spherically Focused Log	Dual Induction-Laterolog	Dual Induction Focused Log	Dual Induction Log
Laterolog-3	Laterolog-3	Focused Log	Guard Log
Dual Laterolog	Dual Laterolog	Dual Laterolog	Dual Guard Log
Microlog	Micro-Electrical Log	Minilog	Contact Log
Microlaterolog	Microlaterolog	Microlaterolog	F_oR_{xo}Log
Proximity Log		Proximity Log	
Microspherically Focused Log			
Borehole Compensated Sonic Log	Borehole Compensated Sonic	Borehole Compensated Acoustilog	Acoustic Velocity Log
Long Spaced Sonic Log		Long Spacing BHC Acoustilog	
Cement Bond/Variable Density Log	Sonic Cement Bond System	Acoustic Cement Bond Log	Microseismogram
Gamma Ray Neutron	Gamma Ray Neutron	Gamma Ray Neutron	Gamma Ray Neutron

Sidewall Neutron Porosity Log	Sidewall Neutron Porosity Log	Sidewall Epithermal Neutron Log	Sidewall Neutron Log
Compensated Neutron Log	Compensated Neutron Log	Compensated Neutron Log	Dual Spaced Neutron
Thermal Neutron Decay Time Log		Neutron Lifetime Log	Thermal Multigate Decay
Formation Density Log	Compensated Density Log	Compensated Densilog	Density Log
Litho-Density Log		Z-Densilog	
High Resolution Dipmeter	Four-Electrode Dipmeter	Diplog	Diplog
Formation Interval Tester		Formation Tester	Formation Test
Repeat Formation Tester	Selective Formation Tester	Formation Multi Tester	Multiset Tester
Sidewall Sampler	Sidewall Core Gun	Corgun	Sidewall Coring
Electromagnetic Propagation Log	Dielectric Constant Log	Dielectric Log	Dielectric Constant Log
Borehole Geometry Tool	X-Y Caliper	Caliper Log	Caliper
Ultra Long Spacing Electric Log			
Natural Gamma Ray Spectrometry		Spectralog	Compensated Spectral Natural Gamma
General Spectroscopy Tool		Carbon/Oxygen Log	
Well Seismic Tool		Borehole Seismic Record	
Fracture Identification Log	Fracture Detection Log		

BIBLIOGRAPHY

Bateman, R. M.: *Log Quality Control,* IHRDC (1985).

Neinast, G. S., and Knox, C. C.: "Normalization of Well Log Data," paper I, *Trans.,* SPWLA, 14th Annual Logging Symposium, Lafayette, May 1973.

Schlumberger Limited: "This is Schlumberger," New York (March 1981).

Schlumberger: "CSU Well Site Products and Calibrations Guide CP 26," M-081027 (Nov. 1983).

Schlumberger: "Log Interpretation Charts," Schlumberger Well Services (1984).

Schlumberger: "Cyber Service Unit Wellsite Products and Calibration Guide CP 26," Schlumberger Well Services (1984).

Schlumberger: "CSU," brochure (no date).

Answers to Text Questions

QUESTION #5.1

$2T \cos 45° = 10,000 \times 0.7071 = 7707$ lb

QUESTION #5.2

250 cu ft or 44.5 bbl

QUESTION #5.3

6774.44 ft

QUESTION #5.4

a.

	Before	After	Change	Tolerance
FFDC	335	337	+2	±14
NFDC	524	529	+5	±22

b. -2.118 kg/m^3 or -2.118×10^{-3} g/cc

METHODS OF ANALYSIS AND APPLICATION OF RESULTS

CHAPTER SIX

THE PHYSICS OF ROCK/FLUID SYSTEMS

Petrophysics is the name given to the study of rock/fluid systems. It is particularly important that the log analyst be aware of the way in which rocks and fluids interact in both static and dynamic situations. Although logging measurements are made under static reservoir conditions, the prediction of reservoir behavior under dynamic flow conditions can only be made if the physics of fluid flow is understood. The objective of this discussion is to equip the formation evaluator with the information needed to relate log response to reservoir performance (which is what really counts) rather than just to static reservoir content. It is hoped that from what follows the reader will develop a feel for why some reservoirs with low water saturations produce with high watercut while other reservoirs with much higher computed water saturations produce water-free hydrocarbons.

THE GENESIS OF RESERVOIR ROCKS

A reservoir rock is one that has both storage capacity and the ability to allow fluids to flow through it; that is, it must possess both porosity and permeability. Porosity can develop as the void space between grains of sediments as they are laid down. Typical of this intergranular porosity would be sandstone reservoirs. Porosity can also develop when chemicals react with rocks after they have been deposited. Typical of this solution type of porosity would be carbonate reservoirs. Porosity can also develop as fractures induced by the stresses of tectonic movement. Porosity per se does not guarantee permeability. Swiss cheese, for example, is highly porous but impermeable.

Porosity

The porosity developed in sedimentary rocks is a function of many variables—such as grain shape, size, orientation, and sorting—broadly grouped as *rock texture.* If all the grains are of the same size, sorting is said to be good. If grains of many diverse sizes are mixed together, sorting is said to be poor. The *packing* of the grains (see fig. 6.1) determines the porosity. For a given sorting, the porosity is independent of the grain size. For example, if spheres of diameter d are packed in a cubic lattice arrangement, the porosity can be calculated in the following manner.

In unit volume, n^3 spheres are packed n to a side. The total volume is $(nd)^3$. The volume of any one sphere is $(4/3)\pi(d/2)^3$, thus the

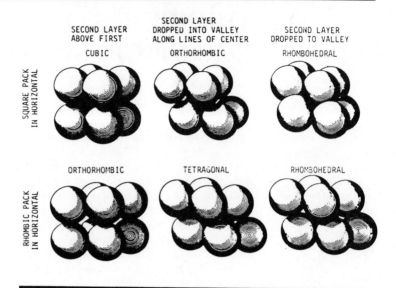

FIGURE 6.1 *Packing of Spheres. Reprinted by permission of the University of Chicago Press from Graton and Fraser 1935.*

volume occupied by n^3 spheres is $(4/3)\pi(nd/2)^3$. The porosity is thus

$$\frac{(nd)^3 - 4/3\pi(nd/2)^3}{(nd)^3},$$

which simplifies to $(1 - \pi/6)$ or 0.4764. Note that the term d cancels out and is not a determining factor.

Cubic packing is not an efficient way to store spheres in a box; and nature seeks more compact packing mechanisms—such as rhombohedral packing, which produces a porosity of 25.95% (as opposed to 47.64% for cubic packing). For a given grain size, porosity will decrease as sorting gets poorer, since intergranular pores may be occupied by ever-smaller grains.

Quite apart from the mechanics of how sand grains are packed, their compaction with depth of burial is another matter. Porosity decreases with increasing depth in a predictable manner. A relationship of the sort

$\phi = \phi_0 e^{-\text{depth}/\alpha}$, where α is a compaction constant for a particular geologic area,

generally fits most normally pressured reservoirs—that is, the log of porosity is linear with depth. For example, if ϕ_0, the porosity at surface, is 45% and depth is in feet, a typical value of α might be 12,000, resulting in a porosity of 12.9% at 15,000 ft and 8.5% at 20,000 ft.

Permeability

While porosity is a *static* property of a rock, permeability is a *dynamic* one. Permeability is a measure of the ability of a rock to allow fluids to flow through it. Darcy's law for flow through a porous medium is

$$q = \frac{kA\Delta p}{\mu L} \, .$$

If flow is laminar, Darcy's relation can be used to define permeability k, in terms of flowrate q, area A, length L, pressure differential Δp, and fluid viscosity μ, such that

$$k = \frac{q\mu L}{\Delta p A} \, .$$

If only one fluid is present in the pore system, this relation defines *absolute* permeability—that is, it is a rock property independent of the fluid flowing through the rock. If q is in cc/s, A is in cm^2, $\Delta p/L$ is in atm/cm, and μ is in cp, then k is in darcys. (The practical unit is the millidarcy, abbreviated md, equal to one one-thousandth of a darcy.)

The relationship between permeability and porosity depends on rock type. In general the log of permeability is linear with porosity for a given rock type; however, the precise relationship is found only through direct measurements of representative rock samples. Figure 6.2 shows some of these trends. Various investigators have, over the years, developed theoretical permeability-to-porosity relationships taking into account textural features such as the size, shape, and distribution of pore channels in the rock. One such is the Kozeny relationship:

$$k = (\text{constant}) \, d^2 \phi^3 / (1 - \phi)^2.$$

For fracture systems, generalized formulas have been developed relating the permeability to the square of the fracture width.

In some reservoirs, permeability is a vector—that is, it takes on directional properties. Depositional effects may tend to align grains along their long axis, thus increasing the permeability in that direction. Also, vertical permeability may be different from horizontal permeability. In fractured reservoirs, permeability is likely to be highly directional depending on the azimuth of the fracture planes.

FLUID DISTRIBUTION IN THE RESERVOIR

Initially, sediments are laid down in water, either in river and lake beds (continental), in deltas and along shore lines (transitional), or on the continental shelves (marine), as illustrated in figure 6.3. (Exceptions to this rule are eolian dune sediments, which are initially deposited in a water-free environment.) Later in geologic time, after the reservoir rock has been buried, hydrocarbons from neighboring

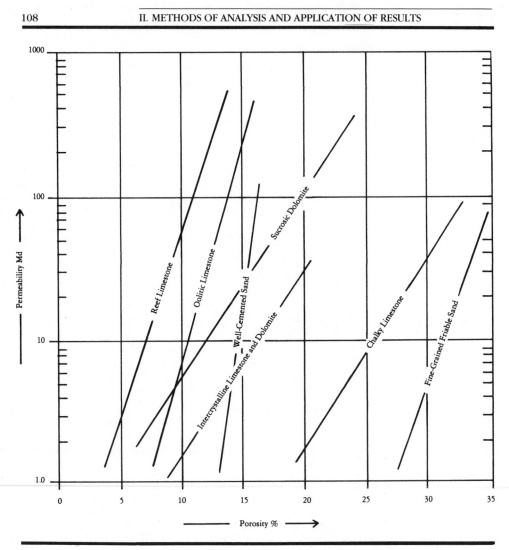

FIGURE 6.2 *Permeability/Porosity Relationships. After Core Labs.*

source rocks migrate into the reservoir. By simple gravity segregation, gas accumulates above oil, which overlies water. In the absence of any rock matrix, gas, oil, and water form distinct layers with sharp contacts between each phase; but in the reservoir, the lines of demarkation between gas and oil and water become blurred. Some common observations in a physics lab will serve to explain why this occurs.

Figure 6.4 shows a simple reservoir containing oil and water. It is divided into three sections. The section at the top is mainly oil, the section at the bottom is all water, and the section in the middle has ever-increasing amounts of water as depth increases. Plotted on the right of the figure is a curve of water saturation, together with a plot of the pressure of the fluids in the pore space. In order to understand the

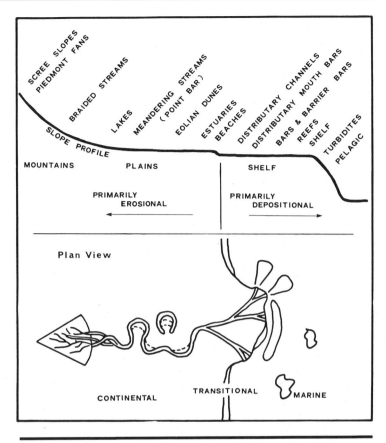

FIGURE 6.3 *Interrelationships of Depositional Environments. Reprinted by permission of the Indonesian Petroleum Association from Goetz et al. 1977.*

shape of the water-saturation curve in the transition zone, consider the case of a small glass tube held in a beaker of water, as in figure 6.5. A capillary tube of radius r will be found to support a column of water of height h. If the density of the air is ρ_a and the density of the water is ρ_w, the pressure differential at the air/water contact is simply $(\rho_w - \rho_a)h$. This pressure differential acting across the cross-sectional area of the capillary is exactly counterbalanced by the surface tension, T, of the water film acting around the inner circumference of the capillary tube. If, at the water/glass interface, the contact angle is θ, then, at equilibrium we have

$$2\pi r T \cos\theta = (\rho_w - \rho_a)h \times \pi r^2$$
$$\text{force} = \text{pressure} \times \text{area}.$$

Simplifying this expression and rearranging we have

$$h = \frac{2T\cos\theta}{r(\rho_w - \rho_a)},$$

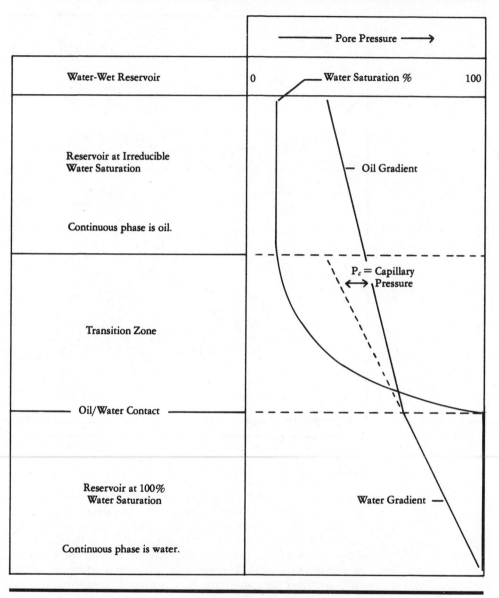

FIGURE 6.4 *Fluid Distribution in a Water-Wet Reservoir.*

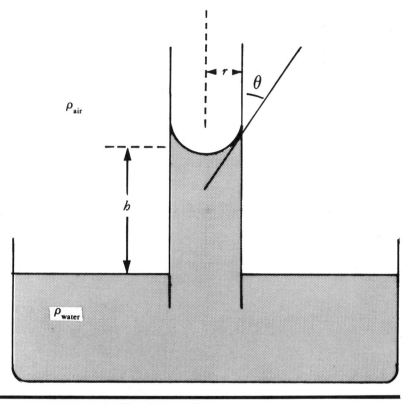

FIGURE 6.5 *Capillary Tube in Water.*

Inspection of this equation shows that the smaller r gets, the larger h gets.

When this laboratory observation is translated into terms of reservoir fluids we see that water can be drawn up into what would otherwise be a 100%-oil column by the capillary effect of the small pores—which act like tiny tubes—in the rock system. We can equate the air in figure 6.5 with oil, water with water, and the tube with pore throats. Thus, the maximum height, h, to which water can be raised by capillary action is controlled by the following factors:

The interfacial tension, T, between the two phases (oil and water in this case)
The contact angle, θ, between the wetting fluid (water in this case) and the rock.
The radius, r, of the pore throats
The density difference between the phases ($\rho_w - \rho_o$ in this case)

Given these factors, it is simple to predict the length of a transition zone in a reservoir. Reservoirs with large pore throats and high permeability will have short transition zones, and the transition zone at a gas/oil contact will be shorter than that at an oil/water contact due

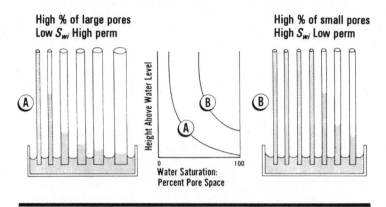

FIGURE 6.6 *Capillary Pressure Effects in Reservoirs. After Core Labs.*

simply to the interphase density differences involved (see fig. 6.6).

Obviously a pore system is made up of a variety of pore sizes and shapes. As a result, no single value of *r* can be assigned to one reservoir. Therefore, depending on the *distribution* on the pore throat radii, as well as their actual size, either many or few of the available pore channels will raise water above the free-water level. The water saturation above the top of the transition zone will thus be a function of porosity and pore-size distribution.

In a water-wet system, water wets the surface of each grain, or lines the walls of the capillary tubes. At the time oil migrates into the reservoir, the effects of the capillary pressure are such that the downward progress of oil in the reservoir is most strongly resisted in the smallest capillaries. There is a distinct limit to the amount of oil that can be expected to fill the pores. Large-diameter pores offer little resistance (capillary pressure, p_c, is low because *r* is large). Small-diameter pores offer greater resistance (p_c is high because *r* is small). For a given reservoir, ρ_o and ρ_w determine the pressure differential an oil/water meniscus can support. Thus, the maximum oil saturation possible is controlled by the relative number of small and large capillaries or pore throats. This maximum possible oil saturation, if looked at in terms of water saturation, translates into a minimum possible water saturation, and this is referred to as the *irreducible water saturation, S_{wi}*. Shaly, silty, low-permeability rocks with their attendant small pore throats tend to have very high irreducible water saturation. Clean sands of high permeability have very low irreducible water saturation. Figure 6.7 illustrates this important concept by comparing capillary pressure curves for four rock systems of different porosity and permeability.

Some rock/fluid systems are preferentially oil-wet rather than water-wet, the distinguishing trait being the contact angle θ at the interface of oil and water on the rock surface (see fig. 6.8). Traditionally, rock/fluid systems exhibiting contact angles less than 75° are considered water-wet, those with more than 105° are considered oil-wet, and systems in between are considered neutral. Note that

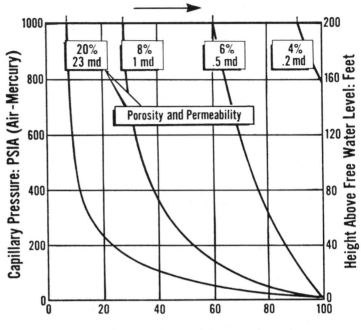

FIGURE 6.7 *Capillary Pressure Curves as a Function of Permeability and Porosity. After Core Labs.*

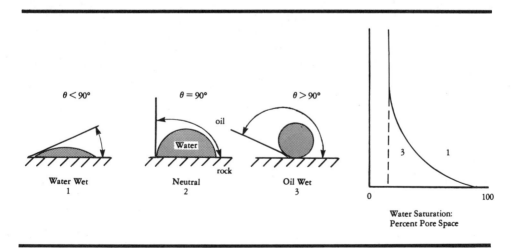

FIGURE 6.8 *Rock/Fluid Wettability and Contact Angle. After Core Labs.*

contact angle is a combined rock/fluid property and studies show it is
dependent on the wetting history of the rock and the concentrations
of polar compounds in the oil. Oil-wet reservoirs have very low S_{wi}
values, since, in effect, the capillary forces assist oil in the
displacement of water.

RELATIVE PERMEABILITY

If only one fluid is present in a pore system, fluid flow is nicely
governed by Darcy's law. If two or more fluids are present together in
a pore system, the dynamic behavior of the individual phases is not
quite so straightforward. Consider the case of oil and water together
in a pore system. *Effective* permeability is defined as the permeability
of the rock to a particular fluid phase at a particular saturation. Thus,
if, under a given pressure gradient, oil and water flow through a pore
system together, we find the effective permeabilities k_o (for oil) and
k_w (for water) are

$$k_o = \frac{q_o \mu_o L}{\Delta_p \times A} \quad \text{and} \quad k_w = \frac{q_w \mu_w L}{\Delta_p \times A} .$$

The total flow rate $Q_t (= Q_o + Q_w)$ is found to be less than the flow
rate either phase would have if it were at 100% saturation. Thus it
appears that the two phases interfere with each other's progress
through the pore system.

Relative permeability, k_r, is the ratio of effective permeability to
one phase to absolute permeability, and it is quoted at some
particular saturation value. Thus,

$$k_{ro} = k_o/k \quad \text{and} \quad k_{rw} = k_w/k.$$

Figure 6.9 shows typical relative permeability curves. Several things
are worth noting. Relative permeability to oil at irreducible water
saturation is 100% or 1; and, as water saturation increases, k_{ro}
decreases until it effectively reaches zero at some high water saturation corresponding to S_{or}, the residual oil saturation. Relative
permeability to water, on the other hand, commences effectively at
zero when the rock is at S_{wi} and thereafter increases as S_w increases.
Note also that for a given S_w, k_{ro} is always less for an oil-wet system
than for a water-wet system. Conversely, k_{rw} is always greater in an
oil-wet system than in a water-wet one. A common way of representing this difference is to plot the relative permeability ratio k_{rw}/k_{ro}
versus water saturation S_w. Figure 6.10 shows that in water-wet
systems the relation is such that if the k_{rw}/k_{ro} ratio is on a log scale
and S_w is on a linear scale, a straight line is obtained. In an oil-wet
system, an S-shaped line is observed.

When plotting relative-permeability curves, a distinction is usually
made between two possible scenarios, *imbibition* and *drainage*.
Imbibition refers to the case where the saturation of the wetting fluid

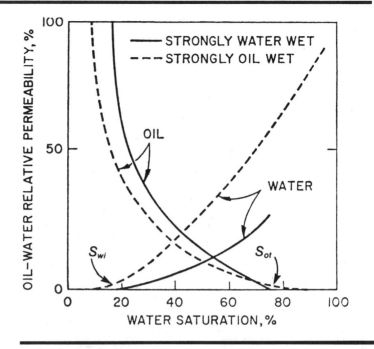

FIGURE 6.9 *Relative-Permeability Curves. Reprinted from Raza et al. 1968.*

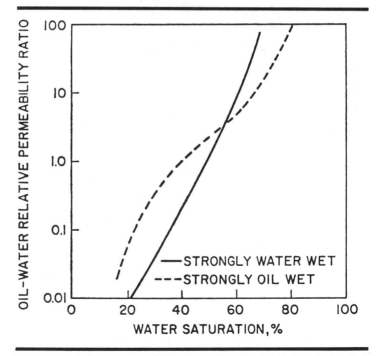

FIGURE 6.10 *Relative-Permeability Ratio Plot. Reprinted from Raza et al. 1968.*

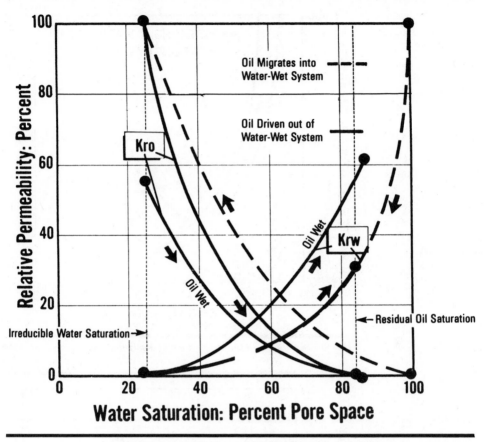

FIGURE 6.11 *Imbibition and Drainage. After Core Labs.*

is increasing. For example, in a water-wet reservoir, a rise in the water table subjects the transition zone to imbibition of water. Drainage refers to the case where the saturation of the wetting fluid is decreasing, as when oil first migrates into a water-wet rock. The difference between the two sets of relative-permeability curves reflects saturation history and the trapping of the nonwetting phase that occurs after it has been imbibed. Figure 6.11 illustrates these different cases.

Many workers in this field have proposed generalized empirical equations to relate k_{ro} and k_{rw} to S_w, S_{wi}, and S_{or}. Particularly worthy of note are those cited in Honarpour et al. (1982), Molina (1983), and Pirson et al. (1964). A commonly used approximation gives

$$k_{ro} = \left(\frac{0.9 - S_w}{0.9 - S_{wi}}\right)^2 \quad \text{and} \quad k_{rw} = \left(\frac{S_w - S_{wi}}{1 - S_{wi}}\right)^3.$$

If a well is completed above the transition zone, which means that the reservoir is at irreducible water saturation (and thus $k_{rw} = 0$), water cannot be produced. If completion is contemplated in the transition zone, it is comforting to know in advance what watercut may be expected. This can be calculated as follows:

The oil flow rate is $\qquad q_o = k_o \Delta p A / \mu_o L.$

The water flow rate is $\quad q_w = k_w \Delta p A / \mu_w L.$

Thus, the water/oil ratio (WOR) is given by

$$WOR = k_w \mu_o / k_o \mu_w.$$

The ratio k_w / k_o is numerically equivalent to k_{rw} / k_{ro}, which can be deduced from measured relative-permeability ratios, or estimated from one of the generalized correlations.

The actual watercut (WC) of the production into the wellbore will be given by

$$WC = q_w / (q_w + q_o)$$

which is equivalent to:

$$WC = WOR/(1 + WOR).$$

Surface watercut will be a function of the formation volume factors (β) of the oil and water, so the complete expression will be

$$WC = WOR\beta_o / (\beta_w + WOR\beta_o)$$

MEASUREMENT OF POROSITY
Porosity may be measured by a variety of methods including:

Borehole gravimetrics
Wireline logging
Core analysis

Each method investigates a different volume of the formation. The borehole gravimeter samples very large volumes on the order of 10^3 to 10^6 ft^3. Wireline logging tools investigate a much smaller volume, on the order of 1 to 10 ft^3, depending on the specific porosity device used. Core analysis investigates much smaller volumes, ranging from 10^{-3} to 10^{-1} ft^3. From one extreme to the other lie nine orders of magnitude, so one should not be needlessly surprised to learn that porosity estimates using different tools and techniques do not always agree. This general problem is enlarged on in chapter 20. Porosity determination from wireline logs is covered in chapter 22 and core analysis in chapter 26.

MEASUREMENTS OF PERMEABILITY

As with porosity, there are many ways to estimate permeability. These include:

Pressure buildup from drillstem tests (chapter 30)
Pressure drawdown and buildup from wireline repeat formation testers (RFT) (chapter 29)
Log analysis
Core analysis (chapter 26)

Again, many orders of magnitude separate the effective radius of investigation of each method. In decreasing order these are:

Method	Approximate radius of investigation, feet
DST	10^2 to 10^4
RFT buildup	10 to 10^2
RFT drawdown	10^{-2} to 10^0
Log analysis	5 to 10^{-1}
Core analysis	8×10^{-2} to 3×10^{-1}

Obviously the different methods of measurement produce different results. Disparate results are also to be expected in a heterogeneous reservoir. Where the drilling process has caused clay swelling in the invaded zone, it is to be expected that measurements made near the wellbore (logs, cores, RFT tests) will reflect permeabilities that are lower than true permeabilities. Disparities can also be expected in the permeability measurements made on cores, since they are influenced by the type of fluid (air or brine) used for the measurement and the pressure and temperature of the sample at the time of the measurement (standard or reservoir conditions).

Many investigators have attempted to correlate rock permeability to measurements made by wireline logging tools. These relationships fall into two categories: those that apply above the transition zone and those that apply only in the transition zone. A few examples are:

$$k = 0.136 \times \phi^{4.4} \times S_{wi}^{-2} \qquad\qquad\qquad \text{Chevron}$$

$$k = 0.105 \times \phi^{4.5} \times S_{wi}^{-2} \qquad\qquad\qquad \text{Timur (1968)}$$

$$k = (250 \times \phi^3 \times S_{wi}^{-1})^2 \quad \text{(oils)} \qquad\qquad \left.\begin{array}{l} \text{Schlumberger} \\ \text{after Wyllie and} \\ \text{Rose (1950)} \end{array}\right.$$

$$k = (79 \times \phi^3 \times S_{wi}^{-2})^2 \quad \text{(gas)}$$

$$k = \phi \times [122/b\,(\rho_w - \rho_o)]^2 \text{ (for O/W)} \qquad \left.\begin{array}{l} \text{Raymer and Freeman} \\ \text{(1984)} \end{array}\right.$$

$$k = \phi \times [140/b\,(\rho_w - \rho_g)]^2 \text{ (for G/W)}$$

where

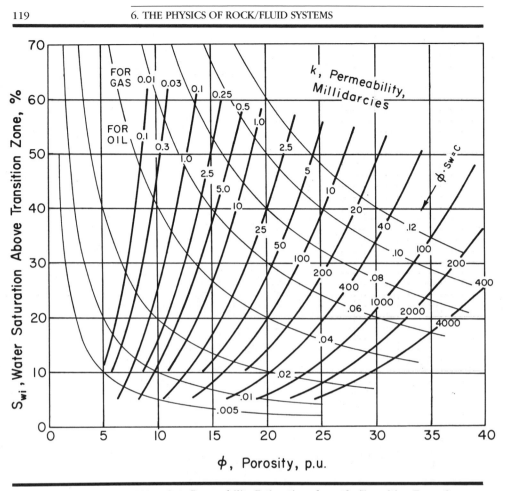

FIGURE 6.12 *Permeability Estimation above the Transition Zone. Courtesy Schlumberger Well Services.*

S_{wi} is fractional irreducible water saturation
ϕ is fractional porosity
h is height in feet from free-water level to the top of the
      transition zone
ρ_w is the water density in g/cc
ρ_o is the oil density in g/cc
ρ_g is the gas density in g/cc

Figure 6.12 shows a graphic representation of the Wyllie and Rose relationship above the transition zone; ϕ and S_{wi} are crossplotted to yield values of k.

In the transition zone, the resistivity gradient is commonly linear—that is, a resistivity log on a linear scale will frequently show a straight line in a transition zone. The resistivity gradient ($\Delta R/\Delta D$) measured in $\Omega \cdot m/ft$ is first normalized by division by R_0 (the wet rock resistivity, see chapter 7) and then related to k provided that the

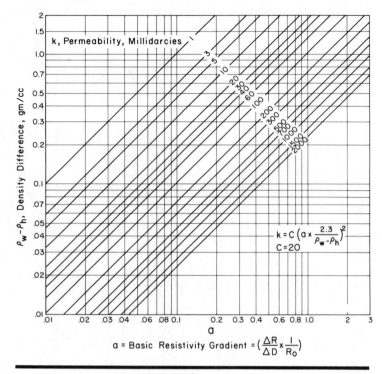

FIGURE 6.13 *Permeability Estimates from Transition Zone Resistivity Gradient. Courtesy Schlumberger Well Services.*

density difference ($\rho_w - \rho_b$) between the wetting and nonwetting phases is known. Figure 6.13 gives a graphic solution to the equation

$$k = c \left(\frac{2.3a}{(\rho_w - \rho_b)} \right)^2,$$

where

$a = \dfrac{\Delta R}{\Delta D} \times \dfrac{1}{R_0}$ (normalized resistivity gradient),

$c = 20$ (a constant to take care of units),

ρ_w and ρ_b are in g/cc,

$\Delta R / \Delta D$ is expressed in Ω/ft, and

k is in md.

MEASUREMENTS OF SATURATION

Fluid saturations for the most part are well measured by log analysis techniques, provided formations are clean and connate waters are saline. Problems arise with shaly formations and fresh formation waters (see chapters 25 and 27). Other methods of saturation

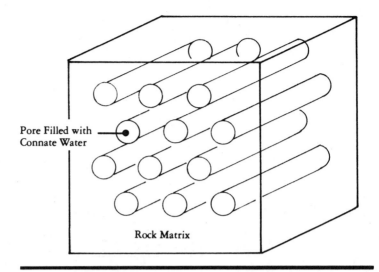

Pore Filled with
Connate Water

Rock Matrix

FIGURE 6.14 *Simplified Model of Water-Filled Rock System.*

determination are available from proper coring and core analysis techniques (see chapters 4 and 26). Mud logging can provide a qualitative measure of oil and gas saturations, as we have seen in chapter 2.

Many similarities exist between the flow of fluids through a rock and the flow of electric current through a rock. The permeability to water, for example, can be equated with the electrical conductivity of a porous system, since both depend on interconnected pores. In the case where both oil and water are present in a pore system there is a parallel between relative permeability to water and electrical conductivity of an oil-bearing sand. Investigation of the electrical properties of wet and oil-bearing rocks was pioneered by Archie. A good starting point for following the development of his experimental observations is the electrical behavior of electrolytes and water-filled rocks.

Water Resistivity R_w

Connate waters range in resistivity from about $1/100$ of an $\Omega \cdot m$ up to several $\Omega \cdot m$ depending on the salinity and temperature of the solution. To find water saturation by quantitative analysis of porosity and resistivity logs, a value of R_w is required.

Chapter 24 is dedicated exclusively to the methods available for finding R_w. Here, it is necessary only to understand that the ability of a rock to conduct electricity is due entirely to the ions in the water found in the pore spaces. Figure 6.14 shows a cube of rock with a system of cylindrical tubes drilled through it. If these cylindrical "pores" are filled with water of resistivity R_w, their total area is A and their length is L, we can estimate the resistivity of the total rock system as proportional to R_w (L/A). If the area A is small, there is a small

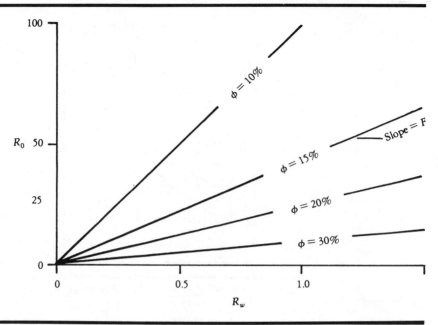

FIGURE 6.15 R_0 *as a Function of* R_w *for Rocks of Different Porosity.*

conductive path of length L and the resistivity of the rock system is high. Conversely, if A is large, the resistivity is low. The resistivity of a rock 100% saturated with water is referred to as R_0. Since A is proportional to the porosity itself, we may write

$$R_0 = f(R_w, \phi),$$

or say that R_0 is related to R_w by some formation factor F such that

$$R_0 = FR_w.$$

Electrical Formation Factor

The method Archie used to arrive at this conclusion was simple. He took a number of cores of different porosity and saturated each one with a variety of brines. He could then measure, at each brine salinity, the resistivity of the water, R_w, and the resistivity of the 100%-water-saturated rock system, R_0. When the results were plotted, he found a series of straight lines of slope F (see fig. 6.15).

F, determined in this manner, is referred to as the electrical formation factor. Archie conducted many experiments in an attempt to link the electrical formation factor to porosity in a predictable manner. (The simple tubular model bears little relationship to the tortuous paths that pores actually take; see fig. 6.16.) The factor L, length of the tubular pore, grows larger as the tortuosity of the pore system increases. It was thus a logical step to propose that F and ϕ had to be related by some inverse power function. Note that *by*

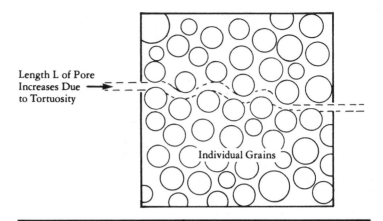

Length L of Pore Increases Due to Tortuosity →

Individual Grains

FIGURE 6.16 *Schematic of Tortuous Pore System.*

definition the formation factor is the ratio of R_0/R_w—that is, the resistivity of a rock sample 100% saturated with water to the resistivity of the water itself. Eventually Archie found that laboratory-measured values of F could be related to the porosity of the rock by an equation of the form

$$F = a/\phi^m$$

Where a and m are experimentally determined constants. a is usually close to 1 and m is usually close to 2 (fig. 6.17). Three commonly used formation factor to porosity relations are:

$$F = 1/\phi^2 \qquad \text{(Carbonates)}$$
$$F = 0.62/\phi^{2.15} \qquad \text{(Humble formula–soft formations)}$$
$$F = 0.81/\phi^2 \qquad \text{(Sands)}$$

In a wet formation we may therefore combine the F to ϕ relationship with the definition of F and arrive at the equation

$$R_w = R_0/F = R_0 \times \phi^m/a$$

Saturation

Archie's experiments showed that the saturation of a core could be related to its resistivity. He found that the fractional water saturation, S_w, was equal to the square root of the ratio of the wet formation resistivity, R_0, to the formation resistivity R_t, that is:

$$S_w^2 = R_0/R_t.$$

In a more generalized form, this equation can be written as:

$$S_w^n = R_0/R_t.$$

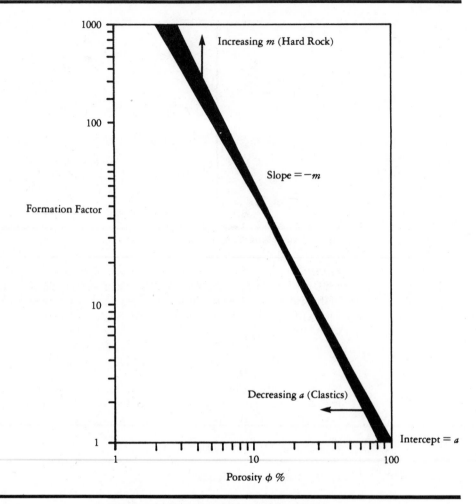

FIGURE 6.17 *Generalized F-to-φ Relationship.*

where n is the saturation exponent and is usually set to 2 (see fig. 6.18).

Because Archie's model considers the electrolyte in the pores as the *only* conductive path, his relationships work well in clean formations; but in shaly formations and where connate water is fresh, they do not work as well.

PRACTICAL PETROPHYSICS

A proper marriage between petrophysics and log analysis can produce some spectacular results. One of the most interesting is the observation that when a formation is above the transition zone, and thus at irreducible water saturation, the product of ϕ and S_w is a constant. Variations of porosity are normal on a local scale, owing to

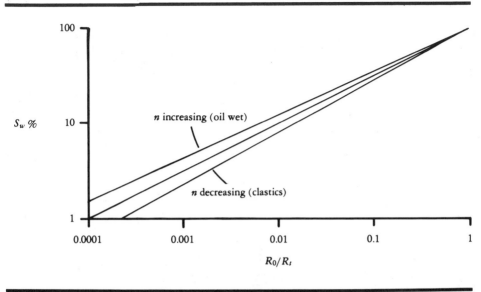

FIGURE 6.18 S_w as a Function of R_0/R_t and n.

changes in depositional environment at the time the sediments were laid down and owing to subsequent diagenesis. If, locally, porosity is reduced, a greater proportion of the pore throats are small ones or there are simply fewer pore throats. Either way, the mean radius r is smaller, thus p_c is larger and more water can be held in the pore maintaining the constant $\phi \times S_{wi}$ product.

This observation has a practical application. After a zone has been analyzed on a foot-by-foot basis for porosity and water saturation, a plot of ϕ versus S_w will reveal the presence or absence of a transition zone. Figure 6.19 illustrates the phenomenon. Note that, on figure 6.19, ϕ and S_w are on linear scales and the points at irreducible saturation fall on a hyperbola. Figure 6.20 shows a similar plot on log-log paper, where points at irreducible saturation plot on a straight line. Points in the transition zone plot to the right of the irreducible line.

Reservoirs may be characterized by the $\phi \times S_{wi}$ product, and this knowledge can be used as a basis for predicting production characteristics. For points not at irreducible saturation, some water production is to be expected depending on the mobility ratio $(k_w\mu_o/k_o\mu_w)$ for the particular fluids present. Figure 6.21 shows the $\phi \times S_{wi}$ product at irreducible saturation for a number of formations. Note that in a low-porosity, low-permeability formation, surprisingly high water saturations can be tolerated without fear of water production. Conversely, in formations with good porosity and permeability, even when they have moderate values of S_w, water production can be expected. This salient fact is all too often overlooked by those who were once taught (1) if S_w is less than 50% (or 60%, or whatever) there are no more troubles and (2) do not

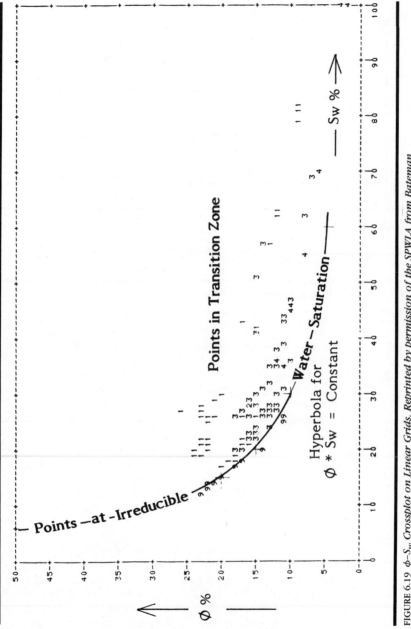

FIGURE 6.19 ϕ–S_w Crossplot on Linear Grids. Reprinted by permission of the SPWLA from Bateman 1984.

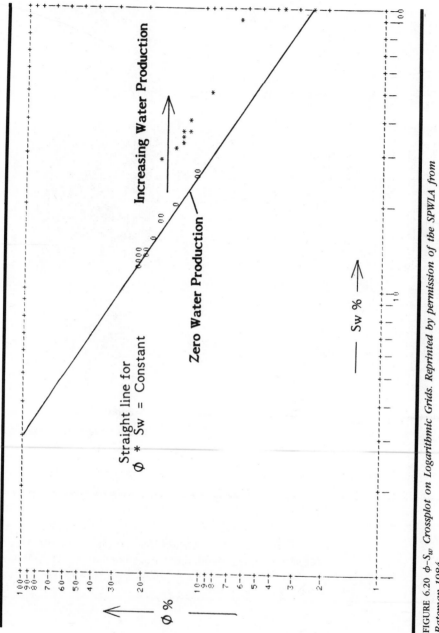

FIGURE 6.20 ϕ–S_w Crossplot on Logarithmic Grids. Reprinted by permission of the SPWLA from Bateman 1984.

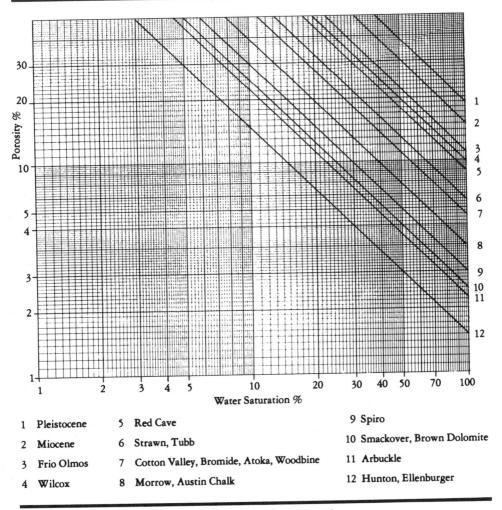

1   Pleistocene   5   Red Cave   9   Spiro

2   Miocene   6   Strawn, Tubb   10 Smackover, Brown Dolomite

3   Frio Olmos   7   Cotton Valley, Bromide, Atoka, Woodbine   11 Arbuckle

4   Wilcox   8   Morrow, Austin Chalk   12 Hunton, Ellenburger

FIGURE 6.21 $\phi \times S_{wi}$ *for Various Formations.*

attempt completions if S_w is greater than so much percent. From those who are now enlightened and emboldened by petrophysics let us see some completions in zones of 65%, 70%, and even 75% S_w, provided permeability is low.

AVERAGING

Averaging is a dangerous pastime. The average depth of the Mississippi is 3 ft, but that does not mean you can try to wade from one side to the other at New Orleans without being prepared either to swim or to drown. So when we talk of *average* porosity, *average* saturation, and *average* permeability we should be careful how we calculate them and how we use them.

For example, a well has two zones. One is at 10% porosity and the other is at 20% porosity. Is the average porosity 15%? Not necessarily. If the 10% zone is 30 ft long and the 20% zone is 50 ft long, obviously the average is skewed in favor of the longer zone. The correct way to average porosities, therefore, is to use

$$\phi_{av} = \frac{\Sigma \phi h}{\Sigma h}$$

where h is the formation thickness. In the case of the above two zones:

$$\phi_{av} = \frac{10\% \times 30 \text{ ft} + 20\% \times 50 \text{ ft}}{30 \text{ ft} + 50 \text{ ft}}$$

$$= \frac{300 + 1000}{80}$$

$$= 16.25\%.$$

Averaging water saturation is even more dangerous. For a 30-ft zone at 15% S_w and a 50-ft zone at 25% S_w we cannot compute an average S_w without also knowing the porosity—that is, $S_{w_{av}}$ is *not* given by:

$$\frac{30 \times 15 + 50 \times 25}{80}.$$

Why? Because figures for the total porosity feet and the total hydrocarbon pore feet are what should be used in the averaging—these must be calculated:

	Feet	Porosity %	Saturation %	$\phi \times Ft$	$\phi(1 - S_w) \times Ft$
	30	10	15	3.00	2.55
	50	20	25	10.00	7.50
Total	80			13.00	10.05

Thus, the two zones combined have 13.00 porosity feet and 10.05 hydrocarbon porosity feet, so the average oil saturation is 77.31% (10.05/13.00) and the average water saturation is 22.69% (2.95/13.00). $S_{w_{av}}$, therefore, is defined as:

$$S_{w_{av}} = \frac{\Sigma \phi S_w h}{\Sigma \phi h}.$$

These sums and averages are most conveniently done using computers working with a digital log base. Permeability averaging may be achieved in a number of ways depending on the intended use of the average itself.

For parallel flow use the *arithmetic average:*
$$\overline{k}_a = (k_1 + k_2 + k_3 \cdots k_n)/n.$$
For series flow use the *harmonic average:*
$$\overline{k}_b = (1/k_1 + 1/k_2 + 1/k_3 \cdots 1/k_n)/n.$$
For random flow use the *geometric average:*
$$\overline{k}_g = (k_1 \, k_2 \, k_3 \cdots k_n)^{1/n}.$$

SUMMARY

The conduction of electric current through a porous rock is conceptually similar to the flow of fluid through the rock. Thus, wireline-tool measurements of formation conductivity are related to formation porosity, permeability, and fluid saturation. By combining the basic relationships established by Archie with the physics of fluid distribution and flow in a reservoir, the analyst may determine the free-water level, the length of the transition zone, and the irreducible water saturation. From these parameters, productivity can be estimated.

BIBLIOGRAPHY

Archie, G. E.: "The Electrical Resistivity Log as an Aid in Determining Some Reservoir Characteristics," *J. Pet. Tech.* (1942).

Bateman, R. M.: "Watercut Prediction from Logs Run in Feldspathic Sandstone with Fresh Formation Waters," paper EE, SPWLA 25th Annual Logging Symposium, New Orleans, June 10-13, 1984.

Bobek, J. E., Mattax, C. C., and Denekas, M. O.: "Reservoir Rock Wettability—Its Significance and Evaluation," *Trans.*, AIME (1958) **213**, 155-160.

Bradley, J. S.: "Fluid and Electrical Formation Conductivity Factors Calculated for a Spherical-Grain Onion-Skin Model," *The Log Analyst* (January–February 1980) **21**, 24-32.

Coates, G. R., and Dumanoir, J. L.: "A New Approach to Improved Log-Derived Permeability," *The Log Analyst* (January–February 1974) **15**, 17-29.

de Witte, L.: "Relations between Resistivities and Fluid Contents of Porous Rocks," *Oil and Gas J.* (August 24, 1950).

Goetz, J. F., Prins, W. J., and Logar, J. F.: "Reservoir Delineation by Wireline Techniques," 6th Annual Convention of the Indonesia Petroleum Association, Jakarta, May 1977.

Graton, L. C., and Fraser, H. J.: "Systematic Packing of Spheres—With Particular Relation to Porosity and Permeability," *J. Geol.* (1935) **43**, 785-909.

Honarpour, M., Koederitz, L. F., and Harvey, A. H.: "Empirical Equations for Estimating Two-Phase Relative Permeability in Consolidated Rock," *J. Pet. Tech.* (December 1982).

Le Blanc, R. J., Sr.: "Distribution and Continuity of Sandstone Reservoirs" (Parts 1 & 2), *J. Pet. Tech.* (July 1977), 776-804.

Leverett, M. C.: "Capillary Behavior in Porous Solids," *Trans.*, AIME (1941).

Molina, Nelson N.: "Systematic Approach Aids Reservoir Simulation," *Oil and Gas J.* (April 1983).

Morris, R. L., and Biggs, W. P.: "Using Log-Derived Values of Water Saturation and Porosity," SPWLA 8th Annual Logging Symposium, June 11–14, 1967.

Pirson, S. J., Boatman, E. M., and Nettle, R. L.: "Prediction of Relative Permeability Characteristics of Intergranular Reservoir Rocks from Electrical Resistivity Measurements," *J. Pet. Tech.* (May 1964) 565–570.

Raymer, L. L.: "Elevation and Hydrocarbon Density Correction for Log Derived Permeability Relationships," *The Log Analyst* (May–June 1981) 3–7.

Raymer, L. L., and Freeman, P. M.: In-Situ Determination of Capillary Pressure, Pore Throat Size and Distribution, and Permeability from Wireline Data," paper CCC, SPWLA 25th Annual Logging Symposium, June 10–13, 1984.

Raza, S. H., Treiber, L. E., and Archer, D. L.: "Wettability of Reservoir Rocks and Its Evaluation," *Producers Monthly* (April 1968).

Rockwood, S. H., Lair, G. H., and Langford, B. J.: "Reservoir Volumetric Parameters Defined by Capillary Pressure Studies," *Trans.,* AIME (1957) **210**.

Schlumberger: "Log Interpretation Principles," 1969.

Schlumberger: "Log Interpretation: Volume I—Principles," 1972.

Schlumberger: "Log Interpretation Charts," 1977.

Timur, A.: "An Investigation of Permeability, Porosity, and Residual Water Saturation Relationship for Sandstone Reservoirs," *The Log Analyst* (July–August 1968) **9**, 8–17.

Tixier, M. P.: "Evaluation of Permeability from Electric-Log Resistivity Gradients," *Oil and Gas J.* (June 16, 1949) **48**, 113.

Wyllie, M. R. J., and Rose, W. D.: "Some Theoretical Considerations Related to the Quantitative Evaluation of the Physical Characteristics of Reservoir Rock from Electrical Log Data," *J. Pet. Tech.* (April 1950) **189**, 105–108.

CHAPTER SEVEN

BASIC CONCEPTS OF LOG ANALYSIS

Successful completion of this section will enable the reader to pick up a suite of logs and perform basic formation evaluation, including:

Identification of porous and permeable reservoir rocks
Porosity estimation
Water saturation calculation
Differentiation between oil- and gas-bearing sections

Before confronting these tasks, it may be profitable to reflect for a moment on the general scope of log analysis and the tasks a log analyst must address. An easy-to-grasp example of the sort of problem with which the log analyst is faced would be to consider one cubic foot of rock of reasonable porosity and containing one gallon of oil (fig. 7.1). The surface area of the grains of sand in that cubic foot of rock might be one acre. Spreading the gallon of oil over an area of one acre results in a layer only 37 millionths of an inch thick. The task of the log analyst is to detect this thin layer of oil and then assist in "producing" it. The fundamental questions to be answered by the analyst are:

What kind of rock is present?
Are any hydrocarbons present and if so should a test be run and the well completed?
What kind of hydrocarbon is it, oil, gas, or condensate?
How much is there? (net pay, porosity, water saturation, etc.)

The parameters needed to answer these questions are:

The porosity and lithology
The water saturation
The bed thickness

and the skill required to find on a log the data necessary for the relevant quantitative calculations.

A generalized "answer" log (plotted on subsea depths) is shown in figure 7.2. This log gives a visual account of the main formation characteristics of interest—porosity, water saturation, rock type, and hydrocarbon type. More sophisticated presentations might include other parameters, and these will be covered later, once a better understanding has been gained of the principles of measurement of the various logging tools available and a more detailed treatment of interpretation has been made.

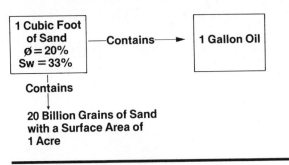

FIGURE 7.1 *Reservoir Perspective.*

On the "answer" log of figure 7.2, the logged column, from top to bottom, shows a shale section over a porous and permeable section containing both a gas cap and an oil column. Below the oil leg, a transition zone leads into a water column. The log is divided into four tracks. From left to right: the depth column shows a gas flag; Track 1 shows shale content and permeability; Track 2 shows the porosity and the distribution of fluids within the pore space; Track 3 shows water saturation and highlights intervals with a water saturation less than 50%; Track 4 gives a picture of the bulk rock volume by splitting it into three components, sand, shale, and pore space. From these four tracks, the analyst may deduce all he needs to know about the logged column. All the necessary information is available for making decisions about completion and testing and also for calculation of the hydrocarbon reserves.

Let us review the steps that went into evolving an answer log of this sort and trace how a log analyst proceeds when studying a suite of logs. Our review will cover all the basic steps involved; the chapters that follow will add many refinements.

LITHOLOGY

Our starting point will be to inspect the section that has been logged and discard from further attention the parts we do not want. Zones of no further interest will be shales and evaporites with zero porosity and permeability (i.e., the nonreservoir rocks). As a rule, this step can be accomplished by taking a marker pen and color coding only those sections that are of interest—that is, those showing porosity and permeability. The criteria used for this visual sifting of the data will be as follows:

Reservoir Rocks	*Nonreservoir Rocks*
Low gamma ray	High gamma ray
Good SP development	Flat SP
Relative separation of resistivity curves	Stacked resistivity curves

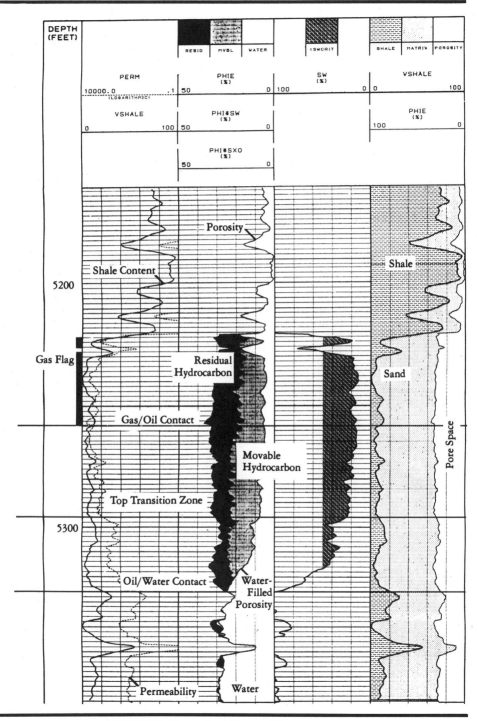

FIGURE 7.2 *Generalized "Answer" Log.*

The logs in figures 7.3 and 7.4 are from a sand–shale sequence. Figure 7.3 shows a dual laterolog–GR and figure 7.4 shows a neutron–density–GR log.

As an exercise in log analysis, it is suggested that the first step should be to delineate the porous and permeable sections on the example log. With a yellow highlighter, mark up the gamma ray log in figure 7.3 or 7.4 to show the clean (shale-free) zones. A practical way to do this is to color from the middle of Track 1 *leftward* to the gamma ray curve. This should give a good visual image of the sand and shale sections. Attention can then be concentrated in the section containing reservoir rock.

POROSITY

The porosity of a formation is defined as the volume of the pore space divided by the volume of the rock containing the pore space. This definition of porosity ignores the question of whether the pores are interconnected or not. Swiss cheese, for example, is quite porous, but is of very low permeability since the void spaces are not interconnected. Intergranular porosity that is interconnected is *effective* porosity. Pores that are blocked in some way (by clay particles, silt, etc.), are *ineffective.* Thus, a preferred definition gives total porosity (ϕ_T) as the volume of the pores divided by the volume of rock and effective porosity (ϕ_e) as the volume of the *interconnected* pores divided by the volume of rock (see fig. 7.5).

In a laboratory, porosity can be measured in a number of ways. One of the simplest is to weigh a sample of rock when it is 100% saturated with water, then remove all the water and reweigh it. Provided the density of the rock matrix (or the volume of the rock sample) is known, the porosity can be calculated.

QUESTION #7.1

A sample of porous sandstone saturated with water is found to weigh 215.5 g. After all the water is removed from the sample it weighs 185.5 g. If the density of the sandstone matrix is 2.65 g/cc and the density of the water is 1 g/cc, what is the porosity of the sample?

The density logging tool provides an in situ bulk-rock density measurement that, if properly rescaled, can be displayed as a porosity trace. The solid curve in Tracks 2 and 3 in figure 7.4 is such a recording. For the time being, the dashed curve (neutron) may be ignored.

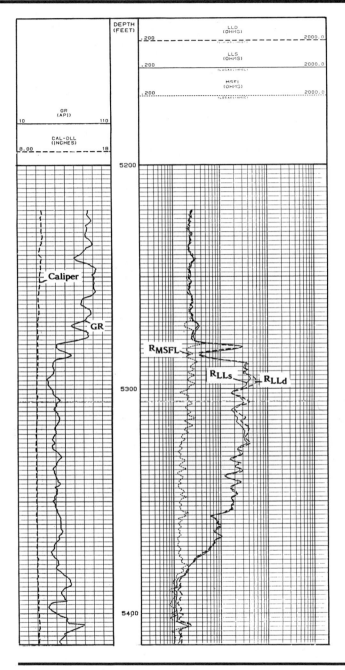

FIGURE 7.3 *A Dual Laterolog Resistivity Log.*

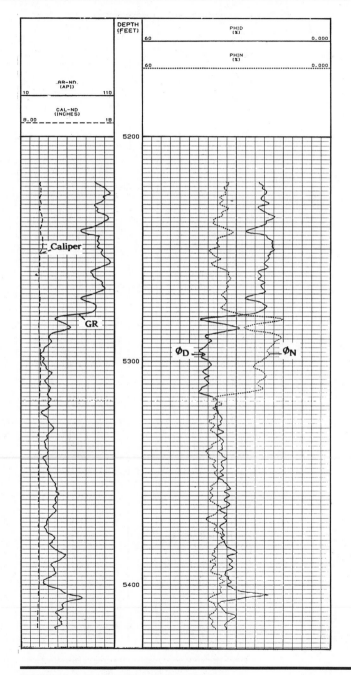

FIGURE 7.4 *A Neutron–Density Porosity Log.*

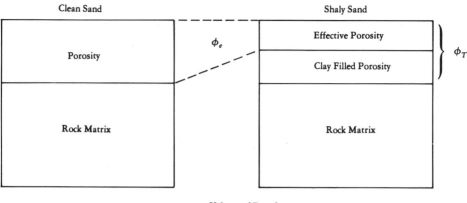

$$\text{Porosity } \phi = \frac{\text{Volume of Pore Space}}{\text{Volume of Rock}}$$

FIGURE 7.5 *Definitions of Total and Effective Porosity.*

QUESTION #7.2
On the example log shown in figure 7.4, the porosity scale is marked
from 0 to 60% (right to left) over Tracks 2 and 3. Using the density/
porosity curve (the solid trace) read the average porosity in the zone
5320 to 5350 ft.

On the example log (fig. 7.4), good porosity development
(33–36%) is seen for the sand section. The shale section at the top of
the log appears to show around 21% porosity, and the gas cap
around 40%. Both of these apparent porosities reflect the effects of
shale and light hydrocarbons and for the present purposes can be
ignored.

WATER SATURATION: S_w
Water saturation is defined as the volume fraction of the pore space
occupied by water (see fig. 7.6). Note that the bulk volume of water
is given by the product ϕS_w; the hydrocarbon saturation is given by
$(1 - S_w)$; and the bulk volume of hydrocarbon is $\phi (1 - S_w)$.
From chapter 6, where the relationship between rock resistivity
and water saturation was defined, we know that

$$S_w = (R_0/R_t)^{1/2}.$$

This equation can be put to good use on the example logs. If, on
figure 7.3, we can read a value for R_0 in the wet zone and a value for
R_t in a hydrocarbon zone, then, if the porosities in each zone are the
same, we can compute a water saturation using Archie's equation.

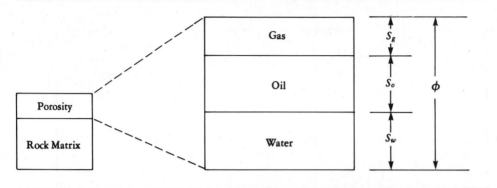

FIGURE 7.6 *Definitions of Porosity and Water Saturation.*

QUESTION #7.3
Read R_0 at 5420 ft.
Read R_t at 5297 ft.
Compute S_w.

In cases where R_0 cannot be read in a wet zone of equal porosity, recourse is made to Archie's equation written in the form

$$S_w^n = F \times R_w/R_t.$$

Solution of this equation requires knowledge of both F, the formation factor, and R_w, the connate water resistivity. Again from chapter 6, we have

$$R_w = R_0/F = R_0 \phi^m/a,$$

which can be used to find R_w from the logs shown in figures 7.3 and 7.4. The oil/water contact in this well is at 5384 ft. Below this level, a value for R_0 can be read from the dual laterolog and a value (ϕ_D) for porosity in the wet zone can be read from the density log. These can be combined to find a value for R_w.

QUESTION #7.4
Read R_0 in the interval 5390–5400 ft.
Read ϕ_D in the same interval.
Assume $F = 0.62/\phi^{2.15}$.
Calculate R_w.

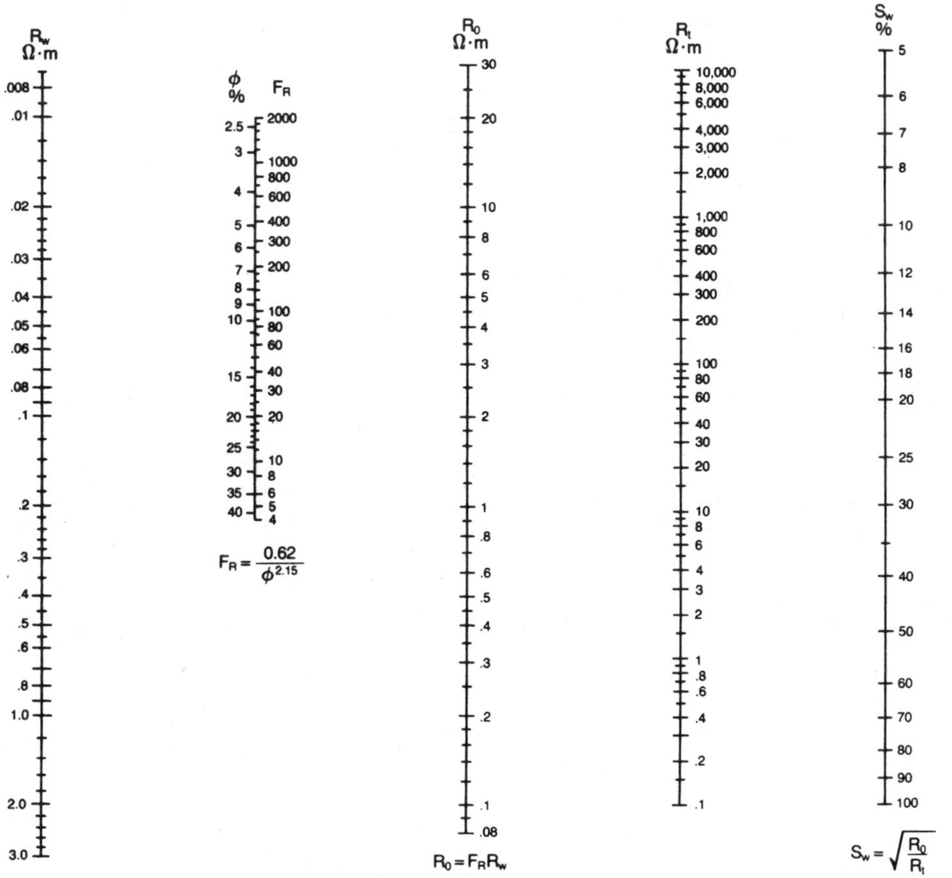

R_w $\Omega \cdot m$

.008 .01 .02 .03 .04 .05 .06 .08 .1 .2 .3 .4 .5 .6 .8 1.0 2.0 3.0

ϕ % F_R

2.5 — 2000
3 — 1000, 800, 600
4 — 400
5 — 300
6 — 200
7
8
9
10 — 100, 80, 60
15 — 40, 30
20 — 20
25
30 — 10, 8
35 — 6
40 — 5, 4

$$F_R = \frac{0.62}{\phi^{2.15}}$$

R_0 $\Omega \cdot m$

30, 20, 10, 8, 6, 5, 4, 3, 2, 1, .8, .6, .5, .4, .3, .2, .1, .08

R_t $\Omega \cdot m$

10,000, 8,000, 6,000, 4,000, 3,000, 2,000, 1,000, 800, 600, 400, 300, 200, 100, 80, 60, 40, 30, 20, 10, 8, 6, 4, 3, 2, 1, .8, .6, .4, .2, .1

S_w %

5, 6, 7, 8, 10, 12, 14, 16, 18, 20, 25, 30, 40, 50, 60, 70, 80, 90, 100

$$R_0 = F_R R_w$$

$$S_w = \sqrt{\frac{R_0}{R_t}}$$

FIGURE 7.7 *Nomogram for Saturation Determination. Courtesy Schlumberger Well Services.*

With R_w defined, Archie's equation may be solved.

QUESTION #7.5
Read the porosity at 5350 ft.
Convert this to an F using $a = 0.62$, $m = 2.15$.
Read R_t at 5350 ft.
Use $R_w = 0.18 \, \Omega \cdot m^2/m$.
Compute S_w at 5350 ft.

A useful chart for solving Archie's equation is shown in figure 7.7. To use this chart (known as a nomogram), draw a first line from R_w through ϕ to find R_0. Then draw a second line from R_0 through R_t to find S_w. Thus, water saturation can be calculated on a point-by-point basis algebraically or by use of a nomogram. It is worth digressing to discuss a wellsite quick-look method for calculating water saturation on a semicontinuous basis.

Effectively, S_w is just a scaled ratio of R_0/R_t. A quick, practical method of finding and scaling this ratio is afforded by the "F" overlay. This method allows comparison of R_t to R_0 at all points on the log. To understand this method and its usefulness we must first modify Archie's equation by taking logs of both sides:

$$S_w^n = F \times R_w/R_t,$$

therefore,

$$n \log S_w = \log F + \log R_w - \log R_t \quad \text{or}$$
$$n \log S_w = \log R_0 - \log R_t.$$

This demonstrates that if a logarithmic scale is used to present R_t, then the distance between a logarithmic R_0 curve and the logarithmic R_t curve is equal to n times the log of S_w.

In practice, the F curve is generated at the time the density log is run by use of a function former that turns the density/porosity value into a formation factor (F).

$$F_D = a/\phi_D^M$$

An example F log (from the well of fig. 7.3) is given in figure 7.8. In practice, the R_t curve is traced onto the F curve after normalizing the F curve to the R_t curve in the wet zone. This effectively turns the F curve into an R_0 curve. The left-to-right adjustment of the F trace relative to the R_t trace, to make them coincide in the wet zone, effectively takes care of adding log R_w to log F, the result being a trace of log R_0. This "F" overlay can now be read using an S_w scaler as described in chapter 21.

HYDROCARBON TYPE

The neutron–density log (see fig. 7.4) can be used to distinguish between oil and gas. For reasons that will be covered later when we discuss the neutron and density tools and their interpretation in chapter 22, we can expect gas to leave its "fingerprint" on the neutron–density combination. The neutron log (dashed curve) will read less than true porosity and the density log (solid curve) will read more than true porosity in gas-bearing formations.

QUESTION #7.6

Inspect figure 7.4.

Color code (red) between the neutron and density logs only where the neutron value is *less* than the density value. Hence, pick the gas/oil contact (GOC).

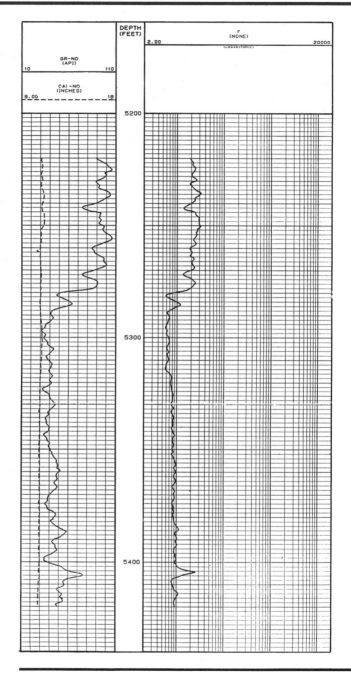

FIGURE 7.8 *F Log.*

PAY COUNTING

Chapter 8 covers reserve calculations in detail. We will take a preview here concerning the various ways to quantify *pay*—that is, the amount of hydrocarbon calculated from logs.

Net pay is measured in feet. For example, the well logged in figure 7.3 has a gross pay section from 5280 ft down to 5384 ft (dual laterolog depths). However, not all of this section will produce hydrocarbon. There are low-porosity shale streaks that will not produce, and there are sections at the base of the interval that will produce water because S_w is too high.

QUESTION #7.7
Find the net pay in feet for the example well.

This net pay can be further classified by a breakdown into so many feet of oil and so many feet of gas.

QUESTION 7.8
Subdivide the net pay into oil and gas pay.

But "*X* feet of pay" is not very informative unless you know how good or bad each foot is. It is necessary to quantify "net pay" with reference to hydrocarbon pore volumes.

Oil-in-place is defined as the fraction of the rock volume occupied by hydrocarbons. For each foot of formation a water saturation and a porosity define the hydrocarbon pore volume (HCPV) by

$$HCPV = \phi \times (1 - S_w).$$

For example, a 35% porous formation at 15% water saturation has a hydrocarbon pore volume of $0.35(1 - 0.15) = 0.3$—that is, 30% of the rock volume is occupied by oil. This is a useful number. If the total reservoir volume is known (as delineated by logs, geological studies, seismic records, mapping, etc.), then the total volume of oil-in-place (OIP) in the reservoir is given by HCPV $\times$ reservoir volume; and the units of oil-in-place will be HCPV $\times$ feet. Since HCPV is a dimensionless fraction, the units of OIP are feet. Because only a fraction of the oil-in-place is recoverable, other factors must be taken into account when speaking of recoverable reserves, and this will be covered in the next chapter.

PERMEABILITY

In chapter 6, the question of permeability estimates from logs was discussed. Two methods were described. One rested on empirical correlations between porosity and irreducible water saturation (S_w

above the transition zone), and another relied on the resistivity gradient in the transition zone itself. These methods can be applied to the example logs in figures 7.3 and 7.4.

QUESTION #7.9
Estimate the permeability of the oil sand at 5345 ft on the example logs of figures 7.3 and 7.4.

QUESTION #7.10
Estimate the permeability of the gas sand at 5296 ft on the example logs of figures 7.3 and 7.4.

QUESTION 7.11
On the dual laterolog of figure 7.3, estimate the resistivity gradient in $\Omega \cdot m/ft$ for the transition zone and then estimate the formation permeability, assuming the density difference $(\rho_w - \rho_{hy})$ is 0.2 g/cc.

SUMMARY
Application of a few basic principles enables the analyst to work from raw logs through to quantitative pay counts. The steps involved included:

Definition of porous and permeable zones
Reading porosity values from a porosity log
Recognizing hydrocarbon-/and water-bearing zones by inspection of a resistivity log
Applying Archie's relationships in order to find R_w and S_w
Making sums of pay thickness and quality
Estimating permeability

Refinements on these clean-formation procedures, to be studied later, will not in any way alter this basic method of formation evaluation from logs.

BIBLIOGRAPHY
Schlumberger: "Log Interpretation Charts" (1984).

Answers to Text Questions

QUESTION #7.1
We need to know two things: the volume of the rock sample including its pores and the volume of the pore space itself. Since the water removed from the sample entirely filled the pore space, it follows that the volume of the water equals the pore volume. Thus, the pore volume (215.5 − 185.5) g at 1 g/cc = 30 cc. The volume of the rock itself is given by its weight divided by its density or 185.5/2.65, which gives 70 cc of dry rock. This total volume of the rock + the pores is thus 100 cc. The porosity of this system is thus 30/100 = 0.3, or 30%.

QUESTION #7.2
$\phi_{av} = 35\%$

QUESTION #7.3
$S_w = 14.4\%$

QUESTION #7.4
$R_w = 0.18$

QUESTION #7.5
$S_w = 24.7\%$

QUESTION #7.6
GOC is at 5314 ft

QUESTION #7.7
72 ft of net pay

QUESTION #7.8
38 ft net oil pay
34 ft net gas pay

QUESTION #7.9
2500 md

QUESTION #7.10
500 md

QUESTION #7.11
0.5 Ω/ft, hence 700 md

RESERVE ESTIMATION

For many end users of log data, the single most important calculation is that of the reserves of hydrocarbons in the reservoir. This chapter will cover the mathematics and philosophy of reserve estimation. It should be noted that accurate reserve figures rely not only on log data and computations from logs but on other data as well. The size and shape of a reservoir are delineated by seismic records and other means independently of log data gathered in one or many wells. Correlation of logs from many wells in a field will undoubtedly help in defining the size, shape, and limits of the subsurface trap. Dipmeter data will also be of assistance. However, in the end analysis, logs only define the concentration of oil- or gas-in-place in terms of bulk-volume fractions. The extrapolation of these figures to reserves requires more than logs alone can provide. In summary, a log analyst can say with a reasonable degree of certainty that, for example, 10% of the volume of a bottle is full of oil. It is up to others to determine the size of the bottle and hence deduce the actual volume of oil available.

OIL- AND GAS-IN-PLACE ESTIMATES

Consider the circumstances depicted in figure 8.1. A cylindrical block of porous rock has a volume V, a porosity ϕ, and a water saturation S_w. The oil-in-place in such a block is $\phi V (1 - S_w)$ in the units of V. If V is expressed in cubic feet or cubic meters then the oil-in-place is expressed in the same units, since ϕ and S_w are merely fractions of 1.

A more general case would be as depicted in figure 8.2. Some reservoir thickness h is delineated to exist over an area A such that $V = h A$. In this case, the units of h and A must be considered in order to produce a volume V. If h is measured in feet and A in acres, the reservoir volume V is expressed in acre-feet. If h is in meters and A in hectares, then V is expressed in hectare-meters.

So far only reservoir units at constant porosity and saturation have been considered. In actual reservoirs, both porosity and saturation vary laterally and vertically. A useful quantity for oil-in-place measurements is therefore the hydrocarbon pore volume, or HCPV, which is defined as

$$\text{HCPV} = \phi(1 - S_w).$$

Thus, at any depth in a well, if both porosity and saturation are deduced from logs the concentration of hydrocarbon in the

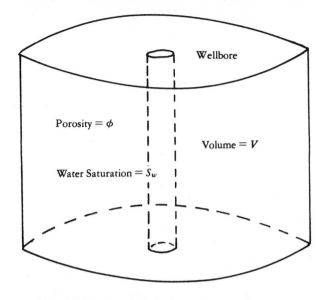

FIGURE 8.1 *Oil-in-Place (Cylindrical Reservoir Unit).*

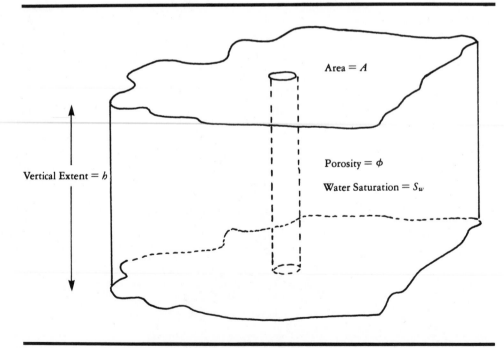

FIGURE 8.2 *Oil-in-Place (Generalized Reservoir Unit).*

reservoir at that depth can be estimated. For example, if porosity is 30% and water saturation is 40%, then HCPV = 0.3(1 – 0.4) = 0.18—18% of the bulk-reservoir volume contains oil. At a neighboring point in the same well, the value of HCPV may be different. Thus, in order to sum the total oil-in-place (OIP), an integration of HCPV with respect to depth and area is called for:

$$OIP = \Sigma\, \phi \times (1 - S_w) \times h \times A.$$

If h is measured in feet and A in acres, the OIP will be expressed in acre-feet. It is common, however, to express the reservoir content as a concentration by leaving out the unknown quantity A. Thus, the total HCPV for the well becomes a fraction times h and is therefore in the same units as h. A length may also be expressed as a volume divided by an area. Thus, once all the individual HCPV values for all depths have been summed, the sum may be expressed in units of barrels/acre or cubic meters/hectares by the simple expedient of a numerical constant dependent on the units chosen.

1 foot = 7757.79 barrels/acre (bbl/acre)
1 meter = 10,000 cubic meter hectare (m^3/ha)

Expressing HCPV in bbl/acre or m^3/ha gives flexibility in the application and use of the numbers found. If, for example, log data show a total HCPV of 10 ft, this may be expressed as 77,578 bbl/acre. If the area of the reservoir is known, the oil-in-place may be calculated directly. If, subsequently, new information becomes available and the mapped area changes, it is a simple matter to recalculate the oil-in-place. It should be noted that the term oil-in-place is used rather laxly since the hydrocarbon could equally well be gas, but none of the expressions change from the mathematical point of view when gas is substituted for oil.

An even more generalized method of expressing the $\phi\,(1 - S_w)\,h$ product summation is in terms of bbl/acre-ft, which is unitless. This, mathematically, is equal to

$$\frac{\Sigma\, \phi\, (1 - S_w)\, h}{\Sigma h}.$$

Once such a number is generated it can be used to extrapolate oil-in-place values for any proposed net pay thickness h or areal extent A.

RESERVE ESTIMATES

So far we have only discussed the volume of hydrocarbon found in a trap. Although this of course is fundamental, the bottom line is the number of stock tank barrels that can be recovered from such a trap. The conversion of oil-in-place to reserves requires two additional

TABLE 8.1 *Units Multiplier C for Reserves*

Use $C =$	Where A is in	h is in	and Hydro-carbon is	To find N in
7.7579×10^{-3}	acres	feet	oil	MMSTB
43.56×10^{-6}	acres	feet	gas	BCF
10×10^{-3}	hectares	meters	oil	MM m^3
10×10^{-3}	hectares	meters	gas	MM m^3

$$MM\ STB = 10^6 \quad \text{stock tank barrels}$$
$$BCF = 10^9 \quad \text{standard cubic feet}$$
$$MM\ m^3 = 10^6 \quad \text{cubic meters}$$

pieces of data: the recovery factor, r, and the formation volume factor, B. Neither of these can be estimated from logs. The recovery factor is a function of the type of reservoir and the drive mechanism, and the formation volume factor is a function of the hydrocarbon properties. The reserves, N, in terms of stock tank volumes are thus expressed as

$$N = \frac{C \times \Sigma \phi (1 - S_w) h}{B} \times A \times r,$$

where C is a constant that takes into account the system of units used for h and A. Table 8.1 lists the value of C for English and metric units in both oil and gas reservoirs.

QUESTION #8.1
From logs, it is determined that a productive formation has a bed thickness of 20 ft, a porosity of 25%, and a water saturation of 30%. Recovery factor is assumed to be 30%. The oil formation volume factor is 1.3. The areal extent of the reservoir is 200 acres. Find

a. HCPV expressed in feet.
b. oil-in-place expressed in bbl/acre.
c. reserves expressed in stock tank barrels.

RECOVERY FACTORS
A number of studies have been published that attempt, either by theoretical means or by empirical studies of historical data, to relate the primary recovery factor to formation parameters that can be deduced from logs. Among these, work of Poston (1984) and Muskat (1949) is useful. Approximate recovery factors may be calculated using the following equations for water-drive (WD) and solution-gas-drive (SG) reservoirs, respectively:

$$r_{WD} = 54.9 [\phi (1 - S_w)/B_o]^a (k \mu_w/\mu_o)^b S_w^c (p_i/p_a)^d$$

where
r_{WD} = recovery factor for water-drive reservoirs,
  a = 0.0422,
  b = 0.077,
  c = −0.1903,
  d = −0.2159,
  p_i = initial reservoir pressure, psia,
  p_a = abandonment reservoir pressure, psia,
  μ_w = water viscosity, cp,
  μ_o = oil viscosity, cp;

and

$$r_{SG} = 41.82 \, [\phi \, (1 - S_w)/B_{ob}]^a \, (k/\mu_{ob})^b \, (S_w)^c \, (p_b/p_a)^d \,,$$

where

r_{SG} = recovery factor for solution-gas-drive reservoirs,
B_{ob} = oil formation volume factor at bubble-point pressure,
μ_{ob} = oil viscosity at bubble-point pressure,
p_b = bubble-point pressure,
  a = 0.1611,
  b = 0.0979,
  c = 0.3722,
  d = 0.1741.

For both equations, k is in darcys and ϕ and S_w are fractions.

QUESTION #8.2
Calculate the primary recovery in a water-drive reservoir given the following data:

ϕ = 20%
S_w = 35%
k = 250 md
B_o = 1.25
μ_w = 0.5 cp
μ_o = 1.5 cp
p_i = 2000 psia
p_a = 750 psia

FORMATION VOLUME FACTORS
Because of dissolved gas present in oil at reservoir conditions, the volume of oil recovered at surface is less than the volume of oil in the formation. The shrinkage is quantified by the oil formation volume factor, B_o. It is a function of the gas gravity, γ_g, the

solution-gas/oil ratio, R_{sb}, and the formation temperature, T. The factors required are not always to hand in a wildcat situation, and oil and gas samples may need to be collected and submitted for PVT analysis.

Gas formation volume factors depend on formation pressure and temperature and the supercompressibility factor Z for the particular gas in question. In general, B_g will be a small fraction of 1—that is, many tens or hundreds of times as much gas will be recovered at surface as exist at formation conditions. For further details, the reader is referred to Bateman (1984).

DEPTH AND NET PAY MEASUREMENTS

When summing pay, it is normal practice to compute a number of useful items such as:

Gross section
Net porous and permeable section
Gross pay section
Net pay section

Referring to figure 8.3:

The gross section represents the distance A–D.

The net porous and permeable section is the gross section less shale or other low-porosity, low-permeability sections such as *a, b,* and *c.*

The gross pay is represented by A–C less *a* and *b* (i.e., gross pay includes all hydrocarbon-bearing sections from the top of the formation down to the oil/water contact, less any shale sections, etc.).

Net pay is represented by A–B less *a* and *b* (i.e., it only includes sections above the transition zone likely to produce water-free hydrocarbons).

Net pay figures are the ones to use for h in the HCPV, oil-in-place, and reserve calculations.

SPECIAL CASES FOR DEVIATED
WELLS AND DIPPING BEDS

So far, only vertical wells with horizontal beds have been considered. However, in the case where the wellbore is deviated or the beds are dipping, or both, it is important to make allowances for these effects.

Offshore exploration and development requires the drilling of multiple deviated wells from fixed platforms. As a result, the thickness of a formation as measured from a log made in a deviated hole reflects neither its true stratigraphic thickness nor its true

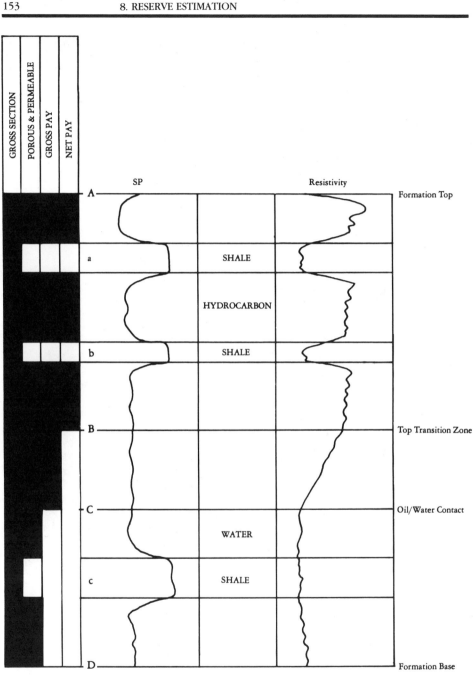

FIGURE 8.3 *Gross and Net Pay.*

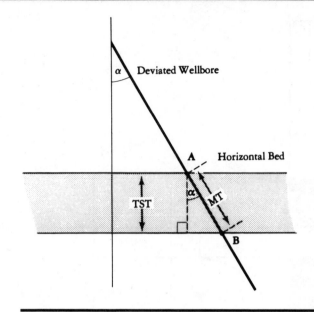

FIGURE 8.4 *Deviated Well and Horizontal Bed. Reprinted by permission of the SPWLA from Bateman and Konen 1979.*

vertical thickness. The geologist who tries to trace a given bed across the field finds it difficult to deduce whether variations in bed thickness measured from a log are due to legitimate bed thickness variations or whether they are geometrical distortions resulting from hole deviation and bed-dip effects. The engineer trying to compute reserves is faced with the same type of problem. Measured bed thickness can be less than true vertical bed thickness in some cases and greater in other cases. A need for a quick and simple solution to the problem is evident, and a method will be presented to satisfy the demands of both the explorationist and the engineer.

Figure 8.4 illustrates the case of a well deviated $\alpha°$ from the vertical crossing a horizontal bed of measured thickness MT. If the well enters the bed at **A** and leaves the bed at **B**, the measured thickness is **AB** and the true stratigraphic thickness (**TST**) is related to the measured thickness by

$$TST = MT \cos \alpha. \tag{8.1}$$

Figure 8.5 illustrates the case of a straight hole crossing a bed of measured thickness MT dipping $\beta°$ from the horizontal. If the hole enters the bed at **A** and leaves the bed at **B**, the measured thickness is **AB**, and the true stratigraphic thickness is given by

$$TST = MT \cos \beta. \tag{8.2}$$

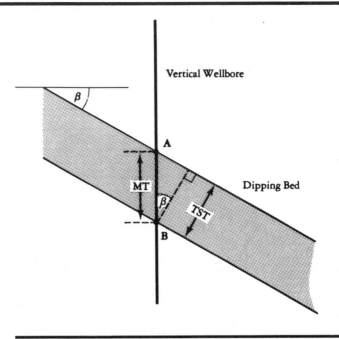

FIGURE 8.5 *Vertical Well and Dipping Bed. Reprinted by permission of the SPWLA from Bateman and Konen 1979.*

Note that where the well is vertical, the azimuth of the dipping bed is immaterial and that where the bed is horizontal, the well azimuth is likewise not required for calculating TST.

Figure 8.6 illustrates the case of a well deviated $\alpha°$ from vertical crossing a bed dipping $\beta°$ from the horizontal. Note that both the bed azimuth and the well azimuth are the same for this particular illustration. Again, the measured thickness is **AB**. It is of interest to note that the true vertical depth difference between **A** and **B** is **AC** where **AC** = **AB** cos α. However, the engineer is interested in the true vertical thickness (TVT) given by **AE**, and the geologist is seeking the true stratigraphic thickness given by **BD**.

For the case where both well and bed dip are oriented in the same azimuth, it is a trivial exercise to deduce TVT (distance **AE** on figure 8.6) and TST (distance **BD** in figure 8.6).

$$\text{TST} = \text{MT} \cos (\alpha + \beta), \tag{8.3}$$

$$\text{TVT} = \frac{\text{TST}}{\cos \beta} = \frac{\text{MT} \cos (\alpha + \beta)}{\cos \beta}. \tag{8.4}$$

Note that equations (8.3) and (8.4) hold only for the case where bed-dip azimuth and well-deviation azimuth are *exactly* the same. In the general case, the well azimuth is different from the bed azimuth, and the simple relationships deduced for the unique case illustrated in figure 8.6 do not hold.

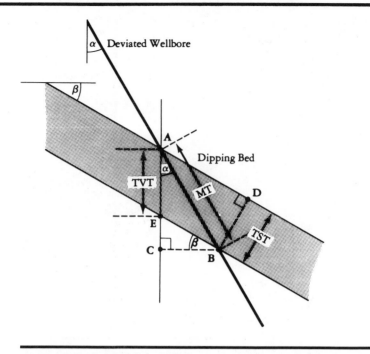

FIGURE 8.6 *Deviated Well and Dipping Bed. Reprinted by permission of the SPWLA from Bateman and Konen 1979.*

The practicing log analyst who has had to deal with dip-meter-related problems will appreciate the similarity between structural dip subtraction and the solution described here. The basic idea is to consider the well as one dip vector and the bed as another and perform a dip rotation. In this method the bed is brought back to the horizontal and the well deviation and azimuth assume new values, in the geographical frame of reference, while maintaining the same orientation relative to the bed itself. The reader may better understand the principle by imagining a pencil (representing the well) stuck through a slab of cheese (representing the dipping bed). If the cheese is initially dipping to north and the pencil transverses the cheese deviated to the west, then the result of placing the cheese on a horizontal surface will be to move the azimuth of the pencil from west to somewhere north of west and, at the same time, to increase the deviation of the pencil from vertical.

An alternative approach would be to allow the pencil to rotate until it was vertical, in which case the cheese slab would dip at an increased angle of dip to some point west of north.

Either of these rotations can be made by treating it as a structural dip subtraction. The manipulation can be made graphically by using a stereo net with a rotating overlay, or it can be made mathematically using the cosine law of spherical triangles. Once the bed is rotated back to horizontal (or the hole back to vertical), the simple relationships illustrated in figures 8.4 and 8.5 can be used. Since the new azimuth of the bed or the hole is not required, the only parameter needed is the new bed dip or the new hole deviation, depending on which approach is used.

For the purposes of illustration, we will rotate the bed back to horizontal. Thus, if

$\alpha°$ = well deviation from vertical,
$\beta°$ = bed dip from horizontal,
$HAZ°$ = hole azimuth,
$DAZ°$ = dip azimuth, and
MT = measured bed thickness,

then the well deviation from vertical after the bed has been rotated to horizontal is given by α', where

$$\cos \alpha' = \cos \alpha \cos \beta - \sin \alpha \sin \beta \cos (HAZ - DAZ). \tag{8.5}$$

If the bed is now horizontal, the true stratigraphic bed thickness is given by

$$TST = MT \cos \alpha' \tag{8.6}$$

or

$$TST = MT [\cos \alpha \cos \beta - \sin \alpha \sin \beta \cos (HAZ - DAZ)] \tag{8.7}$$

and the true vertical thickness of the bed is given by

$$TST = \frac{TST}{\cos \beta} \tag{8.8}$$

or

$$TST = MT [\cos \alpha - \sin \alpha \tan \beta \cos (HAZ - DAZ)]. \tag{8.9}$$

QUESTION #8.3
From the log the following data are collected:

Top bed = 5642 ft
Bottom of bed = 5878 ft
Well is deviated 30° to azimuth 128°
Bed is dipping 25° to azimuth 45°

Find

a. Measured thickness.
b. TVD difference between top and bottom of bed.
c. TST of bed.
d. TVT of bed.

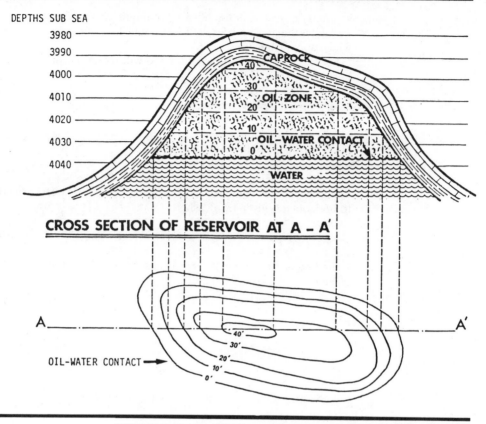

FIGURE 8.7 *Net Pay Thickness Isopach and Structure Cross Section. Courtesy Schlumberger Well Services.*

It is assumed that the true stratigraphic thickness is the same at the point the well enters the bed as it is when the well leaves the bed. It is also assumed that the well deviation and azimuth do not change as the well crosses the bed. If the bed is highly dipping, it is possible that the well will, in fact, enter the bed from the bottom side and exit from the top side. If this occurs, the answers for TST and TVT will appear as negative numbers.

RESERVOIR VOLUMES

The gross rock volume enclosed within a hydrocarbon-bearing structure cannot be deduced from a single log. Even with the aid of seismic mapping, many uncertainties remain. However, after sufficient wells have been drilled, it is usually possible to prepare maps showing how many net feet of pay above the oil/water (or gas/oil or gas/water) contact exist. Such maps are known as *isopachs*. Figure 8.7 shows a cross section through an oil-bearing structure, together with an isopach map contouring the net feet of pay above the oil/water contact.

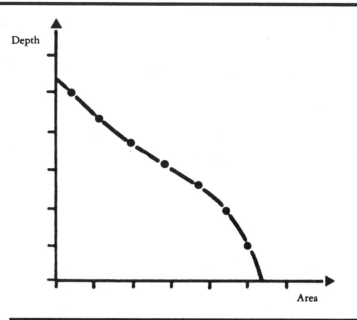

FIGURE 8.8 *Area within Net Pay Contour vs. Depth. Courtesy Schlumberger Well Services.*

Once such a map is drawn, the area enclosed within each contour may be found using a planimeter. A plot may then be made showing the area contained within each contour as a function of depth as shown in figure 8.8. The volume is then calculated as an integration of $A \cdot dh$ either by planimeter or by numerical integration.

The volume of the frustum of a pyramid between two contours is given by

$$V_B = \frac{h}{3} \left[A_n + A_{n+1} + \sqrt{A_n (A_{n+1})} \right]$$

and the total volume of a pyramid is then given by

$$V_{TP} = \sum_{i}^{n} V_{bi}.$$

The volume of a trapezoid section between two contours is given by

$$V_B = \frac{h}{2} (A_n + A_{n+1}),$$

and the total volume of a series of trapezoids is, therefore

$$V_{TT} = \frac{h}{2} (A_0 + 2 A_1 + 2 A_2 + \cdots + 2 A_{n-1} + A_n) + t_{av} A_n,$$

where t_{av} is the average thickness. The total volume required is the average of the two volume estimates:

$$V = \frac{(V_{TP} + V_{TT})}{2}.$$

If areas are in acres and contour thicknesses in feet, then volumes will be in acre-feet.

APPENDIX 8A: CONVERSION FACTORS IN RESERVE ESTIMATION

The following conversions between English and metric units may be helpful.

1	acre-foot	=	43,560	cu ft	=	7,758 barrels =	1233.5 m^3
1	hectare-meter	=	353,147	cu ft	=	62,893 barrels =	10,000 m^3
1	acre-foot	=	0.1233	m-ha			
1	hectare-meter	=	8.107	acre-ft			
1	barrel	=	5.615	cu ft	=	0.159 m^3	
1	cubic meter	=	35.31	cu ft	=	6.289 barrel	
1	Bcf	=	178.1	MMB	=	28.32 MMm^3	
1	MMB	=	5.615	MMcf	=	159 Mm^3	
1	hectare	=	2.471	acres			

BIBLIOGRAPHY

API: "A Statistical Study of Recovery Efficiency," Bulletin D 14 (October 1967).

Banks, R. B.: "New Thoughts on an Old Topic: Reservoir Integration (Volumetrics)," SPE paper 11339, presented at the 1982 Production Technology Symposium, Hobbs, NM, Nov. 8–9, 1982.

Bateman, R. M.: "Cased-Hole Log Analysis and Reservoir Performance Monitoring," IHRDC, Boston (1984), ch. 3.

Bateman, R. M., and Konen, C. E.: "Finding True Stratigraphic Thickness and True Vertical Thickness of Dipping Beds Cut by Directional Wells," *The Log Analyst* (March–April 1979).

Muskat, Morris: *Physical Principles of Oil Production,* IHRDC, Boston (1949; reprinted 1981), ch. 14.

Poston, S.W.: "Numerical Simulation of Reservoir Sandstone Models," SPE paper 13135, presented at the 59th Annual Technical Conference and Exhibition, Houston, Sept. 1984.

Schlumberger: "Reservoir and Production Fundamentals" (1980).

Answers to Text Questions

QUESTION #8.1
a. HCPV   = 3.5 ft
b. Oil-in-place = 27,153 bbl/acre
c. Reserves = 1.253 MMSTB

QUESTION #8.2

Oil-in-place term	$= 54.9 \ [0.2 \ (1 - 0.35)/1.25]^{0.0422} \ =$	49.99
Mobility term	$= \ [(0.25 \times 0.5)/1.5]^{0.077} \qquad =$	0.826
Water saturation term	$= 0.35^{-0.1903} \qquad\qquad\qquad =$	1.221
Pressure ratio term	$= (2000/750)^{-0.2159} \qquad\qquad =$	0.809
Hence r_{WD}	$= 49.99 \times 0.826 \times 1.221 \times 0.809 =$	0.408
or	40.8% of the oil-in-place will be recovered	

QUESTION #8.3
a. MT   = 236 ft
b. TVD difference = 204.38 ft
c. TST   = 179.16 ft
d. TVT   = 197.68 ft

PART III

OPEN-HOLE
LOGGING MEASUREMENTS

CHAPTER NINE

THE SP LOG

HISTORY OF THE SP

The spontaneous potential (SP), one of the first logging measurements ever made, was discovered by accident. It appeared in the borehole as a DC potential that caused perturbations to the old electric logging systems. Its usefulness was soon realized, and it has remained one of the few well log measurements to have been in continuous use for over 50 years.

The SP log has a number of useful functions, including:

Correlation
Lithology indication
Porosity and permeability indications
A means to measure R_w and, hence, formation water salinity

Figure 9.1 shows a typical SP log, where it is represented in Track 1 as a solid curve and shows departures to the left from a base line (or shale line) reading on the right, to a sand line on the left in the "cleanest" nonshale zones. The scale of the log is in millivolts, abbreviated mV. Notice that there is no absolute scale in mV, only a relative scale of so many mV per division.

QUESTION #9.1
On the log shown in figure 9.1, read the maximum SP deflection from the shale line to the sand line:

SP =   mV.

The way in which this measurement can be used quantitatively will be explained later in this section.

RECORDING THE SP

The SP can be recorded very simply by suspending a single electrode in the borehole and measuring the voltage difference between the electrode and a ground electrode, which usually takes the form of a "fish," making electrical contact with the earth at the surface. A generalized illustration of the SP recording system is shown in figure 9.2. Such SP electrodes are built into nearly all logging tools. For example, the SP can be recorded using an induction log, a laterolog, a sonic log, a sidewall core gun, etc.

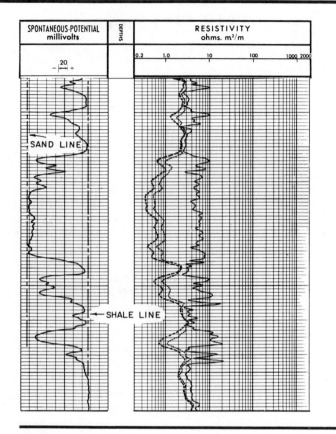

FIGURE 9.1 *Typical SP Log with Resistivity Recording. Courtesy Schlumberger Well Services.*

However, the SP cannot be recorded in oil-based muds since there is no conductive path through the mud, and a conductive path is essential for generation of a spontaneous potential.

THE SOURCE OF THE SP

The SP is a sort of self-calibrated indicator of formation water salinity. To understand how the SP can be used to find R_w, a little should be known about the origin of the SP.

If two sodium chloride solutions of different concentrations are separated by a permeable membrane, then ions from the most concentrated solution will tend to migrate into the least concentrated solution (see fig. 9.3). But Na$^+$ and Cl$^-$ ions do not move with the same alacrity—Cl$^-$ ions move faster than Na$^+$ ions. The result is a conventional current flowing from the weaker solution to the more concentrated solution. This is known is a *liquid junction effect*. In terms of the solutions present in a formation, mud filtrate can be substituted for the weak solution and formation water for the more

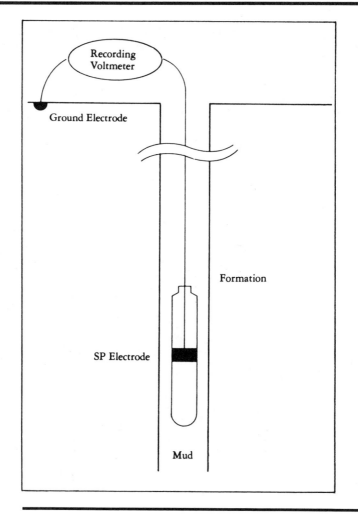

FIGURE 9.2 *Recording the SP. Courtesy Dresser Atlas, Dresser Industries, Inc.*

concentrated solution (see fig. 9.4). The potential is referred to as the *liquid-junction potential* (E_{lj}). The greater the contrast in salinity between the mud filtrate and the formation water, the larger this potential.

A similar "battery" arises in the formation from the molecular construction of shale beds. Shales are permeable to Na^+ ions but not so permeable to Cl^- ions. A shale thus acts as an ionic sieve. This phenomenon occurs because of the crystalline structure of clay minerals whose exterior surfaces have exchange sites to which cations may temporarily cling. (This same surface conductance effect manifests itself in the electrical behavior of shaly sands—see chapter 27.) Since Na^+ ions effectively manage to penetrate the shale bed from the saline formation water to the less saline mud column,

Net current flow

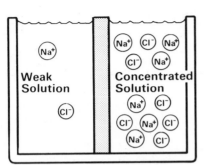
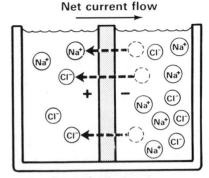

Original Conditions   Dynamic Conditions

FIGURE 9.3 *Liquid-Junction Effects.*

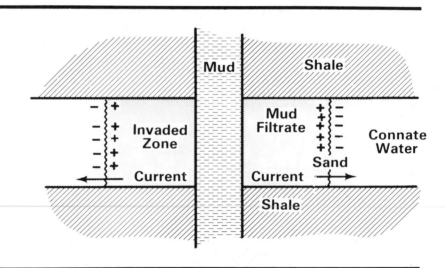

FIGURE 9.4 *Liquid-Junction SP.*

a so-called membrane potential (E_m) is created (see fig. 9.5). The total electrical potential can now be appreciated as the sum of the two components:

$$E_{total} = E_{lj} + E_m = SP.$$

This SP can be measured in the borehole by an electrode. Figure 9.6 shows the total SP picture. It should be noted that the potential measured may include yet another component, the electrokinetic potential (see later under "Factors Affecting the SP"), and to distinguish the liquid junction and membrane potentials they are referred to as the electrochemical components of the SP. When

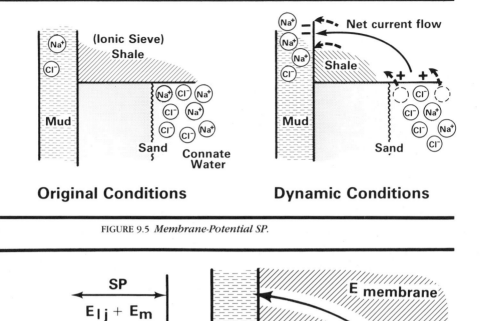

FIGURE 9.5 *Membrane-Potential SP.*

FIGURE 9.6 *Total SP Picture.*

mud-filtrate salinities are lower than connate-water salinities (i.e., when $R_{mf} > R_w$), the SP deflects to the left (the SP potential is negative). This is called a *normal SP*. When the salinities are reversed (i.e., salty mud and fresh formation waters, $R_{mf} < R_w$), the SP deflects to the right. This is called a *reverse SP*. Other things being equal, there will be no SP at all when $R_{mf} = R_w$.

It is quite common to find fresh water in shallow sands and increasingly saline water as depth increases. Such a progression is shown in figure 9.7, where the SP appears deflected to the left deep in the well but is reversed nearer the surface. In sand A, R_w is less than R_{mf} (i.e., formation water is saltier than the mud filtrate). In sand B, the SP deflection is less than in sand A, indicating a fresher formation water. In sand C, the SP is reversed, indicating formation water that is fresher than the mud filtrate so that R_w is greater than R_{mf}. It is a safe guess that somewhere in the region of 7000 ft R_{mf} and R_w are equal.

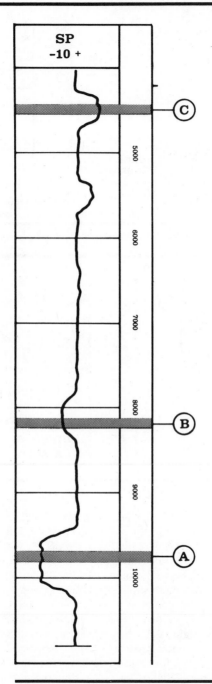

FIGURE 9.7 *Normal and Reverse SP Deflections.*

R_w FROM THE SP

In order to perform quantitative analysis of the SP, the relationship between the SP and the resistivities of the mud filtrate and the formation water must be determined. Analysis of a series of complicated paths of logic based on various laws of physical chemistry leads to the equation:

$$SP = -K \log (R_{mf}/R_w)$$

where SP is measured in millivolts and K is a constant that depends on temperature. By inspection, R_w can be found if SP, K, and R_{mf} are known.

The SP can be read in a water-bearing sand provided it is clean (no shale is present) and is sufficiently thick to allow for full development of the potential. K can be estimated from the temperature of the formation. A good approximation is:

$$K = \frac{T + 505}{8}, \quad \text{where } T \text{ is the formation temperature in } °F.$$

R_{mf} can be estimated from direct measurement on a sample of mud filtrate prepared by placing a circulated mud sample in a mud press. These data are usually entered on the log heading. However, care should be taken when using these values. Logging engineers have been known to take shortcuts and quote R_{mf} as some fraction of R_m, usually $0.75 R_m$, which may be a fair estimate but is not necessarily correct. Likewise, circulated mud samples are not always collected by rig personnel in the correct manner. Even if they are properly collected, the samples are not always representative of the mud in the hole at the time a particular formation was drilled. (Mud engineers love to sprinkle sundry additives into the system daily in order to maintain the appearance of being busy.)

Experiments of the sort reported by Williams and Dunlap (1984), where R_m and R_{mf} were measured on a daily basis as a well was drilled, tend to support the contention that R_{mf} is the least well-defined parameter in SP log analysis. A comparison between the values of R_m and R_{mf} as reported on log headings with the actual values measured on a daily basis show some alarmingly large differences. Figure 9.8 shows such a plot. As can be seen, both R_m and R_{mf} reported on the log heading for this well were low by a substantial factor.

In the absence of any reported value for R_{mf}, and given a value for R_m, a value for R_m can be estimated from figure 9.9, which also serves for estimation of R_{mc}. A fit of this empirical chart gives

$$R_{mf} = (R_m)^{1.065} \times 10^{[(9 - W)/13]} \quad \text{and}$$
$$R_{mc} = (R_m)^{0.88} \times 10^{[(W - 10.4)/7.6]},$$

where W is the mud weight in lb/gal. Another statistical approximation for predominantly NaCl muds is

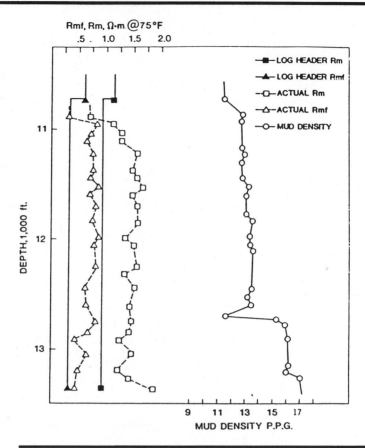

FIGURE 9.8 *Log Header R_m and R_{mf} vs. Short-Term Variations in R_m, R_{mf}, and Mud Density. Reprinted by permission of the SPWLA from Williams and Dunlap 1984.*

$$R_{mf} = 0.75R_m \qquad \text{and} \qquad R_{mc} = 1.5R_m.$$

In all cases, direct measurement on a sample of mud filtrate is preferred. However, even after values for SP, K, and R_{mf} have been determined, there are still some minor wrinkles to be smoothed out. The equation

$$SP = -K \log (R_{mf}/R_w)$$

does not explain adequately the true electrochemical behavior of salt solutions. The actual development of SP is controlled by the relative *activity* of the formation water and mud filtrate solutions. Thus, the SP equation should read

$$SP = -K \log (A_w/A_{mf}),$$

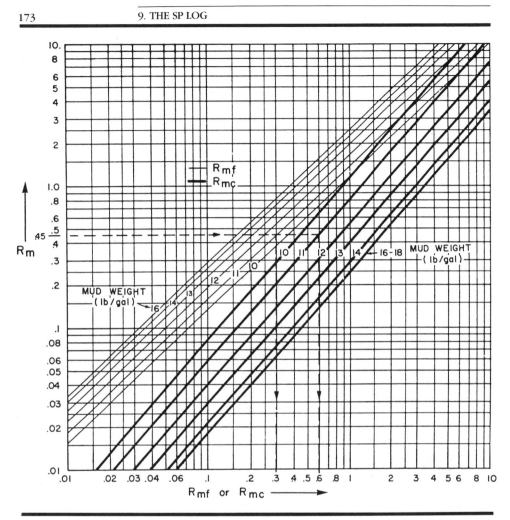

FIGURE 9.9 *Estimation of R_{mf} and R_{mc} from R_m. Courtesy Schlumberger Well Services.*

where A_w and A_{mf} are the activities of the connate water and the mud filtrate, respectively.

The resistivity of a solution is roughly proportional to the reciprocal of its activity at low salt concentrations, but at high concentrations there is a marked departure from this rule. A way to compensate for this departure is to define *effective* or *equivalent* resistivities (R_{we}) for salt solutions, which are, by definition, inversely proportional to the activities ($R_{we} = 0.075/A_w$ at $77°F$). A conversion chart is then used to go from an equivalent resistivity (R_{we}) to an actual resistivity (R_w). For further details refer to Schlumberger (1972). The SP equation can then be rewritten to the strictly accurate formula

$$SP = -K \log (R_{mfe}/R_{we}).$$

The procedure for using this equation is as follows:

1. Establish the formation temperature.
2. Find the value of R_{mf} at formation temperature.
3. Convert R_{mf} at formation temperature to an R_{mfe} value.
4. Compute the R_{mfe}/R_{we} ratio from the SP.
5. Compute R_{we}.
6. Convert R_{we} at formation temperature to an R_w value.

Step 1: Formation Temperature

Formation temperature may be estimated from surface and bottom-hole temperatures by simple extrapolation. Usually log header data are sufficient. Alternatively, formation temperature may be calculated using

$$T_{form} = T_{surf} + \text{depth} \times \text{temperature gradient.}$$

Figure 9.10 plots temperature (in both °F and °C) against depth for a variety of assumed linear temperature gradients and assumed mean annual surface temperatures. A good average case is to assume a surface temperature of 70°F and a gradient of 1°F per ft. (The equivalent metric case would be 20°C and 1.8°C/100 m.)

QUESTION #9.2

According to the log heading, bottomhole temperature at a depth of 11,304 ft was 200°F. If a surface temperature of 70°F is assumed, then:

a. Find the temperature gradient in °F/100 ft.
b. Find formation temperature at 9565 ft.

Step 2: Conversion of R_{mf} to R_{mf} at Formation Temperature

The resistivity of any ionic solution decreases in a predictable fashion as temperature increases. Thus, R_{mf} as measured at surface temperature needs to be converted to downhole conditions at formation temperature before it can be used as a parameter in the SP equation. Sodium chloride solutions behave as shown in figure 9.11, where solution resistivity in Ω·m is plotted against temperature for a number of solutions of different concentration (ppm NaCl).

QUESTION #9.3

R_{mf} is reported on the log heading to be 0.08 at 75°F. Use figure 9.11 to find:

a. R_{mf} at 180°F.
b. The salinity of the solution in ppm NaCl.

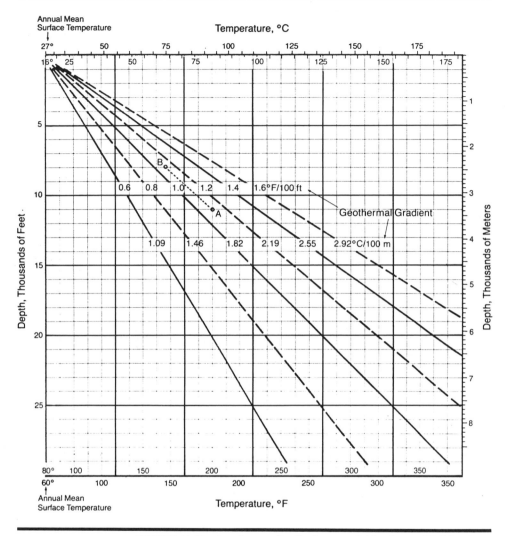

FIGURE 9.10 *Temperature vs. Depth. Courtesy Schlumberger Well Services.*

An approximation to the chart in figure 9.11 is given by the Arps formula, which states:

$$R_2 = \frac{R_1(T_1 + 7)}{(T_2 + 7)} \ ,$$

where R_1 is the solution resistivity at temperature T_1 ($^\circ$F). For example, the data of question #9.3 could be applied to the Arps formula as follows:

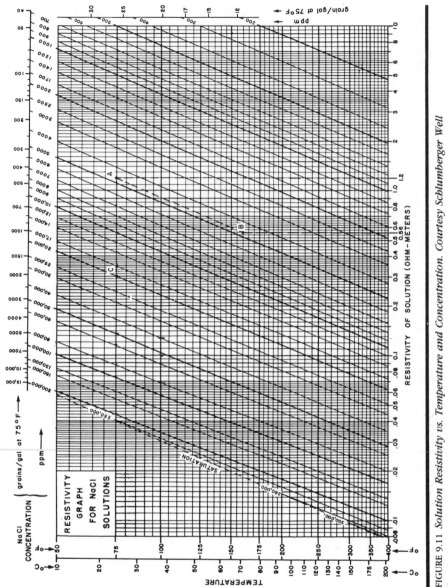

FIGURE 9.11 *Solution Resistivity vs. Temperature and Concentration. Courtesy Schlumberger Well Services.*

176

$$R_{mf180} = R_{mf75}(75 + 7)/(180 + 7)$$

$$= 0.08 \times 82/187$$

$$= 0.0351.$$

A more complex but more accurate approximation to the data of figure 9.11 is given by Hilchie (1984). The Hilchie equation states

$$R_2 = R_1(T_1 + x)/(T_2 + x),$$

where $x = 10 \exp (0.6414 - 0.3404 \log R_1)$.

Step 3: Conversion of R_{mf} to R_{mfe}

Figure 9.12 conveniently takes care of the conversion of R_{mf} to R_{mfe}. It is also applicable to step 6 where an R_{we} needs to be converted back to an R_w. Note that figure 9.12 is constructed for average formation waters containing some magnesium and calcium cations in addition to sodium chloride. Mud filtrates may not contain the same ionic mix; when the mud is fresh, it is normal to treat the conversion of R_{mf} to R_{mfe} by using the relationship $R_{mfe} = 0.85\ R_{mf}$ whenever R_{mf} is greater than 0.1 $\Omega \cdot$ m at 75°F.

QUESTION #9.4
R_{mf} at formation temperature is 0.036 $\Omega \cdot$m. Use figure 9.12 to find R_{mfe}.

Step 4: R_{mfe}/R_{we} Ratio from SP

The SP equation may be used directly to compute the R_{mfe}/R_{we} ratio, provided a calculator is available:

$$R_{mfe}/R_{we} = 10^{-SP/K},$$

where $K = (T°F + 505)/8$. Alternatively, figure 9.13 may be used.

QUESTION #9.5
The SP is −100 mV and formation temperature is 180°F. Use either a calculator or figure 9.13 to find the value of R_{mfe}/R_{we}.

Step 5: Computing R_{we}

R_{we} is found by dividing the known value of R_{mfe} by the R_{mfe}/R_{we} ratio found from the SP.

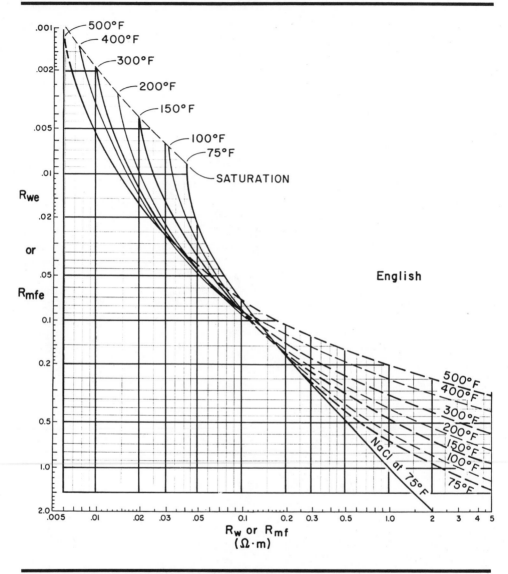

FIGURE 9.12 *Conversion of Solution Resistivities to Equivalent Resistivity. Courtesy Schlumberger Well Services.*

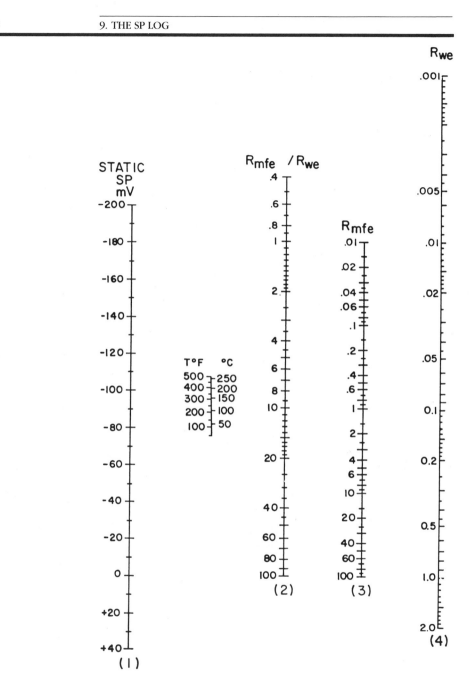

FIGURE 9.13 *Conversion of SP_mV to R_we. Courtesy Schlumberger Well Services.*

QUESTION #9.6

If $R_{mfe} = 0.031$ and $R_{mfe}/R_{we} = 3.84$, find R_{we} using either a calculator or figure 9.13.

Step 6: Converting R_{we} to R_w

The conversion of R_{we} to R_w is a simple matter of reversing the process used to convert R_{mf} to R_{mfe} in step 3. Figure 9.12 is used.

QUESTION #9.7

R_{we} has been found to be 0.0083 at a formation temperature of 180°F.

a. Use figure 9.12 to convert this to an R_w value.
b. Estimate the connate water salinity from figure 9.11.

In order to place in perspective the steps used for calculation of the actual resistivity of formation water from the SP, figure 9.14 is offered as a "road map." For those inclined toward mechanization of the log interpretation process, the flow chart from Bateman and Konen (1977) is reproduced here as figure 9.15.

FACTORS AFFECTING THE SP

Readings of spontaneous potential are usually accurately and easily measured. There are, however, some circumstances where SP readings need careful handling.

1. In oil-based muds: Owing to a complete lack of an electrical path in the mud column, no SP will be generated in oil-based muds.
2. Shaly formations will reduce the measured SP: This phenomenon permits the formation shaliness to be determined if a clean sand with the same water salinity is available for a legitimate comparison.
3. Hydrocarbon saturation will reduce SP measurements: Thus, only water-bearing sands should be selected for R_w determination from the SP.
4. Unbalanced mud columns, with differential pressure into the formation, can cause "streaming" potentials that augment the SP: This effect is noticeable in depleted reservoirs. There is no way to handle it quantitatively. (This phenomenon is called the electrokinetic SP.)
5. In hard formations: Resistivities may be very high in hard formations, except in the permeable zones and in the shales. These high resistivities affect the distribution of the SP currents and, therefore, the shape of the SP curve. This is illustrated in figure 9.16, where the SP currents flowing from shale bed Sh_1 toward permeable bed P_2 are largely confined to the borehole between Sh_1 and P_2 because of the

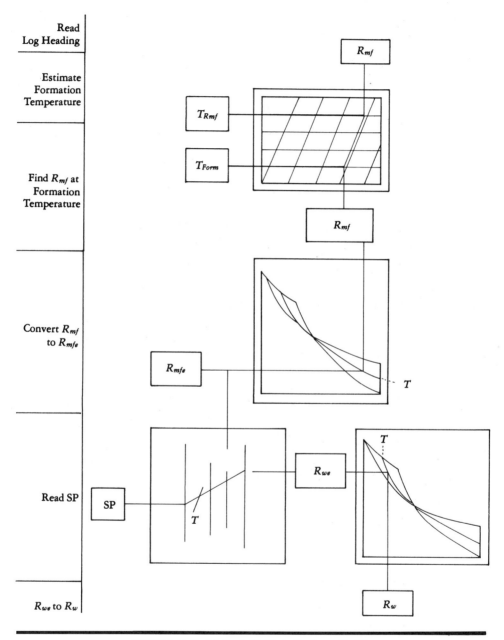

FIGURE 9.14 *Quick-Look Guide to R_w from the SP.*

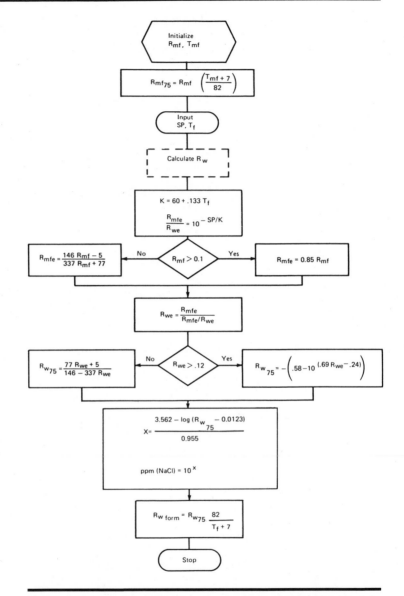

FIGURE 9.15 *Logic Flow Diagram for R_w from the SP. Reprinted by permission of the SPWLA from Bateman and Konen 1977.*

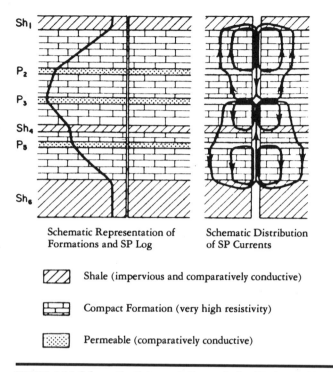

Schematic Representation of   Schematic Distribution
Formations and SP Log   of SP Currents

| ⬚ Shale (impervious and comparatively conductive) |
| ⬚ Compact Formation (very high resistivity) |
| ⬚ Permeable (comparatively conductive) |

FIGURE 9.16 *Schematic Representation of the SP in Highly Resistive Formations. Courtesy Schlumberger Well Services.*

very high resistivity of the formation in this interval. Accordingly, the intensity of the SP current in the borehole in this interval remains constant. Assuming the hole diameter is constant, the potential drop per foot will be constant and the SP curve will be a straight line. In such formations, SP current can leave or enter the borehole only opposite permeable beds or shales, and the SP curve will show a succession of straight portions with a change of slope opposite every permeable interval (with the concave side of the SP curve toward the shale line). The boundaries of the permeable beds cannot be located with accuracy by use of the SP in highly resistive formations.

6. Bed thickness; diameter of invasion; R_{xo}/R_m ratio; neighboring shale resistivity, R_{sh}; hole diameter, dh; and mud resistivity, R_m, are all disturbing factors: Bed thickness can affect the SP measurement quite dramatically. In thin beds, where R_{xo}/R_m is high and invasion is deep, the SP does not fully develop. Figure 9.17 illustrates the factors involved in SP reduction. In the terminology used here, SP refers to the observed SP deflection on the log and SSP (static SP) to the value it would have had if all disturbing influences were removed. Many SP corrections charts are available in the literature, some more complex than others. It is virtually impossible to include on one chart all the possible variables involved in making necessary corrections. Figure

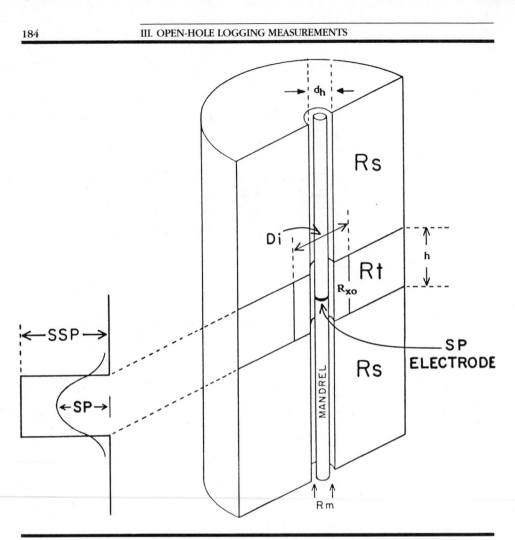

FIGURE 9.17 *Factors Affecting SP Reduction. After Segesman 1962 by permission of the SEG.*

9.18 shows one of the more practical charts. Most of the variables required (d_i, R_{xo}/R_m, and h) are usually known or guessable.

QUESTION #9.8
The SP deflects 25 mV from the base line in a sand 4 ft thick. $R_{xo}/R_m = 50$ and $d_i = 30$ in.

a. Find by what percent the SP has been reduced.
b. Compute the corrected value for the SP.

7. The use of potassium chloride muds affects the derivation of R_w from the SP: Cox and Raymer (1976) cover this subject in detail. A

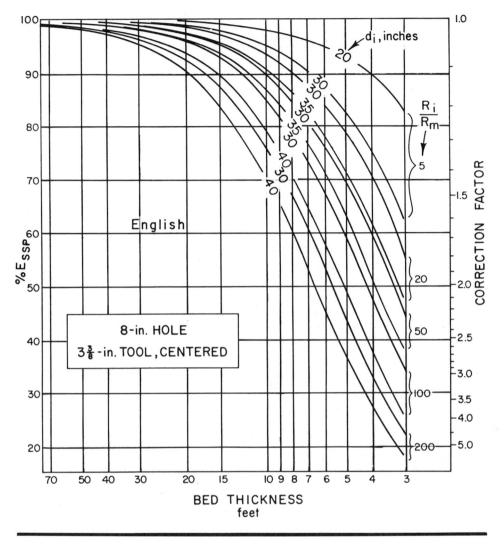

FIGURE 9.18 *SP Correction for Bed Thickness. Courtesy Schlumberger Well Services.*

quick solution to the KCl mud perturbation is simply to take the observed SP deflection, subtract 25 mV, and then treat the computation as a NaCl mud case. The R_{mf} to R_{mfe} relationship is slightly different for KCl filtrates than for NaCl filtrates. Again, a quick rule of thumb is to add 30% to the measured R_{mf} and proceed as for NaCl filtrate.

THE SP AS A SHALE INDICATOR

The presence of shale in an otherwise clean sand will tend to reduce the SP. This fact can be used in estimating the shale content of a formation. If SP_{sand} is the value observed in a clean, water-bearing sand and SP_{shale} is the value observed in a shale, then any intermediate value of the SP may be converted into a value for the shale volume (V_{sb}) by the relationship

$$(V_{sb})_{SP} = \frac{SP - SP_{sand}}{SP_{shale} - SP_{sand}} .$$

QUESTION #9.9
In figure 9.19, assume sand A is clean and wet. Use the SP deflection in sand B to determine the shale content. Find

$$(V_{sb})_{SP} = ?$$

SP DEPOSITIONAL PATTERNS

Quite apart from water salinity variations, SP deflections also respond to depositional changes. In particular, channels, bars, and other depositional sequences, where either sorting, grain size, or cementation change with depth, produce characteristic SP shapes. These are variously referred to as *bells* or *funnels*. Figure 9.20 illustrates some of these patterns.

SP QUALITY CONTROL

Although SP logs are prone to errors, these errors can easily be detected in the field.

1. A poor ground can cause the SP baseline to move. For example, placing the SP ground in a mud pit whose level is changing or on the ocean floor where currents can drag it back and forth can cause problems.
2. Cyclical or saw-tooth SP profiles can indicate a magnetized cable winch drum or intermittent SP circuit contacts. A log of this type is useless.
3. Sometimes the polarity of the SP is reversed as a result of incorrect galvanometer connections. If the SP moves in the wrong direction, incorrect connections could be the cause.
4. Some tools carrying SP electrodes have two dissimilar metals (steel and bronze, for example) in close proximity. Since the two metals have different electrochemical potentials, the interface via the mud column acts as a small battery. This condition is known as bimetalism and could show itself as an extraneous potential added to the SP. It can be cured by wrapping insulating tape over the offending metal parts of the sonde.
5. Baseline shifts can become quite pronounced in some geographical areas. As the tool is raised up the hole, the SP baseline will gently shift to the left by an amount that, typically, is a few millivolts per thousand

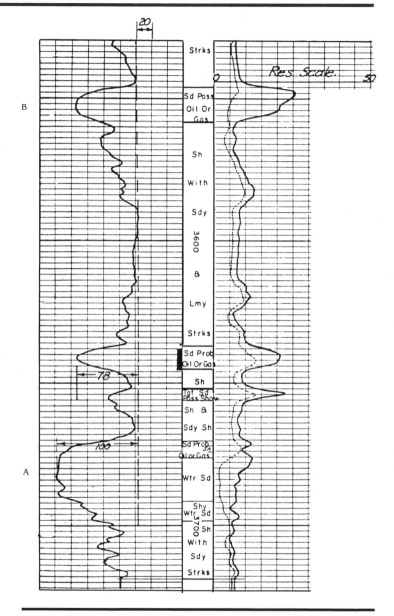

FIGURE 9.19 *Estimation of V_{sh} from SP.*

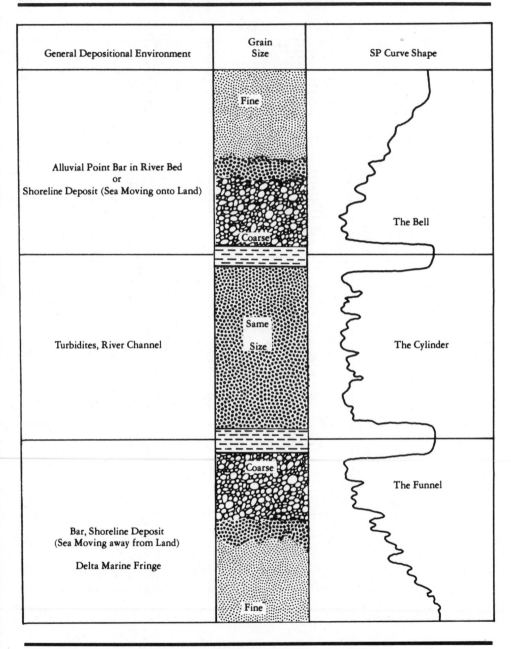

FIGURE 9.20 *SP Shapes in Different Depositional Sequences.*

feet. Eventually, the SP will disappear off the left track edge and a backup trace will appear from the right. This is normal. If a downhole ground is used (cable armor), the baseline will move very rapidly near the casing shoe. Some service company engineers will "ride" the SP millivolt box and purposely keep the baseline straight. If this technique is used, a notation should be made on the log heading.

BIBLIOGRAPHY

Bateman, R. M., and Konen, C. E.: "The Log Analyst and the Programmable Pocket Calculator," *The Log Analyst* (September–October 1977).

Cox, J. W., and Raymer, L. L.: "The Effect of Potassium Salt Muds on Gamma Ray and Spontaneous Potential Measurements," SPWLA, 17th Annual Logging Symposium, New Orleans, June 1976.

Doll, H. G.: "The Invasion Process in High Permeability Sands," *Pet. Eng.* (January 1955).

Doll, H. G.: "The SP Log: Theoretical Analysis and Principles of Interpretation," *Trans.,* AIME (1949) **179**, 146.

Doll, H. G.: "The SP Log in Shaly Sands," *J. Pet. Tech.* (1949) 2912.

Dresser Atlas: "Log Review 1," Dresser Industries Inc., REP 03/81, sec. 4 (1974).

Evers, J. F., and Iyer, B. G.: "A Statistical Study of the SP Log in Fresh Water Formations of Northern Wyoming," SPWLA, 16th Annual Logging Symposium, June 1975.

Gondouin, M., Tixier, M. P., and Simard, G.L.: "An Experimental Study on the Influence of the Chemical Composition of Electrolytes on the SP Curve," *Trans.,* AIME (1957) **210**, 58.

Hilchie, D. W.: "A New Water Resistivity versus Temperature Equation," *The Log Analyst* (July–August 1984) 20.

Overton, H. L., and Lipson, L. B.: "A Correlation of the Electrical Properties of Drilling Fluids with Solids Content," *Trans.,* AIME (1958) **213**.

Schlumberger: "Log Interpretation Charts," Schlumberger Limited (1972).

Schlumberger: "Log Interpretation Charts," Schlumberger Well Services (1984).

Schlumberger: "Log Interpretation Principles," vol. I (1972), ch. 2 and 13.

Segesman, F.: "New SP Correction Charts," *Geophysics* (December 1962) **27** .

Segesman, F., and Tixier, M. P.: "Some Effects of Invasion on the SP Curve," SPE Annual Meeting, Oct. 1958.

Silva, P., and Bassiouni, Z.: "A New Approach to the Determination of Formation Water Resistivity from the SP Log," SPWLA, 22nd Annual Logging Symposium, Mexico City, June 1981.

Williams, H., and Dunlap, H. F.: "Short Term Variations in Drilling Parameters, Their Measurement and Implications," *The Log Analyst* (September–October 1984) 3–9.

Wyllie, M. R. J.: "A Quantitative Analysis of the Electrochemical Component of the SP Curve," *Trans.,* AIME (1949) **186**, 17.

Answers to Text Questions

QUESTION #9.1
140 mV

QUESTION #9.2
a. 1.15 °F/100 ft
b. 180°F

QUESTION #9.3
a. 0.036 $\Omega \cdot$m
b. 90,000 ppm NaCl

QUESTION #9.4
$R_{mfe} = 0.032$ $\Omega \cdot$m

QUESTION #9.5
$R_{mfe}/R_{we} = 3.84$

QUESTION #9.6
$R_{we} = 0.0083$ $\Omega \cdot$m

QUESTION #9.7
a. $R_w = 0.02$
b. Salinity $= 200,000$ ppm NaCl

QUESTION #9.8
a. Reduced by 60%
b. SSP $= 25/0.6 = 41.7$ mV

QUESTION #9.9
V_{sb} from the SP $= 24$%

CHAPTER TEN

THE GAMMA RAY LOG

Gamma ray logs are used for three main purposes:

Correlation
Evaluation of the shale content of a formation
Mineral analysis

The gamma ray log measures the natural gamma ray emissions from subsurface formations. Since gamma rays can pass through steel casing, measurements can be made in both open and cased holes. In other applications, induced gamma rays are measured (e.g., in pulsed neutron logging), but that will not be discussed in this section.

Gamma ray tools consist of a gamma ray detector and the associated electronics for passing the gamma ray count rate to the surface. Some of the common tool sizes, ratings, and applications are given in table 10.1. A typical gamma ray log (see fig. 1A.8) is normally presented in Track 1 on a linear grid and is scaled in API units, which will be defined later. Gamma ray activity increases from left to right. Modern gamma ray tools are in the form of double-ended subs that can be sandwiched into practically any logging tool string; thus, the gamma ray can be run with practically any tool available.

ORIGIN OF NATURAL GAMMA RAYS

Gamma rays originate from three sources in nature: the radioactive elements in the uranium and thorium groups and potassium. Uranium 235, uranium 238, and thorium 232 all decay, via a long chain of daughter products, to stable lead isotopes (see fig. 10.1). An isotope of potassium, $^{40}_{19}K$, decays to argon, giving off a gamma ray as shown in figure 10.2. Each type of decay is characterized by a gamma ray of a specific energy (wave length) and the frequency of occurrence for each decay energy is different. Figure 10.3 shows the relationship between gamma ray energy and frequency of occurrence. This is an important concept since it is used as the basis for measurement in the natural gamma spectroscopy tools.

ABUNDANCE OF NATURALLY OCCURRING RADIOACTIVE MINERALS

An average shale contains 6 ppm uranium, 12 ppm thorium, and 2% potassium. Since the various gamma ray sources are not all equally effective, it is more informative to consider this mix of radioactive

TABLE 10.1 *Scintillation Gamma Tool (SGT) and Natural Gamma Tool (NGT)*

OD (inches)	Temperature (°F)	Pressure (psi)	Application
3⅝	350	20,000	Openhole (SGT)
2¾	500	25,000	HEL* and through drillpipe
1¹¹⁄₁₆	350	20,000	Through tubing
3⅝	350	20,000	Lithology (NGT)
	275	10,000	Lithology + P_e (CSNG-Z)†
	400	20,000	Lithology (CSNG)

* HEL = Hostile-environment logging.
† Mark of Welex, a Halliburton Company.

materials on a common basis; for example, by reference to potassium equivalents (i.e., the amount of potassium that would produce the same number of gamma rays per unit of time). Reduced to a common denominator, the average shale contains uranium equivalent to 4.3% potassium, thorium equivalent to 3.5% potassium, and 2% potassium. An "average" shale is hard to find. Because it is a mixture of clay minerals, sand, silts, and other extraneous materials, there can be no standard gamma ray activity for shale. Indeed, the main clay minerals vary enormously in their natural radioactivity. Kaolinite has practically no potassium, whereas illite contains between 4 and 8% potassium (Appendix 4), and montmorillonite contains less than 1%. Occasionally, natural radioactivity may be due to the presence of dissolved potassium or other salts in the water contained in the pores of the shale.

OPERATING PRINCIPLE OF GAMMA RAY TOOLS

Traditionally, two types of gamma ray detectors have been used in the logging industry: Geiger-Mueller and scintillation detectors. Today, practically all gamma ray tools use scintillation detectors containing a sodium iodide crystal (fig. 10.4). When a gamma ray strikes the crystal, a single photon of light is emitted. This tiny flash of light then strikes a photocathode made from cesium antimony or silver-magnesium. Each photon, when hitting the photocathode, releases a bundle of electrons. These, in turn, are accelerated in an electrical field to strike another electrode, producing an even bigger shower of electrons. This process is repeated through a number of stages until a final electrode conducts a small current through a measure resistor to give a voltage pulse which signals that a gamma ray has struck the sodium iodide crystal. The system has a very short "dead time" and can register many counts per second without becoming swamped by numerous signals.

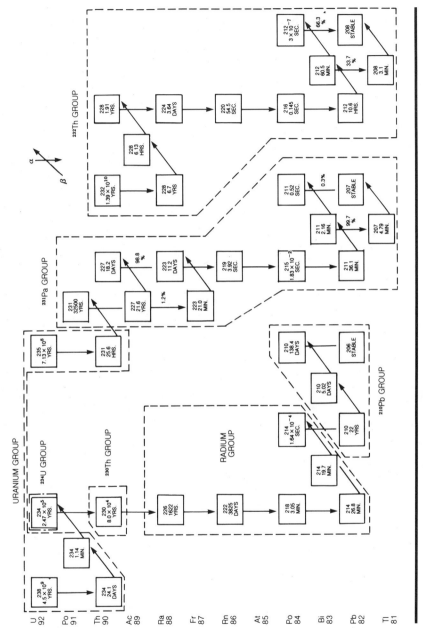

FIGURE 10.1 *Classification of the Radioactive Disintegration Series.*

193

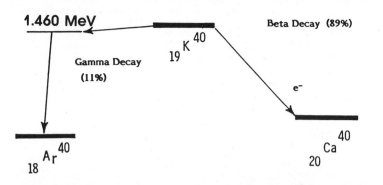

FIGURE 10.2 *Decay Scheme of $^{40}_{19}K$. From Tittman 1956 by permission of the University of Kansas.*

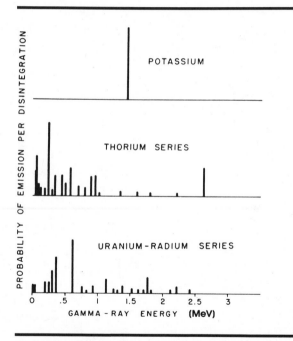

FIGURE 10.3 *Gamma Ray Emission Spectra of Radioactive Minerals. From Tittman 1956 by permission of the University of Kansas.*

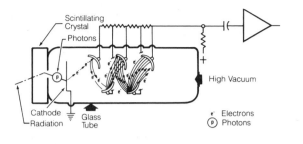

FIGURE 10.4 *Scintillation Counter. Courtesy Gearhart Industries, Inc.*

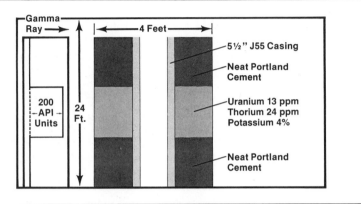

FIGURE 10.5 *API Gamma Ray Standard.*

CALIBRATION OF GAMMA RAY DETECTORS AND LOGS

One of the problems of gamma ray logging is the choice of a standard calibration system, since all logging companies use counters of different sizes and shapes encased in steel housings of varying characteristics. On very old logs, the scale might be calibrated in micrograms of radium per ton of formation. For many reasons, this was found to be an unsatisfactory method of calibration for gamma ray logs, so an API standard was devised. A test pit (installed at the University of Houston) contains an "artificial shale," as illustrated in figure 10.5. A cylinder 4 ft in diameter and 24 ft long contains a central 8-ft section consisting of cement mixed with 13 ppm uranium, 24 ppm thorium, and 4% potassium sandwiched by 8-ft sections of neat portland cement on either side. This 24-ft sandwich is cased with 5½-in. J55 casing. The API standard defines the difference in radioactivity between the neat cement and the radioactive cement mixture as 200 API units. Any logging service company may place its tool in this pit to make a calibration.

Field calibration is performed with a portable jig that contains a radioactive *pill*. The pill is a 0.1-millicurie source of radium 226. When placed 53-in. from the center of the gamma ray detector, it produces a known increase over the background count rate. This increase is equivalent to a known number of API units, depending on the tool type and size and the counter it encloses.

TIME CONSTANTS

All radioactive processes are subject to statistical variations. For example, if a source of gamma rays emits an average of 100 gamma rays each second over a period of hours, the source will emit 360,000 gamma rays per hour (100 × 60 seconds × 60 minutes). However, if the count is measured for 1 second, the actual count might be less than 100 or more than 100. Thus, a choice must be made. Gamma rays can be counted for a very short interval of time, resulting in a poor estimate of the real count rate, or they can be counted for a long time, resulting in a more accurate estimate of the count rate at the expense of an inordinately long time period. In order to average out the statistical variations, various time constants may be selected according to the radioactivity level measured. The lower the count rate, the longer the time constant required for adequate averaging of the variations.

In the logging environment, gamma rays can be counted for a short period of time (e.g., 1 second) with the knowledge that, during that time period, the detector will have moved past the formation whose activity is being measured. Thus, the logging speed and the time interval used to average count rates are inter-related. The following rules of thumb are generally recognized:

Logging Speed (in feet/hour)	Time Constant (in seconds)
3600	1
1800	2
1200	3
900	4

With very slow logging speeds (900 ft/hr = 1.5 ft/s) and long time constants, a more accurate measurement of absolute activity is obtained at the expense of good bed resolution. With high logging speeds and short time constants, somewhat better bed resolution is obtained at the expense of absolute accuracy. At some future time, when the efficiency of gamma ray detectors and their associated electronics improve by one or two orders of magnitude, the use of a time constant will be obsolete except in the cases of very, very inactive formations with intrinsically low gamma ray count rates.

To illustrate the interdependence of logging speed and time constant, the same formation is shown logged at two different speeds in figure 10.6. On the first run, the logging speed was 80 ft/min and the time constant 1 second. On the second run, the speed was 30 ft/min and the time constant 2 seconds. Note the differences in both statistics and bed resolution between the two runs.

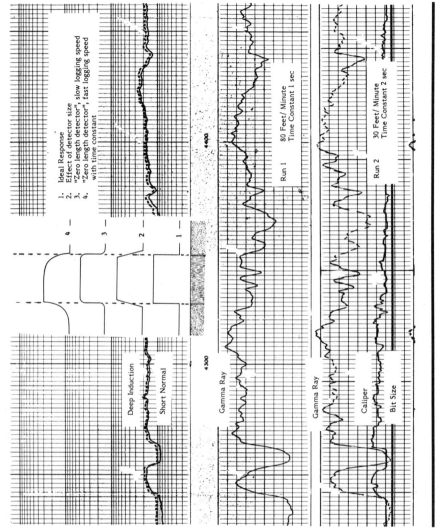

FIGURE 10.6 *Effects of Logging Speed and Time Constant on Gamma Ray Log.*

PERTURBING EFFECTS ON GAMMA RAY LOGS

Gamma ray logs are subject to a number of perturbing effects, including:

Sonde position in the hole (centering/eccentering)
Hole size
Mud weight
Casing size and weight
Cement thickness

Since there are innumerable combinations of hole size, mud weights, and tool positions, an arbitrary set of standard conditions is defined as a 3⅝-in.-OD tool eccentered in an 8-in. hole filled with 10-lb mud. A series of charts is available for making the appropriate corrections. Figure 10.7 applies to logs run in open hole; it corrects for hole size and mud weight. Figure 10.8 applies to logs run in cased hole; it corrects for casing and cement as well.

QUESTION #10.1

Use figure 10.7 to estimate GR_{cor} under the following conditions:

GR_{log} reads 67 API units.
Hole size = 8 in.
Mud weight = 16 lb/gal.
Tool is centered.
GR_{cor} = ?

Note that if a gamma ray log is run in combination with a CNL–FDC tool, it is run eccentered. If it is run with a laterolog or an induction log, it will be centered most of the time.

ESTIMATING SHALE CONTENT
FROM GAMMA RAY LOGS

Since it is common to find radioactive materials associated with clay minerals that constitute shales, it is a commonly accepted practice to use the relative gamma ray deflection as a shale-volume indicator. The simplest procedure is to rescale the gamma ray between its minimum and maximum values from 0 to 100% shale. A number of studies have shown that this is not necessarily the best method, and alternative relationships have been proposed. To further explain these methods, the *gamma ray index* is defined as a linear rescaling of the GR log between GR_{min} and GR_{max} such that:

$$\text{gamma ray index} = \frac{GR - GR_{min}}{GR_{max} - GR_{min}}$$

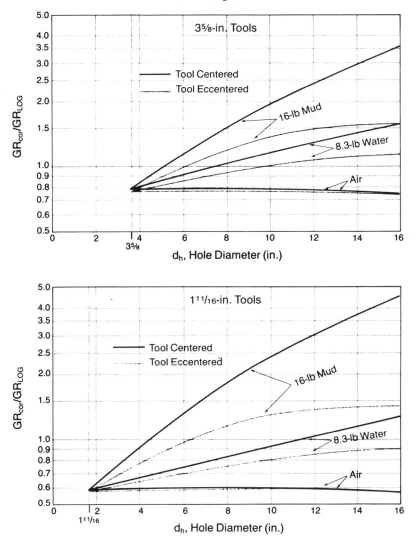

GR$_{cor}$ is defined as the response of a 3⅝-in. tool eccentered in an 8-in. hole with 10-lb mud.

FIGURE 10.7 *Gamma Ray Corrections for Hole Size and Mud Weight. Courtesy Schlumberger Well Services.*

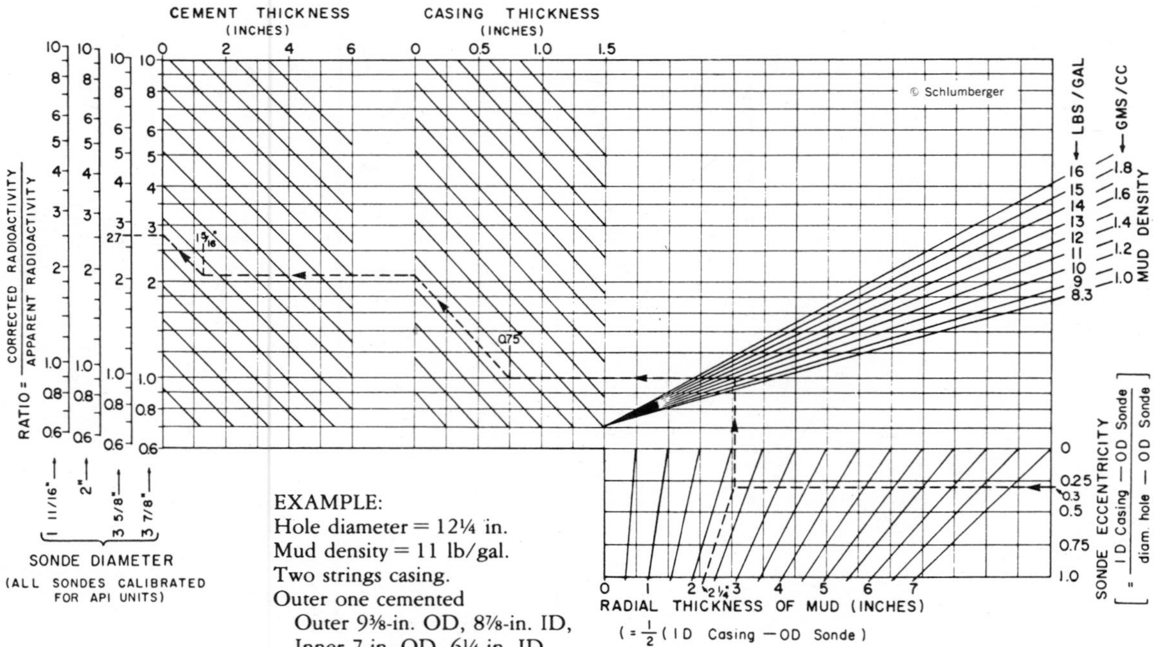

EXAMPLE:
Hole diameter = 12¼ in.
Mud density = 11 lb/gal.
Two strings casing.
Outer one cemented
Outer 9⅝-in. OD, 8⅞-in. ID,
Inner 7-in. OD, 6¼-in. ID.
Sonde OD = 3⅝-in.

From which:
Cement thickness = ½ (12¼ − 8⅝) = 1$\frac{5}{16}$ in.
Casing thickness = ½ (9⅝ − 8⅞ + 7 − 6¼) = ⅝ in.
Mud thickness = ½ (8⅞ − 7 + 6¼ − 3⅝) = 2¼ in.
Sonde eccentricity = (6¼ − 3⅝)/(12¼ − 3⅝) = 0.30.

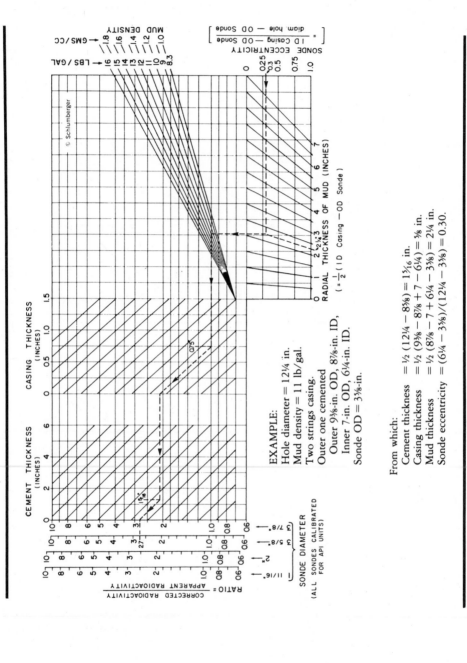

FIGURE 10.8 *Gamma Ray Corrections for Cased Holes. Courtesy Schlumberger Well Services.*

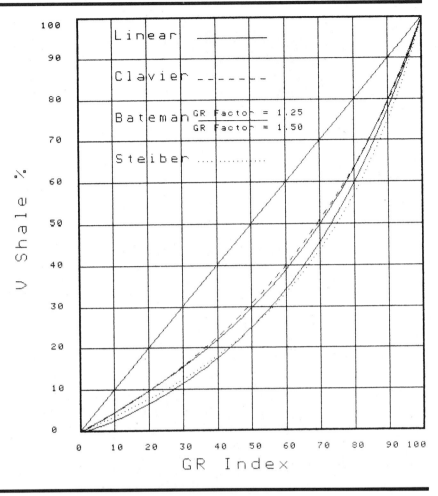

FIGURE 10.9 V_{sh} *as a Function of the Gamma Ray Index.*

If this index is called X, then the alternative relationships can be stated in terms of X as follows:

Relationship	Equation
Linear	$V_{sh} = X$
Clavier	$V_{sh} = 1.7 - \sqrt{3.38 - (X + 0.7)^2}$
Steiber	$V_{sh} = \dfrac{0.5X}{(1.5 - X)}$
Bateman	$V_{sh} = X^{X + GR\,\text{factor}}$

where the GR factor is a number chosen to force the result to imitate the behavior of either the Clavier or the Steiber relationship. Figure 10.9 illustrates the difference between these alternative relationships.

QUESTION #10.2

On the gamma ray log shown on figure 10.10, choose a value for GR_{min} and GR_{max} and then compute V_{sh} in sand C using the linear, Clavier, and Steiber methods.

GAMMA RAY SPECTROSCOPY

Each radioactive decay produces a unique gamma ray. These various gamma rays have characteristic energy levels and occur in characteristic abundance, as expressed in counts per time period. The simple method of just counting how many gamma rays a formation produces can be carried a step further to count how many from each gamma ray energy group it produces. If the number of occurrences is plotted against the energy group, a spectrum will be produced that is characteristic of the formation logged.

Such a spectrum is shown in figure 10.11, where energies from 0 to approximately 3 MeV have been split into 256 specific energy bins. The number of gamma rays in each bin is plotted on the Y-axis. This spectrum can be thought of as a mixture of the three individual spectra belonging to uranium, thorium, and potassium. Some unique mixture of these three radioactive families would have the same spectrum as the observed one. The trick is to find a quick and easy method of discovering that unique mixture. Fortunately, on-board computers in logging trucks are capable of quickly finding a best fit and producing continuous curves showing the concentration of U, Th, and K.

A gamma ray spectral log is shown in figure 10.12. Note that in Track 1 both total gamma ray activity (SGR) and a uranium-free version of the total activity are displayed. Units are API. In Tracks 2 and 3 the concentrations of U, Th, and K are displayed. Logging service companies use various units for spectral logs, which may show counts/ppm, or percent.

QUESTION #10.3

In the example shown in figure 10.12, determine which element is responsible for the high activity seen on the total gamma ray intensity curve at the point marked "A."

INTERPRETATION OF GAMMA RAY SPECTRA LOGS

The interpretation of natural gamma ray spectra logs is a new art. As such, it is still developing at this time. Two general techniques are in use. One, the use of the uranium curve as an indicator of fractures, is described by Fertl et al. (1980). Another technique is to apply the U, Th, and K concentrations in combination with other log data to

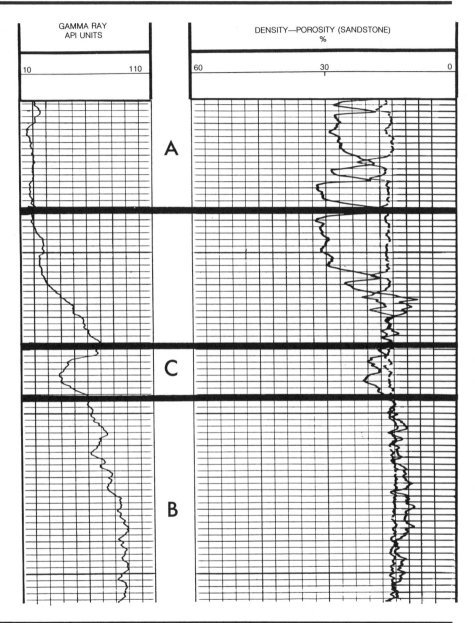

FIGURE 10.10 *Estimation of Shale Content from a Gamma Ray Log.*

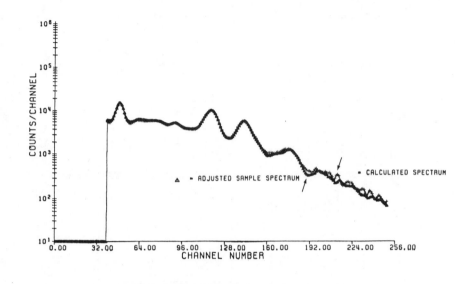

FIGURE 10.11 *Gamma Ray Spectrometry.*

determine lithology and clay type, as described by Marett et al. (1976). Still another approach, described by Hassan et al. (1976), could be called the geochemical method. Figure 10.13 illustrates the variation of the thorium to potassium ratio in a number of minerals ranging from potassium-feldspar to bauxite. A number of radioactive minerals are mapped as a function of their thorium and potassium contents in figure 10.14. If additional data are available, for example, the photoelectric absorption coefficient (P_e) obtained from the litho-density tool, plots of the sort shown in figure 10.15 can be made to assist in mineral identification. Other elemental ratios are also useful indicators. A low U to Th ratio, for example, indicates reduced black shales. Uranium by itself may indicate a high organic carbon content, which in turn may indicate the presence of gas.

Field presentations of gamma ray spectra can assist the analyst in the task of mineral identification by offering curve plots with ratios of the three components (U, Th, and K) already computed. Figure 10.16 gives an example of one such presentation. Track 1 shows a total gamma ray curve together with a uranium-free curve. Track 2 gives three ratios, uranium/potassium, thorium/uranium, and thorium/potassium. Track 3 gives a coded display on which the coded area represents the formations with both the highest potassium and the highest thorium content.

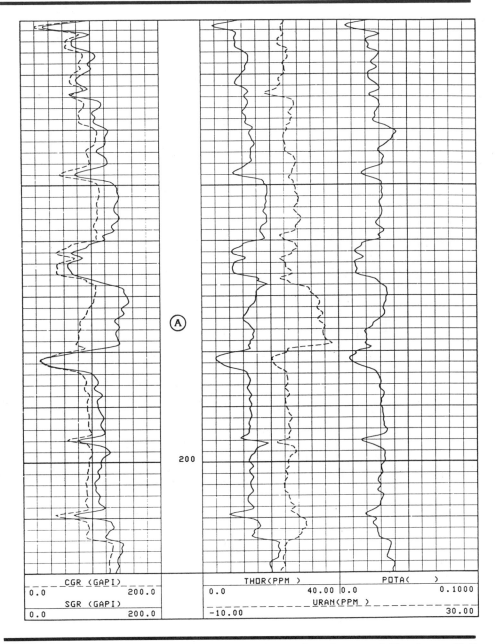

FIGURE 10.12 *Gamma Ray Spectra Log. Courtesy Schlumberger Well Services.*

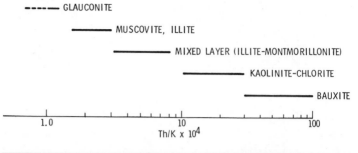

FIGURE 10.13 *Thorium/Potassium Ratios for Various Minerals. Reprinted by permission of the SPWLA from Hassan et al. 1976.*

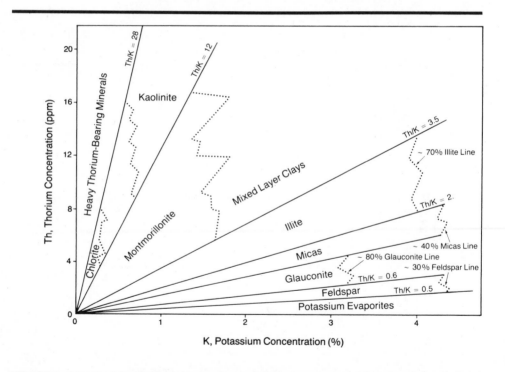

FIGURE 10.14 *Thorium/Potassium Crossplot for Minerals Identification. Courtesy Schlumberger Well Services.*

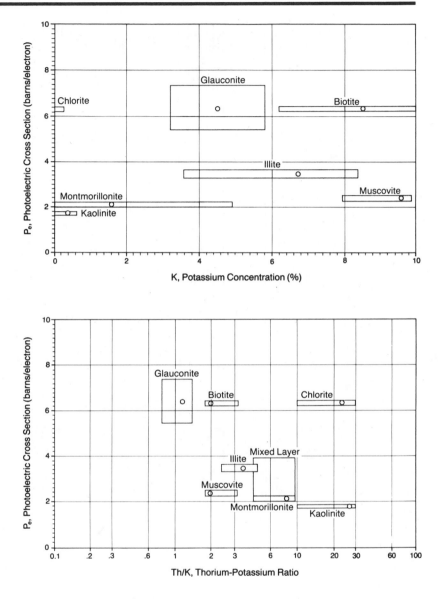

FIGURE 10.15 *Photoelectric Absorption Coefficient (P_e) and Natural Gamma Ray Spectra Crossplots.* (a) P_e *vs. K.* (b) P_e *vs. Th/K. Courtesy Schlumberger Well Services.*

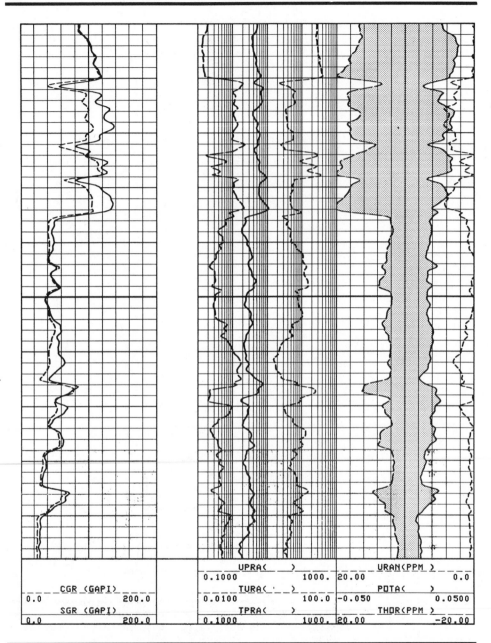

		UPRA( )	URAN(PPM)
		0.1000 1000.	20.00 0.0
CGR (GAPI)		TURA(')	POTA( )
0.0 200.0		0.0100 100.0	-0.050 0.0500
SGR (GAPI)		TPRA( )	THOR(PPM)
0.0 200.0		0.1000 1000.	20.00 -20.00

FIGURE 10.16 *Th, K, and U Ratios Display. Courtesy Schlumberger Well Services.*

SUMMARY

Gamma ray logs are particularly useful to the log analyst because they can be run in both open and cased holes. They are the primary diagnostic for distinguishing between reservoir rocks and shales. In quantitative analysis their measurements are frequently used to compute the shale content of the formation. Gamma ray spectra logs add further to the usefulness of natural gamma ray logs by allowing the analyst to deduce what kind of radioactive rock is present as well as its abundance.

APPENDIX 10A: *Natural Gamma Ray Emitters*

Uranium Series			
Nuclide		Mode of Disintegration	Half-Life
UI	$_{92}U^{238}$	α	4.51×10^4 yr
UX$_1$	$_{90}Th^{234}$	β	24.1 d
UX$_2$	$_{91}Pa^{234m}$	β, IT	1.18 min
UZ	$_{91}Pa^{234}$	β	6.66 hr
UII	$_{92}U^{234}$	α	2.48×10^4 yr
Io	$_{90}Th^{230}$	α	8.0×10^4 yr
Ra	$_{88}Ra^{226}$	α	1620 yr
Rn	$_{86}Em^{222}$	α	3.82 d
RaA	$_{84}Po^{218}$	α, β	3.05 min
RaA'	$_{85}At^{218}$	α, β	2 s
RaA"	$_{86}Em^{218}$	α	1.3 s
RaB	$_{82}Pb^{214}$	β	26.8 min
RaC	$_{83}Bi^{214}$	α, β	19.7 min
RaC'	$_{84}Po^{214}$	α	1.6×10^{-4} sec
RaC"	$_{81}Tl^{210}$	β	1.32 min
RaD	$_{82}Pb^{210}$	β	19.4 yr
RaE	$_{83}Bi^{210}$	α, β	5.01 d
RaF	$_{84}Po^{210}$	α	138.4 d
RaE'	$_{81}Tl^{206}$	β	4.2 min
RaG	$_{82}Pb^{206}$	Stable	

APPENDIX 10A: *(Continued)*

Thorium Series

Nuclide		Mode of Disintegration	Half-Life
Th	$_{90}Th^{232}$	α	1.42×10^{10} yr
MsTh$_1$	$_{88}Ra^{228}$	β	6.7 yr
MsTh$_2$	$_{89}Ac^{228}$	β	6.13 hr
RdTh	$_{90}Th^{228}$	α	1.91 yr
ThX	$_{88}Ra^{224}$	α	3.64 d
Tn	$_{86}Em^{220}$	α	51.5 s
ThA	$_{84}Po^{216}$	α	0.16 s
ThB	$_{82}Pb^{212}$	β	10.6 hr
ThC	$_{83}Bi^{212}$	α, β	60.5 min
ThC'	$_{84}Po^{212}$	α	0.30 μs
ThC'	$_{81}T^{208}$	β	3.10 min
ThD	$_{82}Pb^{208}$	Stable	

APPENDIX 10A: *(Continued) Gamma Ray Lines* in the Spectra of the Important Naturally Occurring Radionuclides*

Nuclide	Gamma Ray Energy, MeV	Number of Photons per Disintegration in Equilibrium Mixture
Bi^{214}(Rac)	0.609	0.47
	0.769	0.05
	1.120	0.17
	1.238	0.06
	1.379	0.05
	1.764	0.16
	2.204	0.05
T4^{208}(ThC')	0.511	0.11
	0.533	0.28
	2.614	0.35
K^{40}	1.46	0.11

* With intensities greater than 0.05 photons per disintegration and energies greater than 100 kev.

APPENDIX 10B: *Thorium-Bearing Minerals*

Name	Composition	ThO_2 Content (%)
Thorium minerals		
Cheralite	$(Th,Ca,Ce)(PO_4SiO_4)$	30, variable
Huttonite	$ThSiO_4$	81.5 (ideal)
Pilbarite	$ThO_2 \cdot UO_3 \cdot PbO \cdot 2SiO_2 \cdot 4H_2O$	31, variable
Thorianite	ThO_2	Isomorphous series to UO_2
Thorite[a]	$ThSiO_4$	25 to 63–81.5 (ideal)
Thorogummite[a]	$Th(SiO_4)_{1-x}(OH)_{4-x};\ x < 0.25$	24 to 58 or more
Thorium-bearing minerals		
Allanite	$(Ca,Ce,Th)_2(Al,Fe,Mg)_3Si_3O_{12}(OH)$	0 to about 3
Bastnaesite	$(Ce,La)Co_3F$	Less than 1
Betafite	About $(U,Ca)(Nb,Ta,Ti)_3O_9 \cdot nH_2O$	0 to about 1
Brannerite	About $(U,Ca,Fe,Th,Y)_3Ti_5O_{16}$	0 to 12
Euxenite	$(Y,Ca,Ce,U,Th)(Nb,Ta,Ti)_2O_5$	0 to about 5
Eschynite	$(Ce,Ca,Fe,Th)(Ti,Nb)_2O_6$	0 to 17
Fergusonite	$(Y,Er,Ce,U,Th)(Nb,Ta,Ti)O_4$	0 to about 5
Monazite[b]	$(Ce,Y,La,Th)PO_4$	0 to about 30; usually 4 to 12
Samarskite	$(Y,Er,Ce,U,Fe,Th)(Nb,Ta)_2O_6$	0 to about 4
Thucholite	Hydrocarbon mixture containing U, Th, rare earth elements	
Uraninite	UO_2 (ideally) with Ce, Y, Pb, Th, etc.	0 to 14
Yttrocrasite	About $(Y,Th,U,Ca)_2(Ti,Fe,W)_4O_{11}$	7 to 9
Zircon	$ZrSiO_4$	Usually less than 1

Source: After Frondel, C., 1956, in Page, L. R., Stocking, H. E., and Smith, H. D., Jr., U.S. Geol. Survey Prof. Papers no. 300.

[a] Potential thorium ore minerals.

[b] Most important commercial ore of thorium. Deposits are found in Brazil, India, USSR, Scandinavia, South Africa, and U.S.A.

APPENDIX 10C: *Uranium Minerals*

Autunite	$Ca(UO_2)_2(PO_4)_2$ 10–12H_2O
Tyuyamunite	$Ca(UO_2)_2(VO_4)_2$ 5–8H_2O
Carnotite	$K_2(UO_2)_2(UO_4)_2$ 1–3H_2O
Baltwoodite	U-silicate high in K
Weeksite	U-silicate high in Ca

APPENDIX 10D: *Potassium, Uranium, and Thorium Distribution in Rocks and Minerals*

	K (%)	U (ppm)	Th (ppm)
Accessory minerals			
Allanite		30–700	500–5000
Apatite		5–150	20–150
Epidote		20–50	50–500
Monazite		500–3000	2.5×10^4–20×10^4
Sphene		100–700	100–600
Xenotime		500–3, 4×10^4	Low
Zircon		300–3000	100–2500
Andesite (av.)	1.7	0.8	1.9
A., Oregon	2.9	2.0	2.0
Basalt			
Alkali basalt	0.61	0.99	4.6
Plateau basalt	0.61	0.53	1.96
Alkali olivine basalt	<1.4	<1.4	3.9
Tholeiites (orogene)	<0.6	<0.25	<0.05
(non orogene)	<1.3	<0.50	<2.0
Basalt in Oregon	1.7	1.7	6.8
Carbonates			
Range (average)	0.0–2.0(0.3)	0.1–9.0(2.2)	0.1–7.0(1.7)
Calcite, chalk, limestone, dolomite (all pure)	<0.1	<1.0	<0.5
Dolomite, West Texas (clean)	0.1–0.3	1.5–10	<2.0
Limestone (clean)			
Florida	<0.4	2.0	1.5
Cretaceous trend, Texas	<0.3	1.5–15	<2.0
Hunton lime, Okla.	<0.2	<1.0	<1.5
West Texas	<0.3	<1.5	<1.5
Clay minerals			
Bauxite		3–30	10–130
Glauconite	5.08–5.30		
Bentonite	<0.5	1–20	6–50
Montmorillonite	0.16	2–5	14–24
Kaolinite	0.42	1.5–3	6–19
Illite	4.5	1.5	
Mica			
Biotite	6.7–8.3		<0.01
Muscovite	7.9–9.8		<0.01
Diabase, Va.	<1.0	<1.0	2.4
Diorite, quartzodiorite	1.1	2.0	8.5
Dunite, Wa.	<0.02	<0.01	<0.01
Feldspars			
Plagioclase	0.54		<0.01
Orthoclase	11.8–14.0		<0.01
Microcline	10.9		<0.01
Gabbro (mafic igneous)	0.46–0.58	0.84–0.9	2.7–3.85

APPENDIX 10D: *(Continued)*

	K (%)	U (ppm)	Th (ppm)
Granite (silicic igneous)	2.75–4.26	3.6–4.7	19–20
Rhode Island	4.5–5	4.2	25–52
New Hampshire	3.5–5	12–16	50–62
Precambrian (Okla.) Minnesota, Col. Tex.)	2–6	3.2–4.6	14–27
Granodiorite	2–2.5	2.6	9.3–11
Colorado, Idaho	5.5	2.–2.5	11.0–12.1
Oil shales, Colorado	<4.0	up to 500	1–30
Peridotite	0.2	0.01	0.05
Phosphates		100–350	1–5
Rhyolite	4.2	5	
Sandstones, range (av.)	0.7–3.8(1.1)	0.2–0.6(0.5)	0.7–2.0(1.7)
Silica, quartz, quartzite, (pure)	<0.15	<0.4	<0.2
Beach Sands, Gulf Coast	<1.2	0.84	2.8
Atlantic Coast (Fla., N.C.)	0.37	3.97	11.27
Atlantic Coast (N.J., Mass.)	0.3	0.8	2.07
Shales			
"Common" shales [range (av.)]	1.6–4.2(2.7)	1.5–5.5(3.7)	8–18(12.0)
Shales (200 samples)	2.0	6.0	12.0
Schist (biotite)		2.4–4.7	13–25
Syenite	2.7	2500	1300
Tuff (feldspathic)	2.04	5.96	1.56

APPENDIX 10E: *Geological Significance of Natural Gamma Ratios*

Ratios	Remarks
Thorium/uranium (Th/U)	In *sedimentary* rocks, Th/U varies with depositional environment
	Th/U > 7: continental, oxidizing environment, weathered soils, etc.
	< 7: marine deposits, gray and green shales, graywackes
	< 2: marine black shales, phosphates.
	In *igneous* rocks, high Th/U indicative of oxidizing conditions by magma before crystallization and/or extensive leaching during postcrystallization history
	Source rock potential estimates of argillaceous sediments (shales)
	Major geologic unconformities
	Distance to ancient shore lines or location of rapid uplift during time of deposition
	Stratigraphic correlations, transgression vs. regression, oxidation vs. reduction regimes, etc.
Uranium/potassium (U/K)	Source rock potential of argillaceous sediments
	Stratigraphic correlations
	Unconformities, diagenetic changes in argillaceous sediments, carbonates, etc.
	Frequent correlation with vugs and natural fracture systems in subsurface formations, including localized correlation with hydrocarbon shows on drilling mud logs and core samples both in clastic and carbonate reservoirs
Thorium/potassium (Th/K)	Recognition of rock types of different facies
	Paleographic and paleoclimatic interpretation of facies characteristics
	Depositional environments, distance from ancient shore lines, etc.
	Diagenetic changes of argillaceous sediments
	Clay typing: Th/K increases from glauconite → muscovite → illite → mixed-layer clays → kaolinite → chlorite → bauxite
	Correlation with crystallinity of illite, average reflectance power, paramagnetic electronic resonance

BIBLIOGRAPHY

Dresser Atlas: "Gamma Ray Spectral Data Assists in Complex Formation Evaluation," REP 06/80 5M 3335, Houston (February 1979).

Dresser Atlas: "Spectralog," 3334 (1981).

Fertl, W. H., and Frost, E., Jr.: "Experiences with Natural Gamma Ray Spectral Logging in North America," SPE paper 11145 presented at the 57th Annual Technical Conference and Exhibition, New Orleans, Sept. 25–29, 1982.

Fertl, W. H., Stapp, W. L., Vaello, D. B., and Vercellino, W. C.: "Spectral Gamma Ray Logging in the Texas Austin Chalk Trend," *J. Pet. Tech.* (March 1980); presented at the 53rd Annual Technical Conference and Exhibition, Houston, Oct. 1–4, 1978.

Hassan, M., Hossin, A., and Combaz, A.: "Fundamentals of the Differential Gamma Ray Log-Interpretation Technique," paper presented at the SPWLA 17th Annual Logging Symposium, Denver, June 9–12, 1976.

Kokesh, F. P.: "Gamma Ray Logging," *Oil and Gas J.* (July 1951).

Marett, G., Chevalier, P., Souhaite, P., Suau, J.: "Shaly Sand Evaluation Using Gamma Ray Spectrometry Applied to the North Sea Jurassic," SPWLA 17th Annual Logging Symposium, Denver, June 1976.

Quirein, John A., Gardner, John S., and Watson, John T.: "Combined Natural Gamma Ray Spectral/Litho-Density Measurements Applied to Complex Lithologies," SPE paper 11143, presented at the 57th Annual Technical Conference and Exhibition, New Orleans, Sept. 25–29, 1982.

Smith, H.D., Jr., Robbins, C. A., Arnold, D. M., and Deaton, J. G.: "A Multi-Function Compensated Spectral Natural Gamma Ray Logging System," SPE paper 12050 presented at the 58th Annual Technical Conference and Exhibition, San Francisco, Oct. 5–8, 1983.

Schlumberger: "Log Interpretation Charts" (1984).

Schlumberger: "Log Interpretation Charts" (1972).

Schlumberger: "Well Evaluation Developments—Continental Europe" (1982).

Tittman, J.: "Radiation Logging Lecture I: Physical Principle" and "Lecture II: Applications," Petroleum Engineering Conference on the Fundamental Theory and Quantitative Analysis of Electric and Radioactivity Logs, University of Kansas, 1956.

Welex: "Open Hole Services," Catalog G 6003.

Answers to Text Questions

QUESTION #10.1
$$GR_{cor} = 100$$

QUESTION #10.2

GR_{min}	=	20
GR_{max}	=	85
In sand C, GR	=	40
GR Index	=	0.3077
V_{sh} linear	=	30.77%
V_{sh} Clavier	=	16.23%
V_{sh} Steiber	=	12.90%

QUESTION #10.3
Uranium

CHAPTER ELEVEN

RESISTIVITY MEASUREMENTS

The very first logging device ever designed was made to measure formation resistivity. It was a modification of a device previously used to detect anomalies of underground resistivity associated with either geologic features or concentrations of metallic ores. This old surface surveying method is illustrated in figure 11.1. A voltage source sent a current through the ground between two widely spaced electrodes. The voltage drop between two other, more closely spaced, electrodes was used as a measure of the ground resistivity. By moving the whole electrode array across the country-side, it was possible to map underground features as shown in figure 11.2. When the entire setup was rotated through a 90° angle and lowered into a borehole, the electric log was born. The first electric log to be run (see fig. 11.3) was in the Pechelbronn field in France in 1927.

Since those early days, a series of improvements has resulted in five main families of resistivity tools. These are:

Electric logs
Induction logs
Laterologs
Microresistivity devices
Dielectric logs

When discussing formation resistivities, it is common to say "this is a 25-ohm sand" rather than "this sand has a resistivity of 25 ohm-meters squared per meter." So the field jargon, when talking about resistivity logs, is to say "ohm" when $\Omega \cdot m^2/m$ is really meant.

Why the need to know the resistivity rather than the resistance? Because resistance is a function not only of the resistivity measured, but also of the geometry of the body of material on which the measurement is being made. The geometry of the body is not of prime interest to the well logger. The measurement that characterizes the rock, as far as fluid content is concerned, is the resistivity and not the resistance. The resistance of a wire stretching across the Pacific Ocean could be high, but the resistivity of the wire itself could be very low.

Conductivity is the reciprocal of resistivity. A substance with infinite resistivity (empty space) has a conductivity of zero, and a substance with low resistivity has high conductivity:

$$\text{conductivity} = \frac{1}{\text{resistivity}} .$$

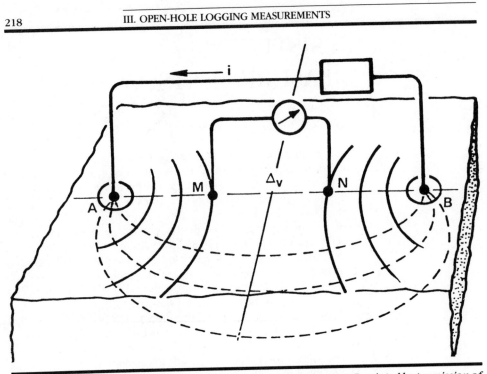

FIGURE 11.1 *Surface Resistivity Measurements. Reprinted by permission of John Wiley and Sons from Alland and Martin 1977.*

The unit of conductivity is the mho/m, one mho/m being the reciprocal of one $\Omega\cdot$m. The practical unit is the millimho, which is the one-thousandth part of a mho.

QUESTION #11.1
Convert 5 ohms to millimhos.
Convert 2000 millimhos to ohms.

TYPICAL RESISTIVITIES
Typical formation resistivities range from 0.5 $\Omega\cdot$m to 1000 $\Omega\cdot$m. Soft formations (shaly sands) range from 0.5 $\Omega\cdot$m to about 50 $\Omega\cdot$m. Hard formations (carbonates) range from 100 $\Omega\cdot$m to 1000 $\Omega\cdot$m. Evaporates (salt, anhydrite) may exhibit resistivities of several thousand $\Omega\cdot$m. Formation water, by contrast, will range from a few hundredths of an $\Omega\cdot$m (brines) to several $\Omega\cdot$m (fresh water). Sea water has a resistivity of 0.35 $\Omega\cdot$m at 75°F.

DEFINITIONS
Ohm's law states that the potential difference, V, between two points on a conductor is equal to the product of the current flowing in the conductor, I, and the resistance of the conductor, R. Practical units of measurement are, respectively, the volt, the amp, and the ohm. Expressed as an equation, this relationship is

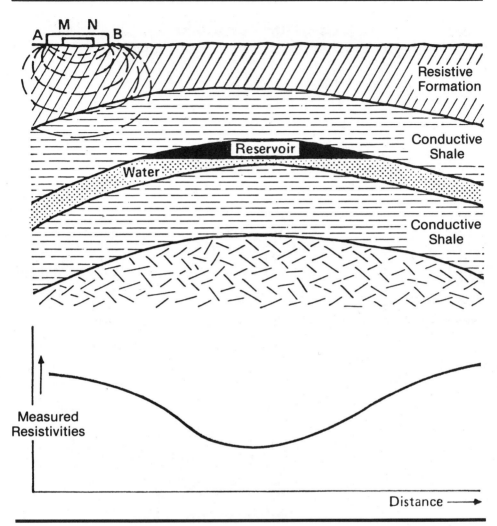

FIGURE 11.2 *Hypothetical Cross Section of an Anticline and Corresponding Resistivity Profile. Reprinted by permission of John Wiley and Sons from Alland and Martin 1977.*

$$V = I \times R$$

volts $=$ amps $\times$ ohms.

In logging, the resistivity of a rock is of more interest than its resistance. *Resistivity* is the specific electrical resistance of a given amount of material. The unit of resistivity is the ohm-meter^2/meter, abbreviated as $\Omega \cdot m^2/m$; that is, the voltage required to cause one amp to pass through a cube one meter long and one meter square (fig. 11.4).

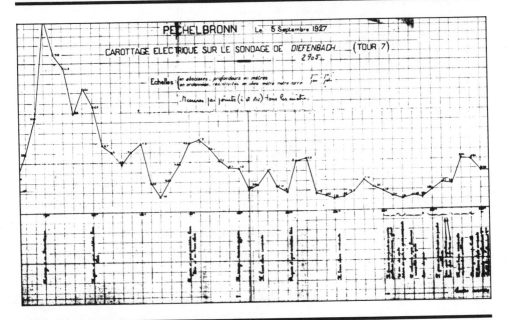

FIGURE 11.3 *This First Electrical Log, a Simple Hand-Plotted Graph, Was a Turning Point in the History of the Petroleum Industry. Courtesy Schlumberger Well Services.*

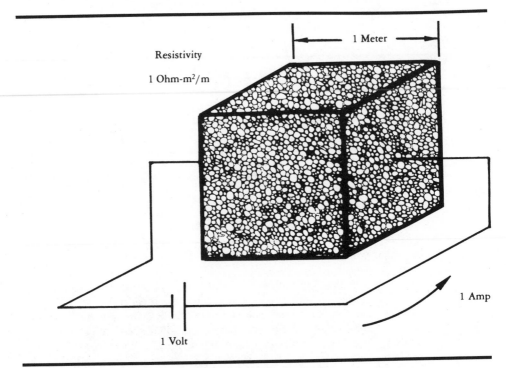

FIGURE 11.4 *Definition of the ohm-meter.*

IDEAL RESISTIVITY MEASUREMENTS

Given an infinite isotropic, homogeneous medium in which there is implanted a spherical electrode emitting a current I radially in a spherical distribution pattern (see fig. 11.5), the voltage drop between any two concentric spherical shells with radii ρ and $\rho + d\rho$ can be determined by

$$dV = I \times dr,$$

where dV is the voltage drop, I is the current, and dr is the resistance between the two shells. If the resistivity of the medium is R,

$$dr = R \times \frac{d\rho}{4\pi\rho^2} \quad \text{and} \quad dV = I \times R \times \frac{d\rho}{4\pi\rho^2}.$$

If we integrate this equation from $\rho = A$ to $\rho = M$, the equation to determine the value for V_m (the voltage measured between an infinite ground and some point M in the formation) becomes:

$$V_m = -I \times \frac{R}{4\pi} \times \left[\frac{1}{\rho}\right]_A^M = \frac{IR}{4\pi AM},$$

where V_m is the measured voltage at some point a distance M from the current electrode A and R is the formation resistivity.

This ideal derivation does not fit the real world for two reasons. First, a borehole is required in order to introduce an electrode into the formation and, second, no formation is infinite and homogeneous. For these reasons, the early resistivity devices evolved into electric logging devices with four electrodes. Two electrodes transmit and receive currents from the formation; two other electrodes measure voltage differences in the borehole. Figure 11.6 illustrates a *normal* resistivity device. Constant current is passed between electrodes A and B. The measure voltage appears between electrodes M and N. The distance *AM* is called the *spacing*. In the 16-in. *short normal* device, electrode A is 16 in. away from electrode M. A *lateral* device is illustrated in figure 11.7. Here, a constant current is sent between the A and B electrodes, and the measure voltage appears between the M and N electrodes. *Inverse* devices utilize another combination of electrode functions and will not be discussed here.

All these old electric devices, which saw great use for many years, were plagued with inherent defects as a result of borehole and adjacent bed effects. Their idiosyncrasies are legion. The student requiring further details to complete a study of old electric logs is directed to a volume published by the Society of Professional Well Log Analysts entitled *The Art of Ancient Log Analysis*. The only survivors from among the early electric logs are the microlog and a device known as the ULSEL, the ultra-long-spaced electric log, which

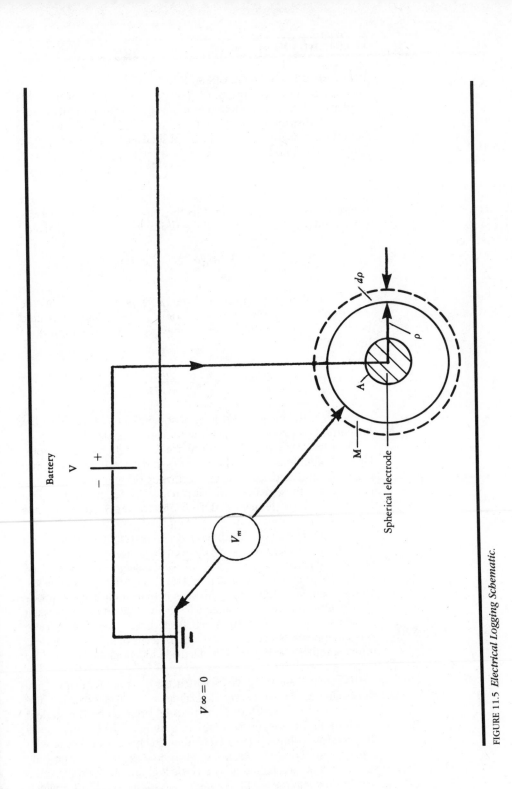

FIGURE 11.5 *Electrical Logging Schematic.*

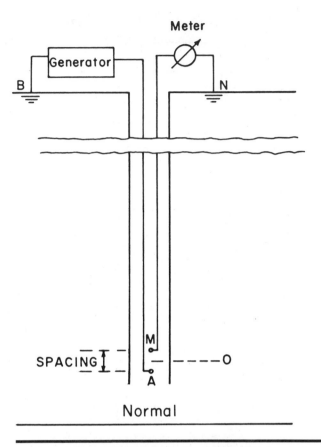

Normal

FIGURE 11.6 *Normal Resistivity Device. Reprinted by permission of the SEG from Moran and Kunz 1962.*

is a normal device with an *AM* spacing of 100 to 1000 ft. It is used for remote sensing of one borehole from another (blowout control) or remote sensing of resistive anomalies such as salt domes.

EVOLUTION OF MODERN RESISTIVITY DEVICES

Allthough the original electric logging principles were sound, their practical embodiments left much to be desired. Efforts to improve the measurement of formation resistivity have been busily pursued for the last 50 years at least. As a result, three main branches of resistivity logging have evolved: focused electric logs, induction logs, and microwave devices.

Focused Electric Logs

Focused electric logs were a logical step up from the early unfocused electric logs. Addition of focusing electrodes to the basic four-electrode array made it possible to steer the current in the right

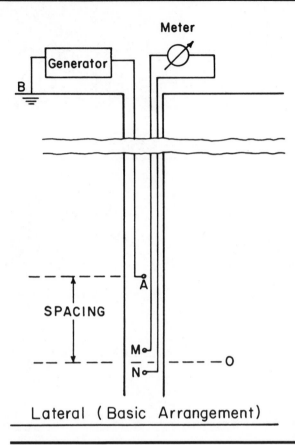

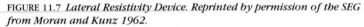

FIGURE 11.7 *Lateral Resistivity Device. Reprinted by permission of the SEG from Moran and Kunz 1962.*

direction. The modern descendant of the early focused electric log is the laterolog (see chapter 13), which uses a multiplicity of focusing electrodes to direct current into the formation and to eliminate most of the detrimental borehole effects that were present on early electric logs. Spherically focused logs also rely on focusing electrodes. Microfocused logging measurements are reviewed in chapter 14.

Induction Logs

Induction logs broke the tradition of using current and voltage-measure electrodes by introducing a system of focused coils to induce currents to flow in the formation, away from the disturbing influence of the borehole and the invaded zone. Induction logs are discussed in chapter 12.

Microwave Devices

More recently still, microwave devices have been used to measure the dielectric constant of the formation. Although, strictly speaking, these do not measure formation resistivity, they are usually classified as resistivity devices since the end use of their measurements is the same—that is, determination of formation fluid saturation. These microwave devices are covered in chapter 15.

PHILOSOPHY OF MEASURING FORMATION RESISTIVITY

Whatever the device used to measure formation resistivity, there are common elements that conspire to confound its efforts. These are:

The effects of the borehole itself
The effects of neighboring beds
The effects of mud filtrate invasion

Although modern resistivity measuring devices represent a considerable improvement over the original unfocused electric log, there is still plenty of room for improvement. Thus, when using a resistivity log, the analyst should always bear in mind that the device used to produce the log is not perfect and the measurement displayed on the log is a composite of the four factors (R_s, R_t, R_m, R_{xo}) illustrated in figure 11.8.

Note that the tool is influenced not only by R_t—the resistivity of the undisturbed zone (which is what we are trying to measure)—but, by its design, also by the resistivity of the borehole, the adjacent beds, and the invaded zone, if present. Thus, there is evident danger in assuming that the reading from a resistivity log truly represents R_t. The device used, the particular circumstances of the well, and the formations logged influence the actual reading, which may be greater or less than R_t. The chapters that follow will give instructions on how to recognize those cases where resistivity measurements depart radically from R_t. For now, suffice it to say that, as a rule of thumb, a big contrast between the resistivity of the bed of interest and the resistivity of either the mud column or the adjacent bed is a danger signal that calls for the use of correction charts. In this context, a "big" contrast could be classified as a factor of 10 or more. Of particular note are conditions where the bed of interest is thin (say 15 ft or less) and/or invasion is deep (d_i greater than 40 in.).

By way of summary: It is safe to assume that a deep resistivity device measures R_t unless:

1. R_t/R_m is greater than, say, 10,
2. R_t/R_s is greater than, say, 10,
3. hole size is greater than, say, 12 in.,
4. the bed is thinner than, say, 15 ft,
5. invasion diameter is greater than 40 in.

If any of these adverse conditions exist, the appropriate correction chart is called for. It will become apparent later that induction logs and focused electric logs (laterologs) behave differently when faced with these problems, and in many cases what is poison to an induction tool is an advantage to a laterolog and vice versa.

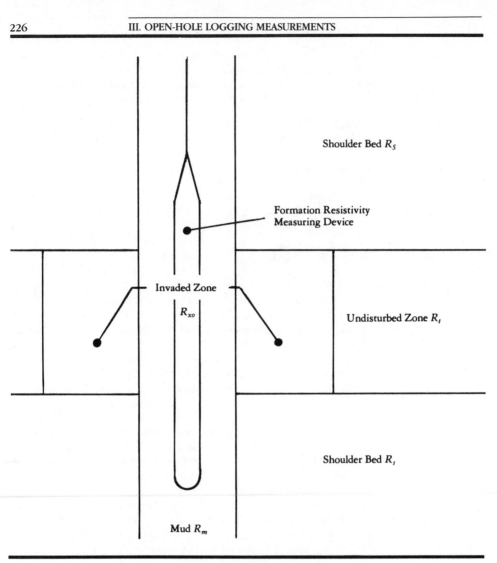

FIGURE 11.8 *Factors Affecting Resistivity Measuring Devices.*

BIBLIOGRAPHY

Alland, L., and Martin, M.: "Schlumberger, the History of a Technique," John Wiley & Sons, New York (1977).

Lynch, E. J.: "Formation Evaluation," Geoscience Series, Harper & Row, New York (1962).

Moran, J. H., and Kuz, K. S.: "Basic Theory of Induction Logging," *Geophysics* (December 1962).

Runge, R. J., Worthington, A. E., and Lucas, D. R.: "Ultra-Long Spaced Electric Log (ULSEL)," *The Log Analyst* (September–October 1969) **10**.

Schlumberger: "Log Interpretation/Principles" (1972).

Schlumberger, C., Schlumberger, M., and Leonardon, E. G.: 3 papers in *Trans.* AIME (1934) **110**. "Some Observation Concerning Electrical Measurements in Anisotropic Media, and their Interpretation," p. 159; "Electrical Coring: a Method of Determining Bottom-Hole Data by Electrical Measurements," p. 237; "A New Contribution to Subsurface Studies by Means of Electrical Measurements in Drill Holes," p. 273.

Answers to Text Question

QUESTION #11.1
200 millimhos
0.5 ohms

CHAPTER TWELVE

INDUCTION LOGGING

Logging systems used before the introduction of induction logging depended on the presence of an electrically conductive fluid in the borehole to transmit electric current to the formation. In most of the rotary-drilled wells, the drilling fluid is a water-based mud that conducts electricity. Some wells are drilled with nonconductive fluids such as oil-emulsion muds, oil-base muds, fresh water, air, and gas. But it is impossible to obtain a satisfactory electrical log using conventional electric logging tools under such conditions.

Induction logging does not depend upon physical contact between the walls of the wellbore and the logging tool. The induction log is similar to a transformer in that the transmitter coil is energized with alternating current, which induces a secondary current in the formation. This current is proportional to the electrical conductivity of the formation and to the cross-sectional area affected by the energizing coil. The higher the conductivity of the formation (the lower the resistivity), the larger the formation current will be. This current in turn induces a signal in a receiver coil, and the intensity of this signal is proportional to the formation current and conductivity. The signal detected by the receiver coil is amplified and recorded at the surface.

The direct measurement made by the induction is one of conductivity. A reciprocator is used to produce a curve on a resistivity scale. Both the conductivity and reciprocated conductivity (resistivity) curves are shown on the log. The deflections of these curves are proportional to formation conductivity. Before making calculations, the resistivity values should be checked with the conductivity values to detect any errors made by the reciprocator. Conductivity is measured in mhos or millimhos per meter. A mho is the reciprocal of an ohm—that is, 1 mho $= 1/\Omega$. A millimho equals $1/1000$ of a mho. Thus, formations having resistivities of 10, 100, or 1000 $\Omega \cdot$m would have conductivities of 100, 10, and 1 mmho/m, respectively.

Induction logging equipment provides a record over a wide range of conductivity values. Accuracy is excellent for conductivity values higher than 20 mmho/m (resistivity values less than 50 $\Omega \cdot$m) and is acceptable in lower conductivity ranges (down to 5 mmho/m). Beyond this limit, the induction log continues to respond to variations in formation conductivity, but with diminished accuracy. There is a small uncertainty of about ± 1 mmho/m on the zero of the present sondes.

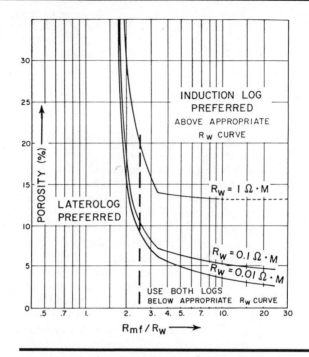

FIGURE 12.1 *Preferred Conditions for Induction Tools. Courtesy Schlumberger Well Services.*

WHEN TO USE AN INDUCTION LOG

Induction logs are recommended for use:

1. where fresh or oil-based mud is employed,
2. where the R_{mf}/R_w ratio is greater than 3,
3. where R_t is less than 150 Ω, and
4. where bed thickness is greater than 30 ft.

They are the only resistivity devices that will work in oil-based mud (where oil is the continuous phase) or air-filled holes. Figure 12.1 illustrates the preferred working conditions for induction tools: R_{mf}/R_w is plotted against porosity; and the top, right portion of the chart shows the conditions in which the induction log should be run.

TOOL AVAILABILITY AND RATINGS

Two commonly used induction devices are the induction spherically focused (ISF) log and the dual induction spherically focused (DISF) log. Each of these tools can be combined with a sonic tool, thereby allowing both sonic and resistivity logs to be recorded simultaneously in one trip into the well. Both tools are rated for 350°F and 20,000 psi. Figure 12.2 shows a typical tool string.

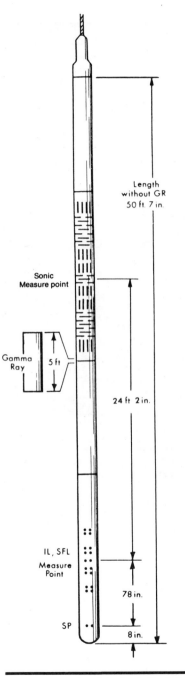

Length
without GR
50 ft. 7 in.

Sonic
Measure point

Gamma
Ray   5 ft

24 ft 2 in.

IL, SFL
Measure
Point

78 in.

SP

8 in.

FIGURE 12.2 *Induction–Sonic Tool Combination. Reprinted by permission of the GCAGS and author from Schuster et al. 1971.*

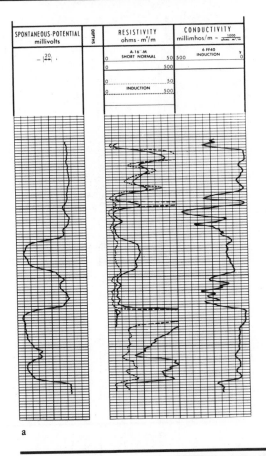

FIGURE 12.3 *Induction Log Presentations.* (a) *Linear.* (b) *Logarithmic.* (c)
Split Grid. Courtesy Schlumberger Well Services.

PRESENTATIONS AND SCALES

Induction logs and combination induction–sonic logs appear on a
variety of scales and presentations. When presented, the primary
measurements of conductivity always appear on a linear scale.
Induction resistivity can appear on a linear scale or a logarithmic
scale. When the sonic curve is also present, a split grid is usually
employed (see fig. 12.3).

THEORY OF INDUCTION DEVICES

Induction tools work like mine detectors or airport security devices
through which the passengers walk. A transmitter coil, with an
alternating current passing through it, sets up an alternating
magnetic field. The alternating magnetic field induces an alternating
potential in the formation. The conductivity of the material in the
formation determines whether a large or a small alternating current

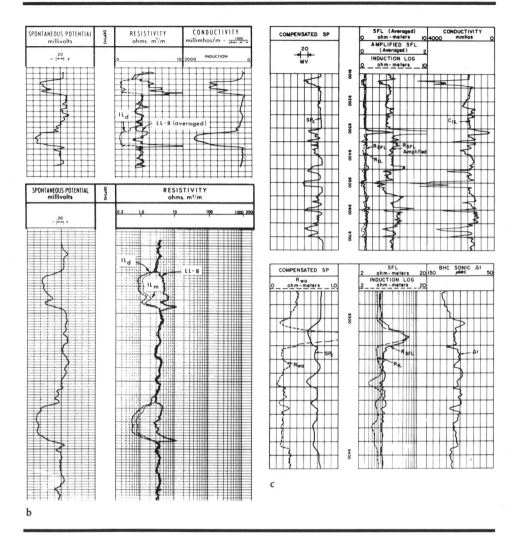

is caused to flow in the formation. The alternating magnetic field associated with this current induces an AC potential in a receiving coil. This potential can be measured and can be used as an indicator of the conductivity of the formation. Figure 12.4 illustrates this concept by reference to a simple two-coil device.

The oscillator attached to the transmitter is held at a constant 20-kHz frequency and with a constant current so that the strength of the magnetic field is also constant. The induced current in the formation is large when the resistivity is low (conductivity high) and therefore, the measured signal (the voltage appearing on the receiver coil) is greatest in low-resistivity formations. In very high-resistivity formations, the measured signal becomes very small and resolution is lost. This loss of resolution in high resistivities

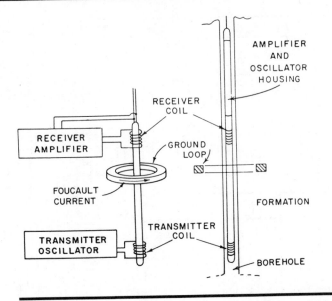

FIGURE 12.4 *Basic Two-Coil Induction Log System. Courtesy Schlumberger Well Services.*

becomes a problem at about 150 $\Omega m^2/m$. Above 250 $\Omega m^2/m$, induction tools are not very useful for quantitative measurements.

An induction tool requires no special borehole fluids and works equally well in air or oil- or water-based muds. In fact, in air-filled holes or oil-based muds (where oil is the continuous phase), it is the only resistivity tool that will work.

Calibration of the system is a two-point method. The sonde is suspended high off the ground well away from conductive materials and is adjusted by zero shift to read zero conductivity (infinite resistivity). A circular ring of metal in the form of a test loop is then placed around the sonde and a scale-sensitivity adjustment is made to the known value of the test loop. Depending on the tool, this will be in the order of the equivalent of a 1- or 2-$\Omega \cdot m$ formation.

When the tool is at the bottom of a 10,000-ft well, there is no way a test loop can be placed around the sonde. An internal calibrator is, therefore, included in the tool. This calibrator will have a nominal value of 1 or 2 $\Omega \cdot m$, and its precise value is determined monthly by reference to the test loop. These internal calibrators shift with aging but behave reasonably well under normal use. A check of the zero-conductivity point when the tool is in the hole is accomplished by simply opening the receiving coil. Any extraneous signal is cancelled out by a zero adjustment.

SKIN EFFECT

To understand the induction response fully, several factors must be considered. In addition to the emf induced by the magnetic field of the transmitter coil, other emfs are induced in the ground loops, and these affect the amplitude and phase of the eddy currents. These additional emfs are produced by linkage of each ground loop with its own magnetic field (a ground loop has self-inductance), and with the magnetic fields of the other nearby ground loops (there is mutual inductance between different ground loops). In this multiple cross-coupled system, the resultant eddy currents will not be quite as predicted on the basis of the simple tool theory already discussed. That is, it cannot be assumed that the individual ground loops are independent of one another.

It can be predicted that, with increasing distance from the source (i.e., from the transmitter coil), there will be attenuation in the amount of transmitted power because:

1. The dissipation of energy by the flow of eddy currents in the region near the source decreases the energy available for transmission to regions farther out.
2. Regions at a distance from the source are shielded from the magnetic field of the transmitter coil by the annulling effect of magnetic fields of opposite sign from the eddy currents in the conductive medium closer to the transmitter. In a sense, the shielding of the outer regions is equivalent to a reflection of the energy back toward the source.

As a consequence of these interactions, there is a reduction of the receiver-coil signal; that is, a reduction of high conductivity. This reduction in the apparent conductivity reading is commonly termed *skin effect*. If σ_g is the conductivity reading that would be observed in a given configuration of media without skin effect and σ_a is the conductivity actually observed, the difference-conductivity, σ_s, is the skin effect, that is,

$$\sigma_s = \sigma_g - \sigma_a.$$

An amount σ_s is added to the observed reading by means of a skin-effect compensating network. The skin effect is nonlinear and can best be illustrated as in figure 12.5. In practical terms, the tool will read a resistivity that is too high unless compensation for the skin effect is working properly. Failure to switch in the compensation can cause an erroneous log.

ENVIRONMENTAL EFFECTS

A practical field tool includes focusing coils (fig. 12.6) in addition to the transmitting and receiving coils of the simple two-coil device (fig. 12.4). These focusing coils make the current ground loop flow as far away from the borehole as possible to eliminate borehole and drilling mud-filtrate invasion effects.

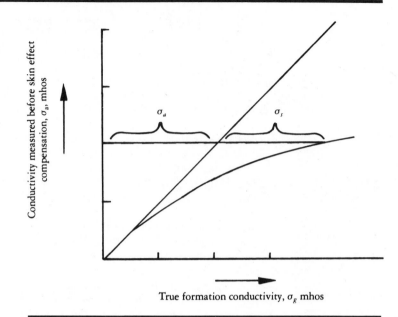

FIGURE 12.5 *Correction of Formation Conductivity for Skin Effect.*

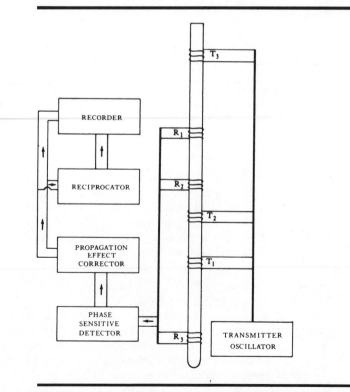

FIGURE 12.6 *Schematic Diagram of a Typical Induction Logging System. Courtesy Dresser Atlas, Dresser Industries, Inc.*

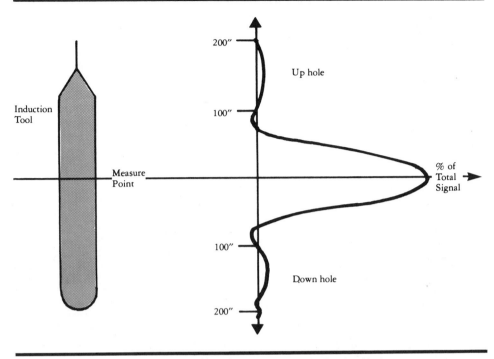

FIGURE 12.7 *Vertical Response of the Induction Tool.*

Bed Thickness Corrections

Unfortunately, a trade-off must be made when designing an induction tool. Good bed resolution can only be obtained with closely spaced transmitter–receiver coil arrangements. But this close spacing results in the radial depth of investigation being relatively shallow. The 6FF40 devices are designed for deep investigation and, therefore, have poor vertical bed resolution. In effect, the signal received is a mixture of signals received from points both above and below the horizon being measured. The surface control equipment tries to offset this poor bed resolution by emphasizing the present interval and playing down the measurement made either side of the horizon being measured. This is achieved by use of a *shoulder-bed resistivity* (SBR) correction routine (see fig. 12.7).

The induction tool, via memorization, can manipulate three measurements in such a way that the reading recorded on the log is equal to some "weight" value A times the value of the interval being measured plus B times the values at points 78 in. above and below the point being measured. The values for A and B are chosen so that $A - 2B = 1$. This is logical since, in a homogeneous formation (where all three measurements are the same), there is no net effect. This scheme assists in correcting the log for the effects of adjacent beds. Even so, it is insufficient to cure completely the problem of poor resolution, and a set of charts (fig. 12.8) is available

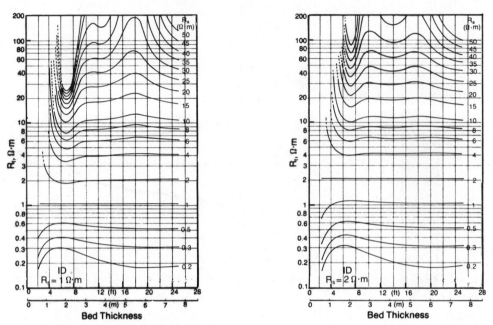

NOTE: These corrections are computed for a shoulder-bed resistivity (SBR) setting of 1 Ω·m. Refer to log heading.

a

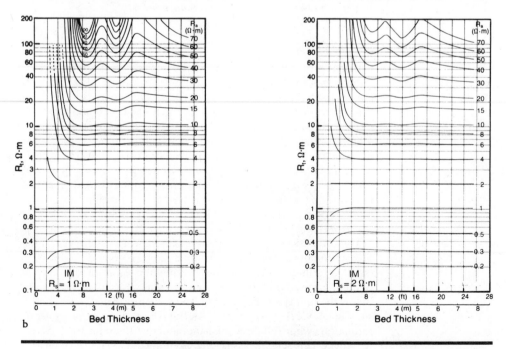

b

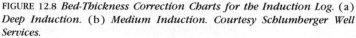

FIGURE 12.8 *Bed-Thickness Correction Charts for the Induction Log.* (a) *Deep Induction.* (b) *Medium Induction. Courtesy Schlumberger Well Services.*

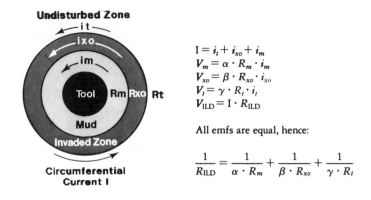

Undisturbed Zone

Tool Rm Rxo Rt

Mud

Invaded Zone

Circumferential Current I

$I = i_t + i_{xo} + i_m$

$V_m = \alpha \cdot R_m \cdot i_m$

$V_{xo} = \beta \cdot R_{xo} \cdot i_{xo}$

$V_t = \gamma \cdot R_t \cdot i_t$

$V_{ILD} = I \cdot R_{ILD}$

All emfs are equal, hence:

$$\frac{1}{R_{ILD}} = \frac{1}{\alpha \cdot R_m} + \frac{1}{\beta \cdot R_{xo}} + \frac{1}{\gamma \cdot R_t}$$

FIGURE 12.9 *Induction Current Paths.*

to finish the job. The symbol R_s refers to the resistivity of an adjacent bed. The bed thickness is plotted against R_t, and the lines on the charts are constant values of the apparent reading R_a, from the log. For example, a bed 18 ft thick with an R_a (log reading) of 30 Ω and adjacent beds of 1 Ω would really represent an R_t of 180 Ω.

Induction-Current Paths

Current loops flow around the borehole in a horizontal plane. The measured signal includes signals from the mud, the filtrate-invaded zone, and the undisturbed zone. The tool "sees" three resistances in parallel (see fig. 12.9). The voltage drop around the loop can be considered constant at V, but in each zone from which signals are being recorded, the current flow will be different. The equations are:

$$V_{mud} = R_m \times i_m \times \alpha$$

$$V_{invaded} = R_{xo} \times i_{xo} \times \beta$$

$$V_{undisturbed} = R_t \times i_t \times \gamma$$

where α, β, and γ are geometric constants dependent on the hole size and diameter of invasion. The total current flowing is

$$I = i_t + i_{xo} + i_m \quad \text{and} \quad V = IR_{ID}.$$

Since all values of V are equal, the equation for conductivity can be written as

$$\frac{1}{R_{ID}} = \frac{1}{\alpha R_m} + \frac{1}{\beta R_{xo}} + \frac{1}{\gamma R_t}.$$

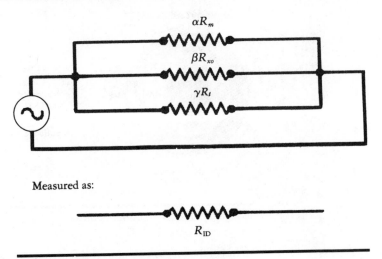

Measured as:

$$R_{\text{ID}}$$

FIGURE 12.10 *An Equivalent Induction Circuit.*

where R_{ID} is the resistivity measured by the deep induction log.

An equivalent circuit for this simplified explanation is shown in figure 12.10. The value of α will depend on hole size, and the values of β and γ will depend on the diameter of the filtrate-invaded zone (d_i) and the contrast between the values of R_{xo} and R_t.

HOLE-SIZE CORRECTIONS

The borehole effects that result from the current loop in the mud can be corrected by using the chart in figure 12.11. The size of the correction is insignificant in fresh (resistive) muds but quite significant in salty (conductive) muds. Instructions for using figure 12.11 are printed beneath the chart. Read them and solve the following example.

QUESTION #12.1

R_a (log reading)	= 20 $\Omega\cdot$m
R_m	= 0.1 $\Omega\cdot$m
SO (standoff)	= 1.5 in.
Hole diameter (caliper)	= 12¼ in.

Find $(R_{\text{ID}})_{\text{cor}}$.

If an SFL is run in conjuction with an induction log, it will also be necessary to make a hole-size correction. Figure 12.12 is provided for this purpose. The R_{SFL}/R_m ratio is plotted against the ratio of $(R_{\text{SFL}})_{\text{cor}}$ to R_{SFL}. The lines on the chart are for constant values of hole size.

Hole Diameter (mm)

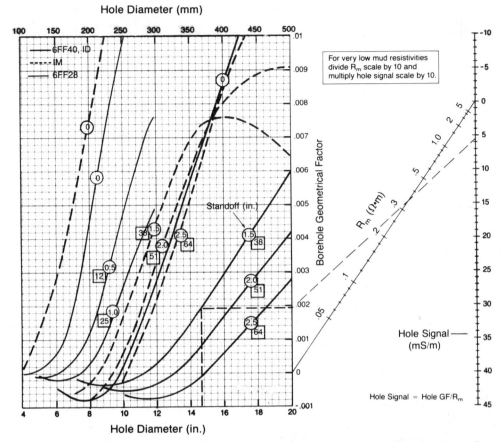

Hole Diameter (in.)

The hole-conductivity signal is to be subtracted, where necessary, from the induction log conductivity reading before other corrections are made.* This correction applies to all zones (including shoulder beds) having the same hole size and mud resistivity.

Rcor-4 gives corrections for 6FF40 or ID, IM, and 6FF28 for various wall standoffs. Dashed lines illustrate use of the chart for a 6FF40 sonde with a 1.5-in. standoff in a 14.6-in. borehole, and $R_m = 0.35$ $\Omega \cdot$m. The hole signal is found to be 5.5 ms/m. If the log reads $R_1 = 20$ $\Omega \cdot$m, C_1 (conductivity) = 50 ms/m. The corrected C_1 is then $(50 - 5.5) = 44.5$ ms/m. $R_1 = 1000/44.5 = 22.4$ $\Omega \cdot$m.

*CAUTION: Some induction logs, especially in salty muds, are adjusted so that the hole signal for the nominal hole size is already subtracted out of the recorded curve. Refer to log heading.

FIGURE 12.11 *Hole Size and Mud Resistivity Correction Chart for the Induction Log. Courtesy Schlumberger Well Services.*

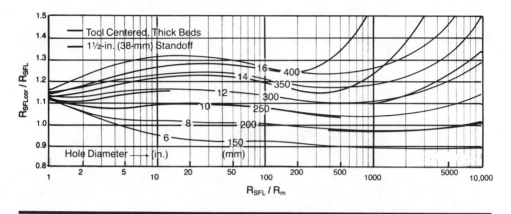

FIGURE 12.12 *Borehole Correction Chart for the Spherically Focused Log (SFL). Courtesy Schlumberger Well Services.*

QUESTION #12.2

If R_{SFL} = 20 $\Omega \cdot$m
   R_m = 0.2 $\Omega \cdot$m
   Hole size   = 14 in.

What is $(R_{SFL})_{cor}$?

Invasion Effects

The radial response of the induction tool is described by the *integrated radial geometric factor*, which is referred to as G. This G factor reveals what fraction of the measured signal comes from what radial distance from the tool. Mathematically, it can be described by the equation

$$\frac{1}{R_{ID}} = \frac{1}{R_{xo}} + \frac{1 - G}{R_t}.$$

If d_i (the diameter of invasion) is small, then G is small and all the signal comes from the undisturbed zone; in this case, R_{ID} will be equal to R_t. If d_i is large, then G also is large, and a large part of the total signal comes from the filtrate-invaded zone. In this case, R_{ID} will read somewhere between R_t and R_{xo}.

In figure 12.13, G is shown as a function of d_i for the deep induction resistivity. This plot can be used to solve the following case. Suppose d_i is 80 in., $R_{xo} = 20$, and $R_t = 10$. What will the induction tool read? From figure 12.13, G for a d_i of 80 in. is found to be 0.4, using the "skin effect included" line. Therefore, the equation given above can be written as

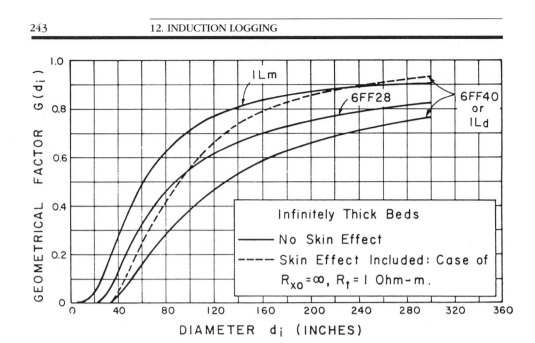

FIGURE 12.13 *Induction Geometric Factor (G). Courtesy Schlumberger Well Services.*

$$\frac{1}{R_{\text{ID}}} = \frac{0.4}{20} + \frac{(1 - 0.4)}{10} = 0.02 + 0.06 = 0.08$$

$$R_{\text{ID}} = 1/0.08 = 12.5.$$

Thus, R_{ID} reads greater than R_t.

The values of R_{xo} and R_t used in this example could easily arise in practice because of mud filtrate invasion of an oil-bearing formation as illustrated in figure 12.14, which shows high filtrate saturation in the invaded zone and low connate-water saturation in the undisturbed zone.

It should be realized that this treatment of the problem is backward—that is, in the field, R_{xo}, R_t, and d_i are not known in advance. The objective is to *find* R_{xo}, R_t, and d_i from the measured R_{ID} value. If only the value of R_{ID} is known, however, there is no solution to the problem. If three unknowns exist, then three known quantities are needed to solve the problem. The solution is the use of the dual induction–SFL combination logging tool. Since the geometric factor (G') for the medium induction log (R_{IM}) is different from the geometric factor (G) for the deep induction log (R_{ID}) at the same d_i, the following three equations can be solved simultaneously:

$$\frac{1}{R_{\text{ID}}} = \frac{G}{R_{xo}} + \frac{1 - G}{R_t},$$

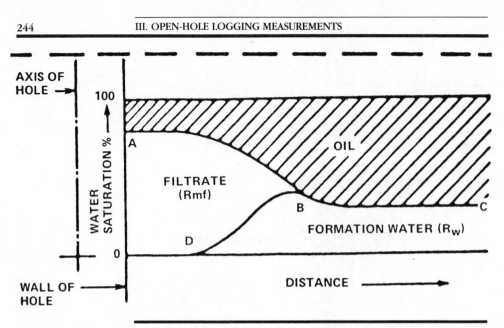

FIGURE 12.14 *Radial Distribution of Fluids in the Vicinity of the Bore-hole, Oil-Bearing Formation (Qualitative). Courtesy Schlumberger Well Services.*

$$\frac{1}{R_{IM}} = \frac{G'}{R_{xo}} + \frac{1 - G'}{R_t},$$

$$R_{SFL} = f(R_{xo}).$$

The solution is provided graphically on figure 12.15.

QUESTION #12.3
Use figure 12.15 to solve the following case:

$$R_{ID} = 10.0 \ \Omega\cdot m$$
$$R_{IM} = 13.8 \ \Omega\cdot m$$
$$R_{SFL} = 65.0 \ \Omega\cdot m$$

Find the values for R_t, R_{xo}, and d_i

CALIBRATION

Induction tools can be calibrated on land at any time. The sonde is placed in a zero-conductivity environment. This is normally done by raising the sonde up in the air well away from metallic objects. This defines a zero point. A calibration loop is then placed around the sonde to give a known conductivity signal, usually 500 mmhos. Calibration is performed monthly. It is almost impossible to perform on an offshore rig owing to the surrounding metal structure. In cases

Thick Beds, 8-in. (203-mm) Hole, Skin-Effect Corrected,
$R_{xo}/R_m \approx 100$, DIS-EA or Equivalent

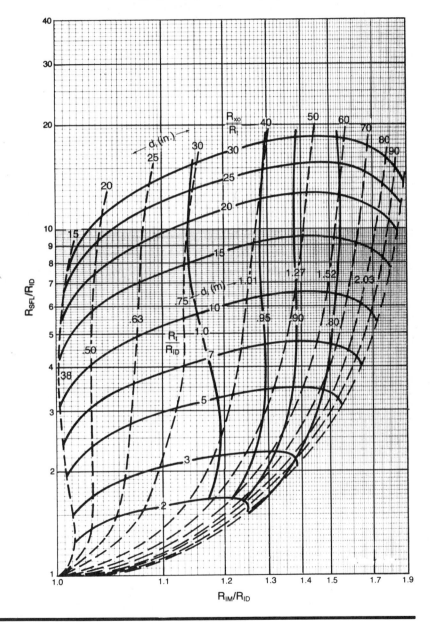

FIGURE 12.15 *Dual Induction "Tornado" Chart. Courtesy Dresser Atlas, Dresser Industries, Inc.*

where it is not possible to set the zero point under controlled conditions at the surface, it is permissible to set it with the tool in the hole opposite a thick very-high resistivity zone (salt, anhydrite, dense low-porosity carbonate, etc.) if such exists. The sonde and its associated electronic cartridge form a matched set and should always be used together.

SUMMARY

In 90% of the cases, it is permissible to assume that the deep induction reading is equal to the true formation resistivity, R_t. Conditions where this assumption is not valid include:

Induction logs run in very large holes
Induction logs run in salt muds
Places where the bed of interest is thin
Where the shoulder-bed resistivity is markedly different from the resistivity of the bed under consideration
Where invasion is abnormally deep

Since all these perturbing effects are a matter of degree, it is suggested that for each log examined, a few quick calculations will show whether any of the sundry correction charts available will introduce any substantial changes in the apparent resistivity. If they do not, it is safe to use R_{ID} as R_t in making an evaluation. If they do, the required corrections should be performed in the following order (see fig. 12.16).

1. Make the hole-size and mud-resistivity corrections.
2. Make bed-thickness corrections.
3. Make invasion corrections.

BIBLIOGRAPHY

Blakeman, E. R.: "A Method of Analyzing Electrical Logs Recorded on a Logarithmic Scale," *J. Pet. Tech.* (August 1962).

Doll, H. G.: "Introduction to Induction Logging," *J. Pet. Tech.* (June 1949).

Dresser Atlas: "Log Review 1" (1974).

Dumanoir, J. L., Tixier, H. P., and Martin, M.: "Interpretation of the Induction-Electrical Log in Fresh Mud," *J. Pet. Tech.* (July 1957).

Moran, J. H., and Kunz, K. S.: "Basic Theory of Induction Logging," *Geophysics* (December 1962).

Schlumberger: "Induction Log Correction Charts" (1962).

Schlumberger: "The Essentials of Log Interpretation Practice" (1972).

Schlumberger: "Log Interpretation Principles" (1972).

Schlumberger: "Log Interpretation Charts" (1984).

Schuster, N. A., Badon, J. D., and Robbins, E. R.: "Application of the ISF/Sonic Combination Tool to Gulf Coast Formations," *Trans.* Gulf Coast Assoc. Geol. Soc., 1971.

Tixier, H. P., Alger, R. P., Biggs, W. P., and Carpenter, B. N.: "Combined Logs Pinpoint Reservoir Resistivity," *Pet. Eng.* (February–March 1965).

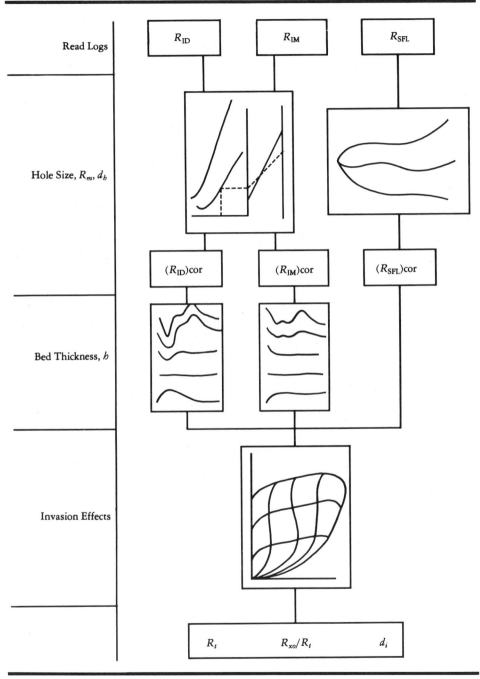

FIGURE 12.16 *Quick Guide to R_t from Dual Induction Logs.*

Answers to Text Questions

QUESTION #12.1
22

QUESTION #12.2
24

QUESTION #12.3

R_{IM}/R_{ID} = 1.38
R_{SFL}/R_{ID} = 6.5
R_t/R_{ID} = 0.9
R_{xo}/R_t = 10
R_t = 9 $\Omega \cdot$m
R_{xo} = 90 $\Omega \cdot$m
d_i = 50 in.

CHAPTER THIRTEEN

THE LATEROLOG

In the 1920s, Conrad Schlumberger put forward the idea of a *guarded electrode*, in an attempt to improve on existing electrical logs, which had undesirable borehole effects. The idea was not put into practice until H. G. Doll designed a working guard-electrode system. From this starting point, laterologs evolved in a number of ways. The laterolog-7, which used small guard electrodes, was later joined by the laterolog-3, which used long guard electrodes (see fig. 13.1). Both operated on the same principle: A constant survey current (i_0) was "forced" into the formation by bucking currents from the guard electrodes. By monitoring the voltage required to maintain the fixed current i_0, the formation resistivity was measured. The conductivity laterolog evolved from these tools. It maintained a constant voltage on the measure electrode while current variations monitored the formation conductivity.

The laterolog tool in most common use today is the 28-ft simultaneous dual laterolog, which is neither a conductivity nor a resistivity laterolog, but rather a hybrid using a constant product of current and voltage (perhaps it should be called a *joule laterolog*). This tool solves many of the problems associated with the earlier laterologs and is now the standard basic resistivity log for salt mud environments.

WHEN TO USE A LATEROLOG
Laterologs should be used when the following conditions exist:

Sea water or brine mud is in the hole.
The R_{mf}/R_w ratio is less than 3.
Hole size is less than 16 in.

The laterolog will produce better results than the induction log when R_t exceeds 150 Ω. It will also give a better estimate of R_t than the induction log when bed thickness is less than 10 ft.

Figure 13.2 should be referred to when there is doubt as to whether the laterolog should be run. This figure shows a plot of the R_{mf}/R_w ratio versus porosity. The laterolog is the tool of choice when the crossplot of R_{mf}/R_w versus ϕ falls in the left side of the chart.

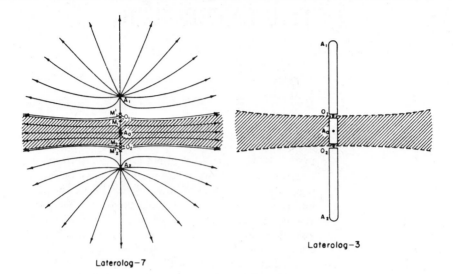

Laterolog-7

Laterolog-3

FIGURE 13.1 *Schematic Diagrams of Early Laterologs. Courtesy Schlumberger Well Services.*

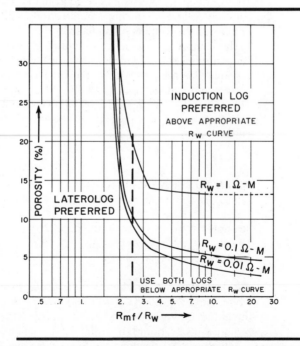

FIGURE 13.2 *Preferred Ranges for Application of Induction Logs and Laterologs. Courtesy Schlumberger Well Services.*

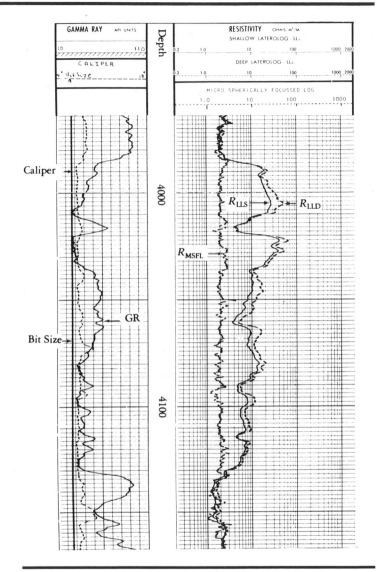

FIGURE 13.3 *Typical Dual Laterolog Presentation.*

THE DUAL LATEROLOG TOOL

The dual laterolog tool (DLL) gives three resistivity curves—the laterolog deep (LLD or LLd), the laterolog shallow (LLS or LLs), and a microspherically focused log (MSFL). Auxiliary curves such as a caliper, gamma ray, and spontaneous potential may also be recorded. The resistivity curves are presented on a standard four-decade logarithmic scale (see fig. 13.3).

Today's dual laterolog tool is the 28-ft simultaneous dual laterolog, which is abbreviated in a number of ways (e.g., DLT, DLL,

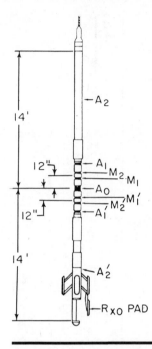

FIGURE 13.4 *Schematic Diagram of the Dual Laterolog-R_{xo} Tool. Reprinted by permission of the SPE-AIME from Suau et al. 1972, fig. 2, p. 3.* © *1972 SPE-AIME.*

DST). The sub carrying the caliper and MSFL devices is optional. When the sub is attached, the tool has a minimum OD of 5¼ in.; when it is not attached, the tool has an OD of 3⅝ in. Temperature and pressure ratings are a standard 350°F and 20,000 psi. Figure 13.4 is an illustration of the tool with its associated measure electrodes.

As this figure shows, the R_{xo} measuring portion of this tool is a pad-type device—once the tool is at the bottom of the well, arms with contact pads on them are extended to fit against the sides of the borehole wall. The mechanics of making both deep and shallow laterolog measurements with a single set of electrodes are handled by circuitry inside the tool. The respective current paths for the LLd and the LLs devices are shown in figure 13.5. The LLd uses long focusing electrodes and a distant return electrode while the shallow laterolog (LLs) uses short focusing electrodes and a near return electrode. The current paths for the MSFL, which has five rectangular electrodes mounted on a pad carried on one of the caliper arms, are shown in figure 13.6.

Under normal conditions found when using a Dual Laterolog, the radial profile of resistivities is as shown in figure 13.7 (i.e., $R_t > R_{xo} > R_m$). Between the invaded zone and the undisturbed formation, there is a transition zone that has a resistivity value between the values of R_t and R_{xo}. Figure 13.8 diagrams a plan view of

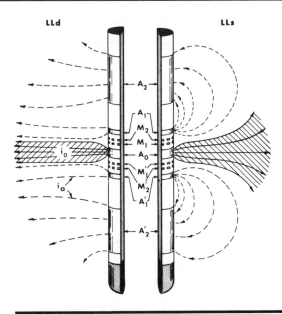

FIGURE 13.5 *Representative Current Patterns for the Deep and Shallow Laterologs. Reprinted by permission of the SPE-AIME from Suau et al. 1972, fig. 3, p. 4. © 1972 SPE-AIME.*

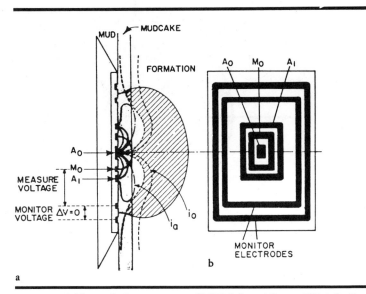

FIGURE 13.6 *Electrode Arrangement of MSFL (a) and Current Distribution (b). Reprinted by permission of the SPE-AIME from Suau et al. 1972, fig. 4, p. 4. © 1972 SPE-AIME.*

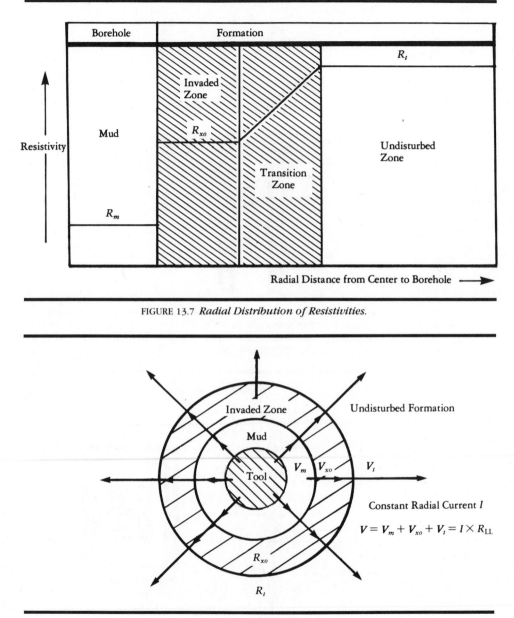

FIGURE 13.7 *Radial Distribution of Resistivities.*

FIGURE 13.8 *Laterolog Current Paths.*

a horizontal slice made through the tool and the formation that surrounds it. This figure shows current flowing radially outward from the tool and passing through the mud, the invaded zone, and the undisturbed formation before arriving at the return electrode. If held constant, the current will thus develop a series of voltage drops across each zone encountered. The relationship between these voltages can be simplistically written as

$$V_{total} \; = \; V_{mud} \; + \; V_{invaded} \; + \; V_{undisturbed}.$$

Each voltage drop is proportional to the product of the current, the resistivity of the zone, and some geometrical constant depending on the size of the zone—that is,

$$V_{mud} \qquad = \; I \times R_m \times \alpha,$$

$$V_{invaded} \qquad = \; I \times R_{xo} \times \beta,$$

$$V_{undisturbed} \; = \; I \times R_t \times \gamma.$$

V_{total} is related to the measured current I and the resistivity (R_{LL}) measured by the tool

$$V_{total} \; = \; IR_{LL}.$$

Thus, the equation for resistivity measured by the laterolog (R_{LL}) can be written as:

$$R_{LL} \; = \; \alpha R_m \; + \; \beta R_{xo} \; + \; \gamma R_t.$$

Logically, α will depend on the hole size, while β and γ will depend on the invasion diameter (d_i) and on the contrast between R_t and R_{xo}. An equivalent circuit for a laterolog measurement is shown in figure 13.9.

The laterolog response in an invaded formation is best described by a *pseudo radial geometric factor, J.* Mathematically, it is possible to partition the amount of total measured resistivity from the invaded and undisturbed zones. This can be done by use of the following equation:

$$R_{LL} \; = \; JR_{xo} \; + \; (1 - J)R_t. \tag{13.1}$$

J is a variable with a value between 0 and 1. At zero d_i, J will be equal to 0, and at infinity d_i, J will equal 1. Each of the two laterolog measurements (deep and shallow) has its own J-d_i relationship, as shown in figure 13.10.

The following is an example of how J factors are used: If the formation has an invaded zone with $R_{xo} = 5 \; \Omega$ and an undisturbed zone with $R_t = 10 \; \Omega$; and the invasion diameter (d_i) is 40 in., figure

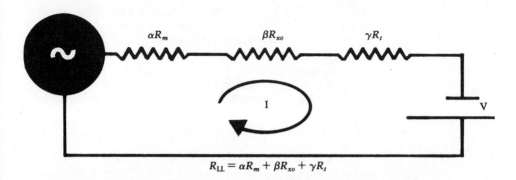

$$R_{LL} = \alpha R_m + \beta R_{xo} + \gamma R_t$$

FIGURE 13.9 *Equivalent Circuit for Laterolog Measurement.*

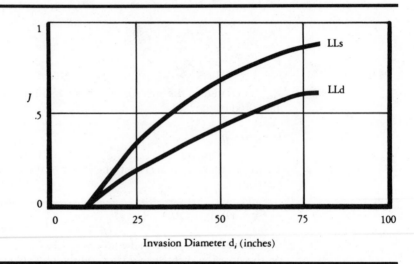

Invasion Diameter d_i (inches)

FIGURE 13.10 *Laterolog Pseudogeometric Factors.*

13.10 can be used to determine that J for the LLd is 0.3. Then, in accordance with equation (13.1), the value of R_{LLd} can be calculated:

$$R_{LLd} = 0.3 \times R_{xo} + (1 - 0.3) \times R_t$$

$$= 0.3 \times 5 + 0.7 \times 10$$

$$= 1.5 + 7$$

$$= 8.5.$$

Obviously, in a practical well logging situation where a resistivity value of 8.5 $\Omega\cdot$m is measured by the log and the value of R_t is desired, the values of R_{xo} and d_i must be known. Since there are three unknowns (R_t, R_{xo}, and d_i), three independent measurements

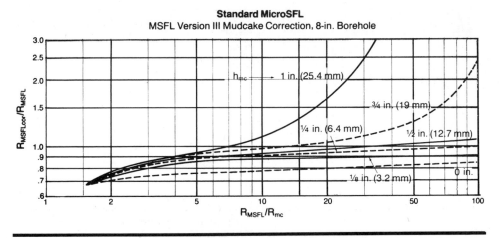

FIGURE 13.11 *Chart for Correcting the Measured Value of* R_{MSFL} *for Mudcake Thickness. Courtesy Schlumberger Well Services.*

are needed. This is accomplished by using the three measurements R_{LLd}, R_{LLs}, and R_{MSFL}. In most cases, R_{xo} can be equated with R_{MSFL}, thus eliminating one of the three unknowns. R_t and d_i are then left to the shallow and deep laterologs that have different values for J at the same d_i (fig. 13.10). For the same d_i, J for the deep measuring tool is less than J' for the shallow measuring tool. The deep measuring tool is more sensitive to R_t, and the shallow measuring tool is more sensitive to R_{xo}.

BOREHOLE AND INVASION CORRECTIONS
Before the three unknowns can be determined, the raw data must be corrected for borehole effects. Charts are available to achieve this.

The MSFL, being a pad-contact device, is sensitive to mudcake thickness (b_{mc}) and mudcake resistivity (R_{mc}). Figure 13.11 is a plot of R_{MSFL}/R_{mc} plotted against $(R_{MSFL})_{cor}/R_{MSFL}$. The lines on the chart represent various values of b_{mc}. The use of this chart corrects for the value of α discussed earlier. The value of b_{mc} is equal to half the difference between the borehole size and the caliper reading, and R_{mc} is the measured value of R_{mc} corrected back to formation temperature.

QUESTION #13.1
Using figure 13.11, solve the following exercise:

$$\begin{aligned}
R_{MSFL} &= 3.5 \ \Omega\text{·m} \\
R_{mc} &= 0.5 \\
b_{mc} &= 1 \text{ in.} \\
(R_{MSFL})_{cor} &= ?
\end{aligned}$$

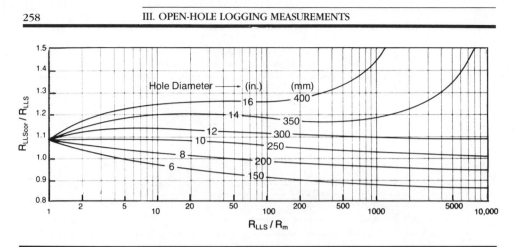

FIGURE 13.12 *Borehole Correction Chart for the Shallow Laterolog. Courtesy Schlumberger Well Services.*

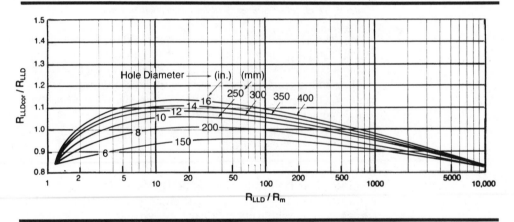

FIGURE 13.13 *Borehole Correction Chart for the Deep Laterolog. Courtesy Schlumberger Well Services.*

The shallow laterolog hole-size correction can be accomplished by using figure 13.12, where R_{LLs}/R_m is plotted versus $(R_{LLs})_{cor}/R_{LLs}$. The lines on the chart are for various values of hole diameter.

QUESTION #13.2

R_{LLs}	=	5.0 $\Omega \cdot m$
R_m	=	0.25 $\Omega \cdot m$
Caliper	=	14.0 in.
$(R_{LLs})_{cor}$	=	?

The deep laterolog hole-size correction is accomplished by using figure 13.13, where R_{LLd}/R_m is plotted against $(R_{LLd})_{cor}/R_{LLd}$. The lines on the chart are for various values of hole diameter.

QUESTION #13.3

R_{LLd} =   50.0 $\Omega \cdot$m
R_m =   0.5 $\Omega \cdot$m
Caliper   =   16.0 in.
$(R_{LLd})_{cor}$ =   ?

In the range of normal interests, where laterolog readings lie in the range of $10 < (R_{LL}/R_m) < 100$, all corrections are within ±10%; but where hole diameters are large, the LL's correction can become intolerably large.

THE BUTTERFLY CHART

After the raw log readings have been corrected for borehole effects, they must be corrected for invasion effects. Figure 13.14 will compensate for the values of β and γ discussed earlier. This correction chart plots the ratio of R_{LLd}/R_{LLs} against the ratio of R_{LLd}/R_{xo}. There are three families of lines on the chart: constant values of R_t/R_{LLd}, constant values of R_t/R_{xo}, and constant values of d_i. In order to use the chart, it is first assumed that $(R_{MSFL})_{cor}$ is equal to R_{xo}. A point is then located on the chart at the coordinates R_{LLd}/R_{LLs} and R_{LLd}/R_{xo}. This point uniquely defines the three unknowns: R_t, R_{xo}, and d_i.

QUESTION #13.4

R_{LLd} =   33 $\Omega \cdot$m
R_{LLs} =   11 $\Omega \cdot$m
R_{MSFL} =   3 $\Omega \cdot$m
R_t =   ?
d_i =   ?

The lower left portion of the chart corresponds to the invasion pattern $R_{MSFL} > R_{LLs} > R_{LLd}$, which usually occurs in water-saturated zones where $R_{mf} > R_w$. Since this area of the chart is rather ill defined on figure 13.14, a better solution can be found using figure 13.15.

QUESTION #13.5

R_{LLd} =   5.0 $\Omega \cdot$m
R_{LLs} =   6.6 $\Omega \cdot$m
R_{MSFL} =   9.0 $\Omega \cdot$m
R_t =   ?
d_i =   ?

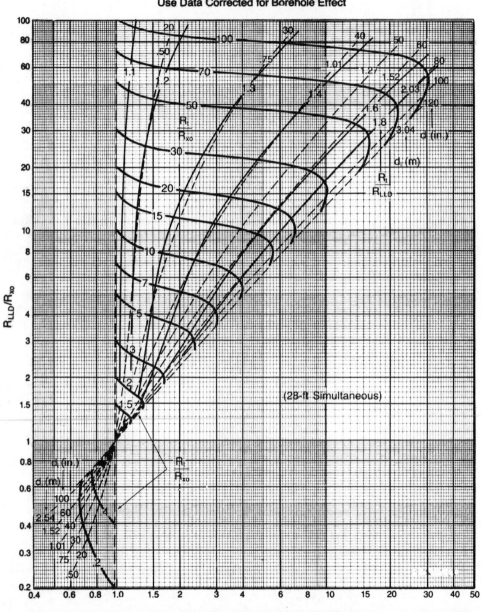

FIGURE 13.14 *Invasion Correction Chart for the Dual Laterolog. Courtesy Schlumberger Well Services.*

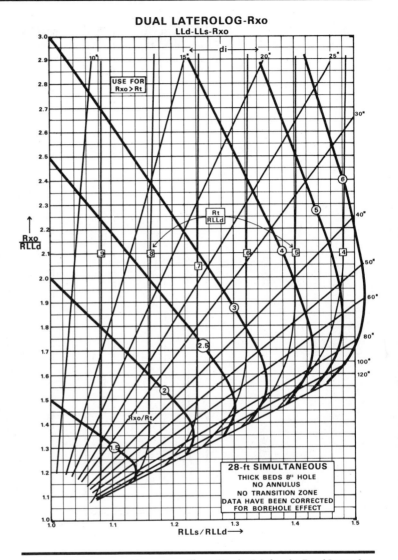

FIGURE 13.15 *Invasion Correction Chart ($R_{xo} > R_t$) for the Dual Laterolog.*

DUAL LATEROLOG "FINGERPRINTS"

The characteristic behavior of the DLL tool in zones with movable hydrocarbons makes quick-look interpretation very simple. The golden rule is that the pattern of $R_{LLd} > R_{LLs} > R_{MSFL}$ (fig. 13.16[a]) is a good indication that hydrocarbons are present. Conversely, the pattern of $R_{MSFL} > R_{LLs} > R_{LLd}$ (fig. 13.16[b]) is a good indication that the zone is wet, i.e., 100% water saturation. Any relative ordering of the curves other than these two cases suggests little or no invasion and indicates that the zone is impermeable (fig. 13.17).

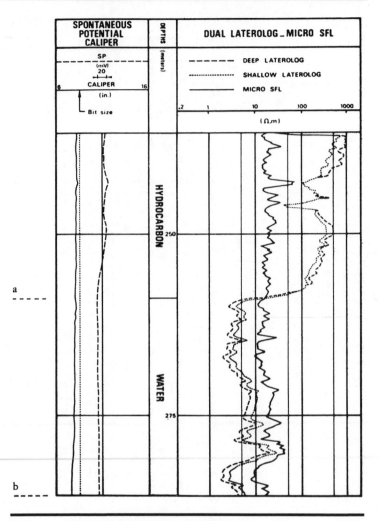

FIGURE 13.16 (a) *Dual Laterolog Oil Pattern*. (b) *Dual Laterolog Wet Pattern. Courtesy Schlumberger Well Services.*

ANOMALOUS BEHAVIOR

The early laterologs were prone to various types of anomalous behavior, which are chronicled here to give some insight into the few anomalies that can still occur even with the dual laterolog.

The Delaware Effect

In the early 1950s in the Permian Basin, logging engineers found that laterologs behaved anomalously when approaching a thick resistive bed, such as the massive anhydrite that overlies the Delaware sand. The effect manifested itself by a gradual increase of apparent resistivity starting when the bridle entered the highly resistive bed.

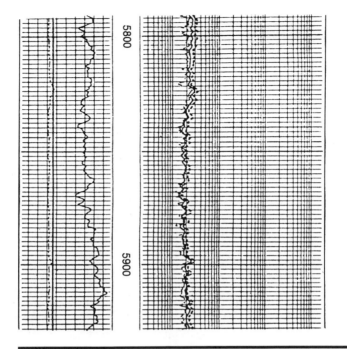

FIGURE 13.17 *Dual Laterolog Shale Pattern.*

Apparent resistivities would climb to as much as 10 times the value of R_t before the sonde itself entered the highly resistive bed. The solution for the laterolog-7 was to place the B return electrode at the surface. For the conductivity laterolog, the solution was not so elegant, since these devices were using a 280-Hz survey current generated in the cartridge. Having the return at the surface did not solve the problem, since skin effect restricted the return current to a sheath around the borehole, thus resulting in the lower part of the cable becoming the effective return electrode (fig. 13.18). Compensation for this effect with the laterolog-3 involved a messy setup with two sondes, one on each side of a cartridge, and a B return on the *bottom* for Delaware situations: Nevertheless, for all practical purposes, the laterolog-3 remained prone to the Delaware effect.

Anti-Delaware Effect

In an attempt to improve on the situation and provide a dual spacing laterolog, a tool was introduced with both deep and shallow devices. This device also exhibited anomalous behavior beneath highly resistive beds, however. The deep laterolog showed a gradient of *decreasing* resistivity, the exact opposite of the Delaware effect. With the B electrode being at surface (effectively, at zero potential), the N electrode acted as the takeoff point for a potential divider formed by the borehole below and above N. The approach-

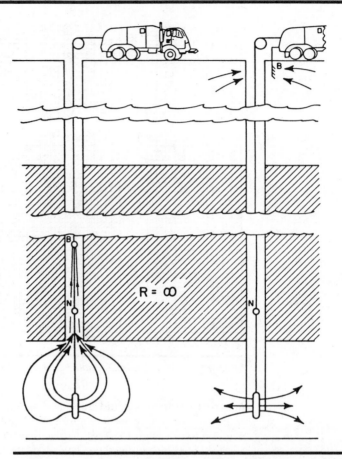

FIGURE 13.18 *Schematic of the Delaware Effect. Reprinted by permission of the SPE-AIME from Suau et al. 1972, fig. A-1, p. 12.* © *1972 SPE-AIME.*

ing sonde, at some positive potential, would then cause N to raise its potential. The anti-Delaware effect would at worst cause a 50% reduction in the deep-laterolog measurements and would only be noticeable within 35 ft of the resistive bed. In fact, the effect had been present on the earlier "B at surface" Delaware-free laterologs, but it had not been noticed because there was no shallow laterolog with which to compare the deep laterolog. The dual laterologs in use today have incorporated into them features that assure virtual freedom from Delaware and anti-Delaware effects.

The Groningen Effect

First noticed during gas-well logging in Holland, the Groningen effect is manifested as a too-high LLd reading when the N electrode enters a highly resistive bed. From a distance of A_0N below the bed boundary (102 ft for the simultaneous DLT), the LLd reading will rise

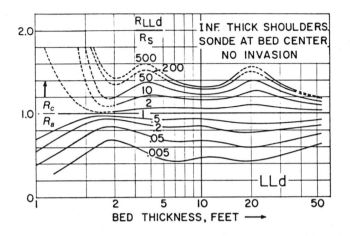

FIGURE 13.19 *Shoulder Bed Correction Chart for the Deep Laterolog. Reprinted by permission of the SPE-AIME from Suau et al. 1972, fig. 8, p. 8.* © *1972 SPE-AIME.*

over a short distance to an anomalously high value, which it will then maintain until the bed is entered. The precise cause of the effect is not yet known, but experiments have indicated that it is dependent on the operating frequency and is only troublesome in low-resistivity formations immediately below a massive salt or anhydrite bed. The Groningen effect will appear (if at all) within 102 ft (31 m) of a resistive bed and will be of interpretive importance only where R_t in the underlying bed is less than 10 Ω. It can still appear even if casing is set to the bottom of the resistive bed.

Even if resistive-bed effects do not distort the dual laterolog measurements, other environmental influences will, simply because a tool cannot be built that is entirely immune to the disturbing effects of the borehole and adjacent beds. For interpretative work, these distorting influences must be taken into account. Hole-size and invasion effects have been covered in the foregoing discussions; the need for shoulder-bed corrections is also worth noting.

Shoulder-Bed Corrections— Squeeze and Antisqueeze

When the sonde is in front of a bed, on either side of which there is a resistive shoulder, current tends to concentrate in the least-resistive path; in other words, the current is "squeezed" between the resistive shoulders into the formation of interest. In order to correct for this effect, the charts in figures 13.19 and 13.20 have been compiled. The correction factor (R_c/R_a) to be applied to the borehole corrected log reading is shown as a function of bed thickness and R_a/R_s, which is the contrast between the apparent reading and the shoulder resistivity. Where R_a/R_s is less than one, a squeeze situation exists and the apparent log reading is too high (lower half of chart). Where

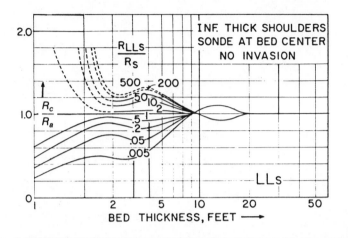

FIGURE 13.20 *Shoulder Bed Correction Chart for the Shallow Laterolog.*
Reprinted by permission of the SPE-AIME from Suau et al. 1972, fig. 8, p. 8.
© *1972 SPE-AIME.*

R_a/R_s is greater than one, the bed is surrounded by a *conductive*
shoulder and the current tends to fan out into the path of least
resistance—the conductive shoulders. (Since this is the reverse of
squeeze it is referred to as *antisqueeze.*) The apparent log readings
are too low in this situation.

Two things are of importance here. First, there is an inversion as
bed thickness increases through 9.5 ft for the LLs (this is not evident
from the chart), and, for formation thicknesses greater than 20 ft, the
LLs is free from shoulder effect. Second, the LLd is much more
affected by squeeze and antisqueeze than is the LLs, even in what
might be considered thick beds (50 ft or more).

When making detailed interpretations, use the shoulder-bed
correction charts for LLd after borehole correction and before any
other step. Invasion corrections may then be made using figure 13.14.
A word of caution is in order. In general, an *ideal* laterolog has a *J* that
behaves logarithmically with d_i, but it is also a function of the contrast
between R_{xo} and R_t. Additionally, the effect of a hole size larger than 8
in. is that part of the R_{xo} zone is replaced by mud; this changes the
effective position of the origin on the *J* versus d_i plot. Use of figure
13.14 thus gives a good answer in 8-in. holes with thick beds. In larger
holes or thin beds, it will be in error. If the effect of antisqueeze is
taken into account (resistive beds, conductive shoulders—e.g., pay
sand with adjacent shaly shoulders), the result is that the effective
depth of investigation of the LLd is reduced. The contrast between LLs
and LLd measurements is therefore reduced. A recent study has
shown that even at a bed thickness of 8 ft, the effective depth of
investigation of the LLd is close to that of the LLs. Under these
conditions, there can be no solution for R_t unless d_i is known. Thus,
separation between LLd and LLs values will not be observed in beds
thinner than 8 ft, even when R_t is greater than R_{xo}.

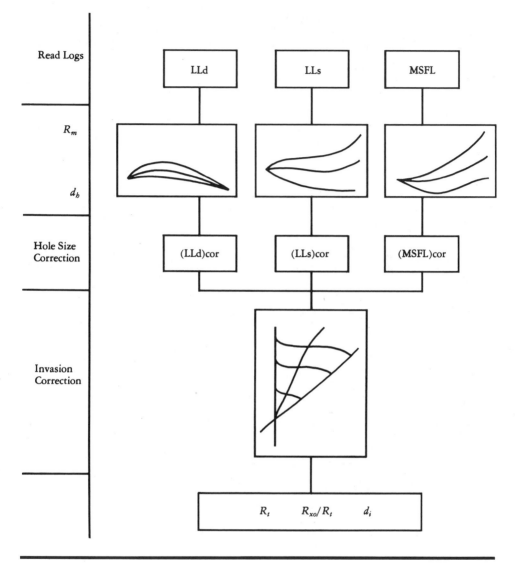

FIGURE 13.21 *Quick Guide to R_t from the Dual Laterolog.*

SUMMARY

In low-resistivity (<10 $\Omega \cdot$m) formations below a massive resistive bed, watch for the Groningen effect (R_{LLd} reads too high within a distance of A_0N of the bed boundary).

When interpreting laterologs, make borehole corrections first and shoulder-bed corrections second, if required, and invasion corrections last (see fig. 13.21).

Use figure 13.14 with care in large holes and when there is a large R_{xo}/R_t contrast in thin formations.

BIBLIOGRAPHY

Doll, H. G.: "The Laterolog," *J. Pet. Tech.* (November 1951).

Horst, G., and Creager, L.: "Progress Report on the Interpretation of the Dual Laterolog-R_{xo} Tool in the Permian Basin," SPWLA, 15th Annual Logging Symposium Lafayette, LA, June 1974.

Schlumberger: "Log Interpretation Principles," (1972).

Schlumberger: "Well Evaluation Conference," (1979).

Schlumberger: "Log Interpretation Charts," (1984).

Suau, J., Grimaldi, P., Poupon, A., Souhaite, P.: "The Dual Laterolog-R_{xo} Tool," SPE paper 4018, presented at the 47th Annual Meeting, San Antonio, Oct. 8-11, 1972.

Answers to Text Questions

QUESTION #13.1
$(R_{MSFL})_{cor} = 3.5 \ \Omega \cdot m$

QUESTION #13.2
$(R_{LLs})_{cor} = 6.0 \ \Omega \cdot m$

QUESTION #13.3
$(R_{LLd})_{cor} = 55.0 \ \Omega \cdot m$

QUESTION #13.4
$R_t = 46.2 \ \Omega \cdot m$
$d_i = 40 \ in.$

QUESTION #13.5
$R_t = 3.0 \ \Omega \cdot m$
$d_i = 40 \ in.$

CHAPTER FOURTEEN

MICRORESISTIVITY MEASUREMENTS

The measurements made by microresistivity tools have a variety of uses and applications such as:

Well-to-well correlation
Flushed-zone saturation S_{xo}
Residual-oil saturation (ROS)
Hydrocarbon movability
Hydrocarbon density (ρ_{hy})
Invasion diameter (d_i)
Invasion corrections to deep resistivity devices

A variety of tools, old and new, are available; each has its own special characteristics. This section will present the names, uses, and idiosyncracies of microresistivity tools.

MICRORESISTIVITY TOOLS

The following list covers the majority of the present-day microresistivity tools:

16-in. SN	short normal log
LL8	laterolog-8
SFL	spherically focused log
MLL	microlaterolog
PL	proximity log
MSFL	microspherically focused log
ML	microlog

These tools can be divided into two main groups: the mandrel tools and the pad-contact tools. The electrodes of the mandrel tools are placed on a cylindrical mandrel that is run in the hole. They do not require physical contact with the formation. The electrodes of the pad-contact tools are embedded in an insulating pad that is carried on a caliper arm, which is forced against the borehole wall. The sizes and ratings of the microresistivity tools are summarized in table 14.1. These devices are usually run in combination with some specific deep-resistivity device as shown in table 14.2.

The microlog is worthy of special mention since it is an underrated device that should be run more frequently than it is. It was one of the first microresistivity devices on the market and has had a spectacular career. Originally, it was used as a pseudoporosity

TABLE 14.1 *Microresistivity Tools: Type and Rating*

Name	OD (inches)		
16-in. SN	2¾ to 3⅞		
LL8	3⅜ & 3⅞	Mandrels	All devices are rated
SFL	3½		for 20,000 psi
MLL	$5^1/_{16}$		and 350°F.
PL	5½	Pads	
MSFL	5¼		
ML	$5^1/_{16}$		

TABLE 14.2 *Microresistivity Tools: Applications*

Micro Tool	Usually Run with	Main Tool
MSFL		dual laterolog
LL8 or SFL		dual induction log
SFL or SN		induction log
MLL or PL		microlog
ML		microlaterolog or proximity log

device. When that function was improved upon with modern porosity devices, the microlog was relegated to the pile of has-beens by many people in the logging industry. Nevertheless, it is still a valuable tool because it offers a superb visual identification of porous and permeable zones.

Figure 14.1 shows a microlog–proximity log presentation. The presence of permeability is indicated wherever the micronormal curve reads higher than the microinverse curve, and the micro-inverse curve reads close to R_{mc}. That is, the microlog, recording two resistivity curves at shallow depths of investigation, detects any contrast in resistivity between the mudcake and the flushed zone. If no porosity or permeability is present in the formation, there is no filtrate invasion and, therefore, no mudcake buildup; hence there is no separation between the two resistivity curves. The micronormal curve is referred to as $R_{2''}$ and the microinverse curve as the $R_{1'' \times 1''}$. Figure 14.2 crossplots $R_{2''}/R_{mc}$ against $R_{1'' \times 1''}/R_{mc}$. The plotted point defines values of R_{xo}/R_{mc} and mudcake thickness (h_{mc}).

QUESTION #14.1
Use figure 14.2 to estimate R_{xo} and h_{mc} if it is known that

$$R_{mc} = 1 \ \Omega\cdot m$$
$$R_{2''} = 3 \ \Omega\cdot m$$
$$R_{1'' \times 1''} = 2.3 \ \Omega\cdot m$$

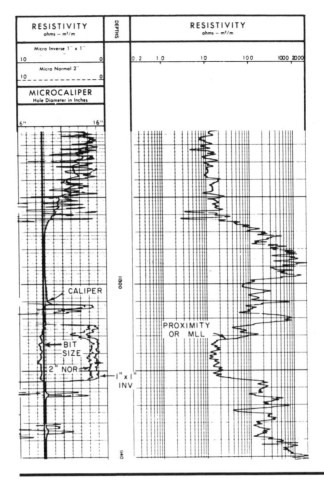

FIGURE 14.1 *Presentation of Proximity Log–Microlog. Courtesy Schlumberger Well Services.*

DEPTH OF INVESTIGATION

Each microresistivity tool has its characteristic depth of investigation. It is important to know what this is for each tool in order to select the one with the right characteristics for the job. When invasion is shallow, a tool with a shallow depth of investigation is needed if the tool is to read R_{xo} without undue influence from R_t. Conversely, in situations where deep invasion exists, a deep investigation tool will assure a reading of R_{xo} free from any effects of R_{mc}.

As with other tools, no single value for the depth of investigation can be used. Rather, a pseudogeometric factor must be used. This factor indicates how much of the total tool signal is received from an annular formation volume represented by distances (expressed in

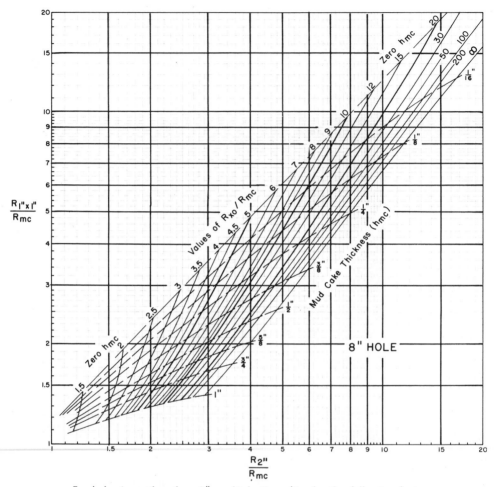

For hole sizes other than 8″, multiply $R_{1''x1''}/R_{mc}$ by the following factors before entering the chart: 1.15 for 4¾-inch hole, 1.05 for 6-inch hole, 0.93 for 10-inch hole.

NOTE: An incorrect R_{mc} will displace the points in the chart along a 45° line. In certain cases this can be recognized when the mud-cake thickness is different from direct measurement by the microcaliper. To correct, move the plotted point at 45° to intersect the known h_{mc}. For this new point read R_{xo}/R_{mc} from the chart and $R_{2''}/R_{mc}$ from the bottom scale of the chart. $R_{xo} = R_{xo}/R_{mc} \times R_{mc}$.

FIGURE 14.2 *Microlog Interpretation Chart. Courtesy Schlumberger Well Services.*

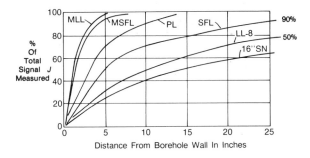

FIGURE 14.3 *Depth of Investigation for Microresistivity Tools.*

TABLE 14.3 *Microresistivity Tools: Depth of Investigation*

Rank	Tool Name	Depth for 90% Response (inches)
1. ML	Microlog	1
2. MLL	Microlaterolog	4
3. MSFL	Microspherically focused log	4½
4. PL	Proximity log	10
5. SFL	Spherically focused log	24
6. LL8	Laterolog-8	49
7. SN	16-in. short normal log	70+

TABLE 14.4 *Microresistivity Tools: Bed Resolution*

Rank	Tool Name	Depth for 90% Response (inches)
1. ML	Microlog	3
2. MLL	Microlaterolog	6
3. MSFL	Microspherically focused log	12
4. PL	Proximity log	12
5. SFL	Spherically focused log	12
6. LL8	Laterolog-8	12
7. SN	16-in. short normal log	24

inches) from the borehole wall (see fig. 14.3). Table 14.3 lists the depth of investigation for microresistivity tools.

BED RESOLUTION

Just as each of the microresistivity tools has its characteristic depth of investigation, so too does each tool have its own characteristic bed resolution; that is, some tools are better than others at distinguishing thin beds. Tools with large-bed resolution values are "blind" to thin shale and/or sandstone layers. For example, 3-in. shale streaks will not be "seen" by a short normal log. Table 14.4 summarizes the bed resolution of each microresistivity device.

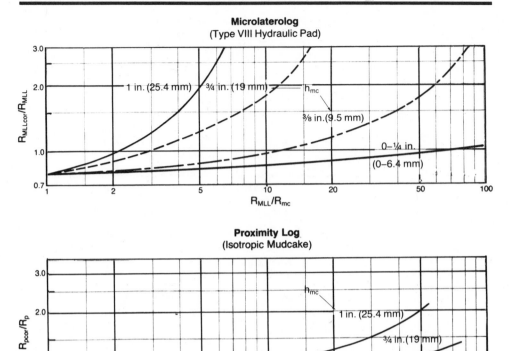

FIGURE 14.4 *Mudcake Correction Chart for Pad-Contact Microresistivity Tools. Courtesy Schlumberger Well Services.*

ENVIRONMENTAL CORRECTIONS

Microresistivity devices of the mandrel type are subject to abber-rations resulting from the size of the wellbore. These effects can be quite severe. The pad-contact tools, however, are only affected by excessive mudcake thickness. If pad contact with the formation is maintained, the pad-contact tools are unaffected by the size of the wellbore.

Mudcake corrections can be made by using figure 14.4 for the microlaterolog and the proximity log, and figure 13.11 for the micro SFL. For each chart, plot the ratio of the log reading to R_{mc} (remember to correct R_{mc} to formation temperature) against mud-cake thickness (h_{mc}) to determine the required correction factor. The lines on each chart are lines of constant mudcake thickness (h_{mc}). Mudcake thickness is calculated as half the difference be-tween the bit size and the measured caliper reading when the caliper reads less than bit size.

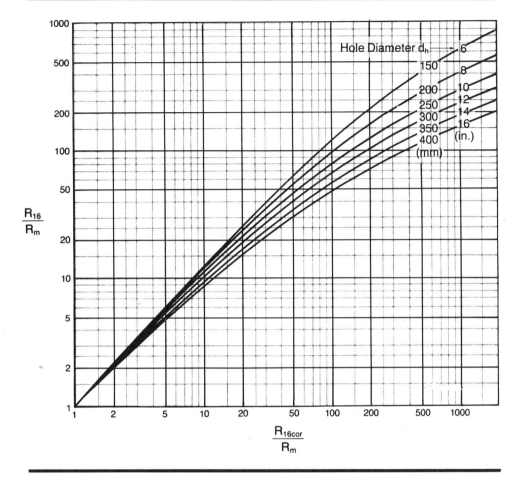

FIGURE 14.5 *Borehole Correction Chart for the 16-in. Short Normal Tool. Courtesy Schlumberger Well Services.*

QUESTION #14.2
Use figure 14.4 to solve the following case:

b_{mc} = ¾ in.
R_{mc} = 1 Ω·m
R_{MLL} = R_{PL} = R_{MSFL} = 10 Ω·m

What is the correct value for the resistivity reading of each tool?

Corrections for mandrel-type tools can be made by using figures 14.5, 14.6, and 12.12 for the 16-in. SN, the LL8, and the SFL tools, respectively. The ratio of the log reading is plotted against the mud

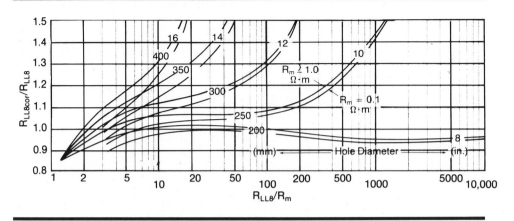

FIGURE 14.6 *Borehole Correction Chart for the Laterolog-8. Courtesy Schlumberger Well Services.*

resistivity (R_M) to determine the correction factor. The lines on the charts are lines of constant hole size in inches.

QUESTION #14.3

R_m	=	0.1 Ω·m
$R_{16''}$	=	7.0 Ω·m
d_b	=	12 in.
$(R_{16''})_{cor}$	=	?

QUESTION #14.4

R_m	=	1 Ω·m
R_{LL8}	=	50 Ω·m
d_b	=	12 in.
$(R_{LL8})_{cor}$	=	?

QUESTION #14.5

R_m	=	0.5 Ω·m
R_{SFL}	=	10 Ω·m
d_b	=	8 in.
$(R_{SFL})_{cor}$	=	?

S_{xo} AND HYDROCARBON MOVABILITY

The water (filtrate) saturation in the flushed zone (S_{xo}) may be estimated by using Archie's equation:

$$(S_{xo})^n = FR_{mf}/R_{xo},$$

where $F = a/\phi^m$. To solve this equation, the values of a, m, n, ϕ, R_{mf}, and R_{xo} must be known. R_{mf} should be converted to formation temperature.

QUESTION #14.6
Apply Archie's equation to the following case:

$$
\begin{aligned}
R_{mf} &= 0.5 \ \Omega\cdot m \ @ \ 75°F \\
T_{form} &= 175°F \\
R_{xo} &= 10 \ \Omega\cdot m \\
F &= 25
\end{aligned}
$$

Find S_{xo} in %, assuming that $n = 2$.

The value of S_{xo} may not reveal much about the amount of oil-in-place, but it will reveal a great deal about whether the oil that is in place is likely to flow or not. The invasion process acts like a miniature waterflood. Invading filtrate will displace not only connate water, but also any movable hydrocarbons. In the undisturbed state at initial reservoir conditions, the fraction of pore volume that is occupied by oil is $\phi(1 - S_w)$. After filtrate invasion has taken place, the fraction of pore volume that is occupied by oil is $\phi(1 - S_{xo})$. The difference between these two values is the fraction of pore volume that contains movable oil (see fig. 14.7). The pore-volume fraction of movable oil is determined by the relationship $\phi(S_{xo} - S_w)$. The percentage of the original oil-in-place that has moved is determined by

$$\% \text{ OOIP moved} = \frac{S_{xo} - S_w}{1 - S_w}$$

QUESTION #14.7
If $S_w = 30\%$ and $S_{xo} = 65\%$, what percentage of the original oil-in-place has been moved?

This index can be used as a measure of the quality of the pay. In formations where the relative permeability to oil is low, S_{xo} is likely to be close to S_w and the index will be low. Such a formation will not be as productive as a formation with the same value of S_w but better relative permeability to oil and, therefore, a higher value of S_{xo}.

HYDROCARBON DENSITY
Computation of the value for hydrocarbon density in a pay zone can be critical when there is a question as to whether the formation

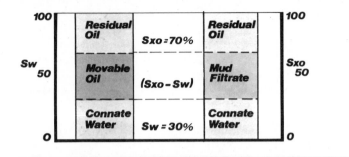

FIGURE 14.7 *Hydrocarbon Movability.*

contains oil, light oil, condensate, or gas. Since the porosity tools make their measurements in the flushed zone, they "see" a bulk volume of hydrocarbon equal to $\phi(1 - S_{xo})$. This leads to the interesting paradox that where hydrocarbons are movable they will have been flushed away from the zone where logging tools can react to them. For this reason, large hydrocarbon effects on porosity tools may be misleading and may really only indicate large volumes of *residual* hydrocarbons. Lack of significant hydrocarbon effects could mean that either movable hydrocarbons are present or the formation is wet. Either way, a good value for S_{xo} is essential for correct evaluation of hydrocarbon density and, therefore, of the type of hydrocarbon present in the formation. A further discussion of this topic will be found in chapter 22.

CALIBRATION AND QUALITY CONTROL
Quality control for microresistivity devices can be summarized by the following maxims:

Beware of washed-out holes because (a) pad-contact tools lose contact with the formation and float in the mud column and (b) mandrel tools give horridly inaccurate readings.

Beware of thick mudcakes because pad-contact tools require large corrections. If hole conditions are bad, forget about trying to measure R_{xo} because either the tool will stick or the pad will tear up. Either way, no usable log reading will be obtained.

BIBLIOGRAPHY
Doll, H. G.: "The Microlog," *Trans.*, AIME (1950) **189**.
Doll, H. G.: "Filtrate Invasion in Highly Permeable Sands," *Pet. Eng. J.* (January 1955).
Doll, H. G.: "The Microlaterolog," *J. Pet. Tech.* (January 1953).
Doll, H. G., and Martin, M.: "Suggestions for Better Electric Log Combinations and Improved Interpretations," *Geophysics* (August 1960).
Schlumberger: "Log Interpretation, Volume 1—Principles" (1972).
Schlumberger: "Log Interpretation Charts" (1984).

Answers to Text Questions

QUESTION #14.1
$$h_{mc} = \text{⅜ in.}$$
$$R_{xo} = 7 \ \Omega\cdot m$$

QUESTION #14.2
$$(R_{MLL})_{cor} = 18 \ \Omega\cdot m$$
$$(R_{PL})_{cor} = 10.5 \ \Omega\cdot m$$
$$(R_{MSFL})_{cor} = 9 \ \Omega\cdot m$$

QUESTION #14.3
$$(R_{16''})_{cor} = 10 \ \Omega\cdot m$$

QUESTION #14.4
$$(R_{LL8})_{cor} = 60 \ \Omega\cdot m$$

QUESTION #14.5
$$(R_{SFL})_{cor} = 10 \ \Omega\cdot m$$

QUESTION #14.6
$$S_{xo} = 75\%$$

QUESTION #14.7
% OOIP moved $= 50\%$

CHAPTER FIFTEEN

DIELECTRIC MEASUREMENTS

The dielectric constant of a material affects the way in which an electromagnetic wave passes through it. Since the dielectric constants of oil and water are different, the behavior of electromagnetic waves in reservoir rocks is of interest. Currently, there are three logging tools for measuring dielectric constants.

EPT   Electromagnetic Propagation Tool (Schlumberger)
DCL   Dielectric Constant Log (Gearhart)
DPT   Deep Propagation Tool (Schlumberger)

The EPT is an ultrahigh-frequency (uhf) tool. The DCL and DPT are high-frequency (hf) tools.

Traditionally, the measurement of the electrical conductivity, or resistivity, of a formation has been one of the main surveys performed in a borehole, primarily to determine water saturation. Development of the first dielectric device, the electromagnetic propagation tool (EPT), made it possible to use another electrical characteristic of the formation—the dielectric permittivity (or dielectric constant)—to estimate formation water saturation. This dielectric permittivity (constant) is not read directly. The basic measurements made by the dielectric tools are of the propagation time and attenuation of an electromagnetic wave as it passes through a specific interval of formation. Since this propagation time is substantially higher in water than in the hydrocarbons or minerals, the dielectric measurement is affected primarily by the water-filled porosity. (The nuclear porosity tools, on the other hand, are influenced by the total porosity.) For a wide range of salinities, the propagation time in water is practically constant, so saturation estimations can be made without water resistivity data. When other open-hole log data are available, it is possible to distinguish between oil, gas, and water in reservoirs with unknown or changing R_w.

PHYSICAL PRINCIPLE

The dielectric constant ϵ is one of the main factors affecting the way an electromagnetic wave propagates through a medium. In general, the electric field E due to an electromagnetic wave can be described by the equation

$$\Delta E = \epsilon\mu \; \frac{\partial^2 E}{\partial t^2} \tag{15.1}$$

Material used in this chapter has in part been abridged from Schlumberger 1982 with their kind permission.

where μ is the magnetic permeability of the medium. In the case of a plane wave passing through a nonferromagnetic material, the general equation has the solution

$$E = E_0 e^{j\omega t - \gamma x} \tag{15.2}$$

where x is the distance traveled by the wave, ω is the frequency of the wave, E_0 is the value of E at time 0, j is the square root of -1, t is time, and γ is the complex propagation constant ($\gamma^2 = -\omega^2 \mu_0 \epsilon$). ϵ is usually a complex number and can be written $\epsilon = \epsilon' - j\epsilon''$, where ϵ' is the permittivity of the medium and ϵ'' is the loss factor. The term γ can be simplified to a form $\alpha + \beta j$, where α is the attenuation factor and β is the phase shift.

If the above relationships are developed, the dielectric permittivity of the medium can be found by measuring the phase shift and attenuation of a single-frequency electromagnetic wave. By relating this permittivity to the dielectric permittivity of free space, ϵ_0, the dielectric constants ϵ_r' and ϵ_r'' can be calculated:

$$\epsilon_r' = \frac{\epsilon'}{\epsilon_0} = \frac{1}{\epsilon_0 \mu_0 \omega^2} (\beta^2 - \alpha^2), \tag{15.3}$$

$$\epsilon_r'' = \frac{\epsilon''}{\epsilon_0} = \frac{1}{\epsilon_0 \mu_0 \omega^2} (2\alpha\beta), \tag{15.4}$$

where the subscript r stands for relative—that is, ϵ_r is the dielectric constant relative to free space—and μ_0 is μ of free space.

Up to this point, the only attenuation of the waveform taken into account is the attenuation resulting from conductivity losses (the loss factor ϵ''). When the electromagnetic wave is measured at a point close to the wave source, spherical waves must be considered—as opposed to the plane-wave approach taken above.

The dielectric permittivity is proportional to the electric dipole moment per unit volume, so formations containing a large number of polar molecules will have a high dielectric constant. Water is one of the few polar substances found in nature, its molecules forming permanent dipoles, and has a greater-than-normal dielectric constant. Table 15.1 lists the relative dielectric permittivities (dielectric constants) of some common geological materials, measured at a frequency of 1.1 GHz.

Dielectric constant logging devices attempt to measure the same physical formation property, although the EPT and DCL tools operate at different frequencies and with different instrumentation. Interpretation of either survey is based on the same principles, although some adjustments are required to suit the idiosyncrasies of the distinct measurements. Detailed discussion of interpretation will be limited to the EPT measurement.

TABLE 15.1 *Relative Dielectric Constants and Propagation Times for Common Minerals and Fluids*

Mineral	Relative Dielectric Constant ϵ_r	Propagation Time t_{pl} (ns/m)*
Sandstone	4.65	7.2
Dolomite	6.8	8.7
Limestone	7.5–9.2	9.1–10.2
Anhydrite	6.35	8.4
Halite	5.6–6.35	7.9–8.4
Gypsum	4.16	6.8
Dry colloids	5.76	8.0
Shale	5–25	7.45–16.6
Oil	2.0–2.4	4.7–5.2
Gas	1.0	3.3
Water	56–80	25.30
Fresh water	78.3	29.5

*ns/m = nano seconds per meter.
Source: Courtesy Schlumberger Well Services.

ELECTROMAGNETIC PROPAGATION TOOL (EPT)

Measurement Principle

The EPT (fig. 15.1) is a pad-type tool with an antenna pad attached rigidly to the body of the tool. A back-up arm has the dual purpose of pressing the pad against the borehole wall and providing a caliper measurement. A standard microlog pad is also attached to the main arm, allowing a resistivity measurement to be made with a vertical resolution similar to the electromagnetic measurement. A smaller arm, exerting less force, is mounted on the same side of the tool as the pad and is used to detect rugosity of the borehole. The borehole diameter is the sum of the measurements from these two independent arms.

Two microwave transmitters and two receivers are mounted in the antenna pad assembly in a borehole-compensation array that minimizes the effects of borehole rugosity and tool tilt (fig. 15.2). The two transmitter-receiver spacings, 8 cm and 12 cm, are chosen so that there is an acceptable depth of investigation while the detected signals will have sufficient amplitude without the possibility of phase wrap-around (see fig. 15.3). A 1.1 GHz electromagnetic wave is sent sequentially from each of the two transmitters; the amplitude and phase shift of the wave are measured at each of the two receivers (fig. 15.4). The absolute values of the amplitude and phase shift are found by comparison with an accurate known reference signal generated in the tool. The phase shift ϕ, the propagation time for the wave, t_{pl}, and the attenuation A over the receiver–receiver spacing, are calculated from the individual measurements. In each case, an average is taken of the measurements derived from the two transmitters. A complete borehole-compensated measurement is made 60 times per second; these

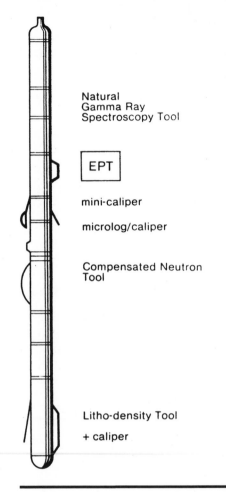

Natural
Gamma Ray
Spectroscopy Tool

EPT

mini-caliper

microlog/caliper

Compensated Neutron
Tool

Litho-density Tool

+ caliper

FIGURE 15.1 *EPT Tool. Reprinted by permission of the SPWLA from Johnson and Evans 1983. Eighth European Formation Evaluation Symposium.*

individual measurements are accumulated and averaged over a formation interval of either 2 in. or 6 in. prior to recording on film and tape.

Because of the close proximity of the receivers to the transmitters, spherical waves are being measured. Therefore, a correction factor is applied to the measured attenuation so that the plane-wave theory can be applied. The increased attenuation resulting from the spherical spreading of the wave is compensated for by applying a spherical loss correction factor, SL. Thus, the corrected attenuation, A_c, is given by $A_c = A - SL$. In air, SL has a value of about 50 dB; but, since the term is porosity dependent, a more exact approach can be taken when correcting downhole measurements:

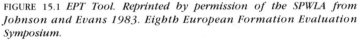

$$SL = 45.0 + 1.3t_{pl} + 0.18t_{pl}^2. \qquad (15.5)$$

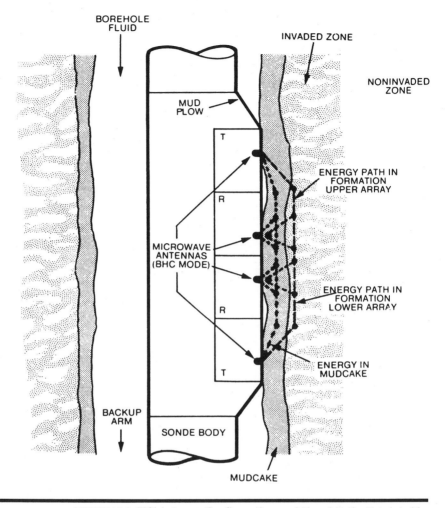

FIGURE 15.2 *EPT Antenna Configuration and Signal Paths. Reprinted by permission of the SPWLA from Johnson and Evans 1983. Eighth European Formation Evaluation Symposium.*

The dielectric parameters for the formation can then be obtained from the log data, since the attenuation factor α is directly proportional to the recorded attenuation A and the phase shift β is proportional to the propagation time t_{pl} ($\beta = \omega t_{pl}$).

The basic data available from the EPT sensors are the electromagnetic, microlog, and caliper measurements. A standard EPT log presentation is shown in figure 15.5 over an interval containing two sandstones (168–179 m and 202–207 m) separated by shale. Track 1 contains the borehole diameter (HD) and the micronormal (MNOR) and microinverse (MINV) resistivity curves. The electromagnetic wave attenuation (EATT) and propagation time (TPL) are

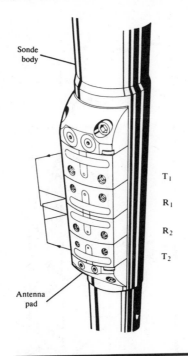

FIGURE 15.3 *Antenna Pad of the EPT Sonde. Courtesy Schlumberger Well Services.*

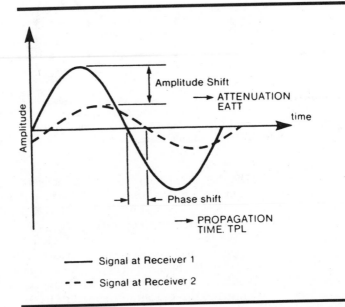

FIGURE 15.4 *Electromagnetic Propagation Signals. Reprinted by permission of the SPWLA from Johnson and Evans 1983. Eighth European Formation Evaluation Symposium.*

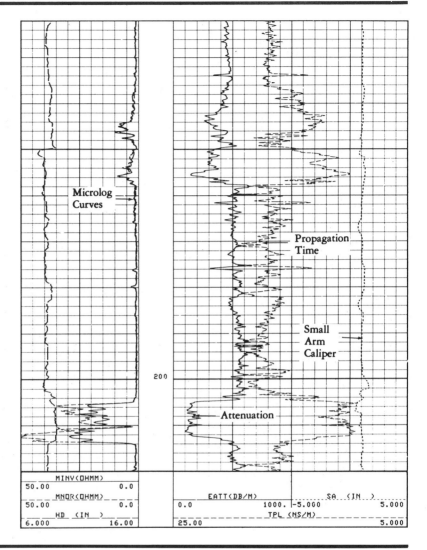

FIGURE 15.5 *Presentation of Data Available from an EPT Survey. Courtesy Schlumberger Well Services.*

recorded in Tracks 2 and 3. The measurement of the smaller caliper arm (SA) can be displayed to monitor the borehole rugosity, and thereby the quality of the EPT data.

Interpretation Methods

The EPT measurement responds mainly to the water content of a formation, rather than to the matrix or any other fluid. The water present in a formation can be the original connate water, mud filtrate, or bound water associated with shales. Because of the shallow depth of investigation of the tool (1–6 in.), it can usually be

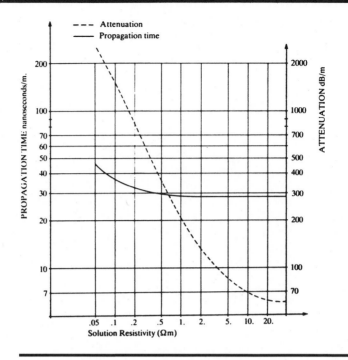

FIGURE 15.6 *Propagation Time and Attenuation as Functions of Water Salinity. Courtesy Schlumberger Well Services.*

assumed that only the flushed zone is influencing the measurement and that the free water is filtrate.

Under normal circumstances, if fresh muds are used, the propagation time of the electromagnetic waves is essentially unaffected by the water salinity (fig. 15.6). An increase in salinity increases the loss factor ϵ'' and decreases the permittivity ϵ', but the effects tend to cancel out. If salt-saturated fluids are encountered, the loss factor increases to the extent that the electromagnetic waves are highly attenuated, and therefore measurements are more difficult. EPT measurements are unaffected by mudcake up to a thickness of about 0.4 in., but rugosity can result in spurious readings as mud comes between the antenna pad and the formation. The situation deteriorates further in boreholes filled with air or oil when even a thin film of the fluid results in the tool responding only to the fluid and not to the formation.

Measurements performed on various samples have produced an empirical relationship for the response of the EPT readings:

$$\gamma = \phi \, \gamma_f + (1 - \phi)\gamma_{ma}, \tag{15.6}$$

where γ is the complex propagation constant of the formation, γ_f and γ_{ma} are the propagation constants for the fluid and the matrix, and ϕ is the porosity of the formation. Using the transforms of equations 15.2 and 15.6, further equations can be derived that relate porosity to the logged parameters:

$$t_{pl} = \phi t_{pf} + (1 - \phi) t_{pma}, \tag{15.7}$$

$$A_c = \phi A_{cf}, \tag{15.8}$$

where t_{pl}, t_{pf}, and t_{pma} are the propagation times for the total formation, the fluid, and the matrix. A_c and A_{cf} are the corrected attenuations for the formation and the fluid. Normally, the matrices are lossless, so there is no corresponding term A_{cma}.

Wellsite Interpretation: t_{po} Method

The t_{po} interpretation method is used in Schlumberger's CSU program during the recording of an EPT log to determine porosity from the measured values of attenuation and propagation time. The principle behind the method is that all values used in the computations are treated as if they were measured in a lossless formation. The measured data must, therefore, be related back to lossless conditions by applying a correction factor that is a function of the attenuation of the electromagnetic waves in the lossy medium. If the measured propagation time is t_{pl}, then the propagation time in the lossless formation, t_{po}, is given by

$$t_{p0}^2 = t_{pl}^2 - \frac{A_c^2}{3604} \tag{15.9}$$

A similar correction can be applied to transform the water propagation time to lossless conditions, t_{pw} to t_{pw0}. The attenuation A_c is computed from the logged attenuation by applying a constant correction of -50 dB. Hydrocarbons and matrices are lossless media, and therefore $t_{pma0} = t_{pma}$ and $t_{ph0} = t_{ph}$. The lossless propagation time of water is obtained from the temperature-related equation

$$t_{pw0} = 20 \times \frac{710 - T/3}{444 + T/3}, \tag{15.10}$$

where T is the temperature in degrees Fahrenheit.

Porosity is computed assuming a clean water-bearing formation:

$$\phi_{EPT} = \frac{t_{po} - t_{pma}}{t_{pwo} - t_{pma}}. \tag{15.11}$$

If the presence of hydrocarbons is included in the response equation, again assuming lossless conditions, the relationship takes the form

$$S_{xo} = \frac{1}{\phi_T} \frac{t_{po} - t_{pma}}{t_{pwo} - t_{pma}} = \frac{\phi_{EPT}}{\phi_T}. \tag{15.12}$$

A comparison of the EPT measurement of porosity with the total porosity measured by the neutron, density, and sonic tools allows a quick-look determination of the water saturation in the flushed zone. Figure 15.7 compares sonic porosity with EPT porosity. The sonic porosity (SPHI) and EPT porosity (EMCP) are displayed in Tracks 2 and 3, and the computed gamma ray (CGR) and total gamma ray (SGR) from the NGT (natural gamma ray tool) survey are recorded in Track 1. There is a change of lithology at 245 m, with a limestone above this depth and a sandstone with calcareous cement below. The matrix parameters for the sonic and EPT porosity calculations were selected accordingly, using the data in table 15.1. The limestone and lower section of the sandstone are water bearing, and the hydrocarbon content of the upper section of the sand is clearly indicated by the separation of the two porosity curves. The original oil/water contact is at 267 m, while the present contact is at 261 m. Generally, the EPT porosity will read the same as a nuclear-derived porosity in water-bearing zones and shales. In hydrocarbon-bearing intervals, however, the EPT porosity will be less than either the total porosity or the density porosity. In gas zones, the separation between the neutron porosity and the EPT porosity will not be so apparent. Figure 15.8 illustrates these differences.

The value of the matrix propagation time to be used in the porosity computation is chosen by one of several available methods. If a simple known lithology is being dealt with, the values in table 15.1 can be used directly. In a dual mineral formation containing two of the most common matrices, the chart of figure 15.9 can be used. If any other mineral is known to predominate in the formation, its matrix parameters can similarly be entered in the chart.

DIELECTRIC CONSTANT LOG (DCL)

The DCL uses a lower operating frequency (16 to 30 MHz) than the EPT and a much longer spacing (±3 ft) between transmitter and receiver (see fig. 15.10). As a result, the DCL measures formation properties beyond the invaded zone. This tool can, therefore, be used for monitoring enhanced recovery projects where plastic pipe has been set. Figure 15.11 shows the progress of a waterflood through repeat logs on different dates.

DEEP PROPAGATION TOOL (DPT)

The DPT operates at 20 MHz with long spacing. Figure 15.12 shows a schematic of the tool and a sample log. The DPT, which is currently rated as experimental, provides measurement of both the dielectric constant and formation resistivity. It is limited in use to small (8-in.) boreholes, muds with R_m greater than 0.5 $\Omega \cdot m$, and formation resistivity greater than 5 $\Omega \cdot m$. Measurements are currently claimed to be accurate to within 20%.

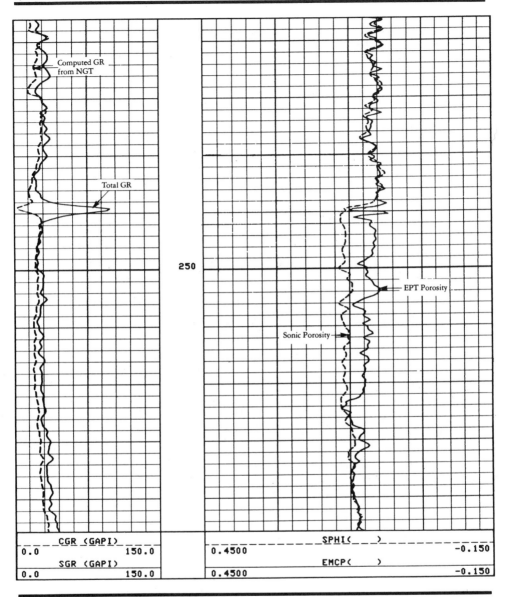

FIGURE 15.7 *Quick-Look Identification of Hydrocarbon-Bearing Intervals through Comparison of Sonic and EPT Porosity. Courtesy Schlumberger Well Services.*

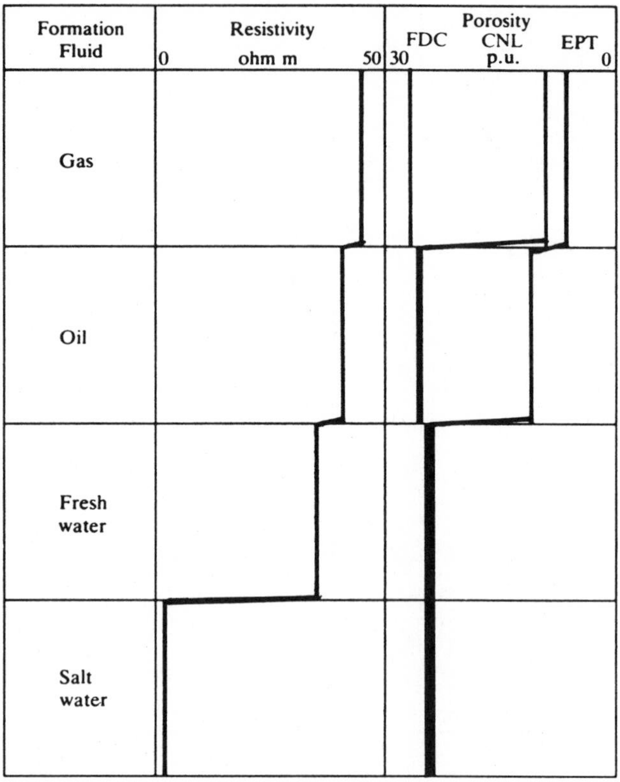

Formation Fluid	Resistivity		FDC	Porosity CNL p.u.	EPT
	0 ohm m 50		30		0
Gas					
Oil					
Fresh water					
Salt water					

FIGURE 15.8 *Variation of Log Readings for Water and Hydrocarbons. Courtesy Schlumberger Well Services.*

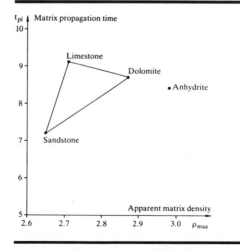

FIGURE 15.9 *Determination of Matrix Propagation Time in a Two-Mineral Formation. Courtesy Schlumberger Well Services.*

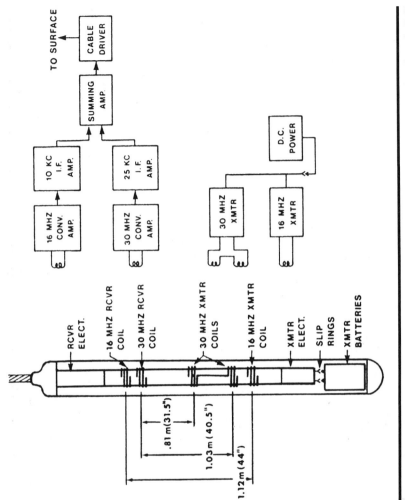

FIGURE 15.10 *Block Diagram and Sonde Arrangement for 16-MHz Dipole–30-MHz Quadrapole DCL System. Reprinted by permission of the SPE-AIME from Meador and Cox 1975, fig. 7. © 1975 SPE-AIME.*

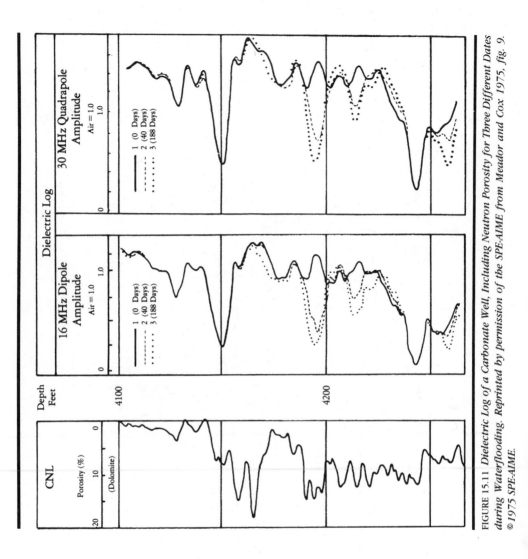

FIGURE 15.11 *Dielectric Log of a Carbonate Well, Including Neutron Porosity for Three Different Dates during Waterflooding. Reprinted by permission of the SPE-AIME from Meador and Cox 1975, fig. 9.* © 1975 SPE-AIME.

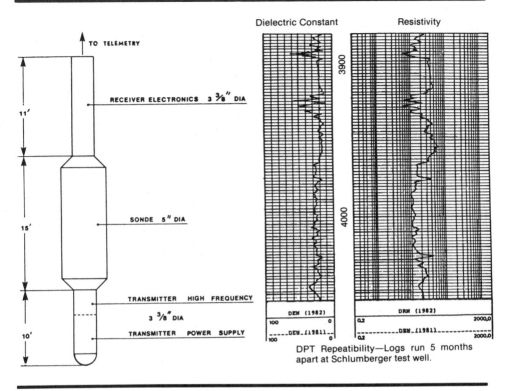

DPT Repeatibility—Logs run 5 months apart at Schlumberger test well.

FIGURE 15.12 *DPT and Sample Log. Reprinted by permission of the SPE-AIME from Huchital et al. 1981, figs. 3 and 4. © 1981 SPE-AIME.*

BIBLIOGRAPHY

Chew, W. C.: "Response of the Deep Propagation Tool in Invaded Bore-holes," SPE paper 10989, presented at the 57th Annual Technical Conference and Exhibition, New Orleans, Sept. 25–29, 1982.

Berry, W. R., Head, M. P., and Mougne, M. L.: "Dielectric Constant Logging a Progress Report," SPWLA paper VV, 20th Annual Logging Symposium, Tulsa, June, 1979.

Eck, M. E., and Powell, D. E.: "Application of Electromagnetic Propagation Logging in the Permian Basin of West Texas," SPE paper 12183, presented at the 58th Annual Technical Conference and Exhibition, San Francisco, Oct. 5–8, 1983.

Geng Xiuwen, Yong Yizu, Lu Da, and Shao Shufang: "Dielectric Log A Logging Method for Determining Oil Saturation," *J. Pet. Tech.* (October 1983).

Huchital, G. S., Hutin, R., Thoraval, Y., and Clark, B.: "The Deep Propagation Tool (A New Electromagnetic Logging Tool)," SPE paper 10988, presented at the 56th Annual Technical Conference and Exhibition, San Antonio, Oct. 5–7, 1981.

Johnson, R., and Evans, C. J.: "Results of Recent Electromagnetic Propagation Time Logging in the North Sea," paper presented at the 8th European Formation Evaluation Symposium, London, March, 1983.

Kenyon, W. E., and Baker, P. L.: "EPT Interpretation in Carbonates Drilled with Salt Muds," SPE paper 13192, presented at the 59th Annual Technical Conference and Exhibition, Houston, Sept. 16–19, 1984.

Mazzagatti, R. P., Dowling, D. J., Sims, J. C., Bussian, A. E., and Simpson, R. S.: "Laboratory Measurements of Dielectric Constant near 20 MHz," SPE paper 12097, presented at the 58th Annual Technical Conference and Exhibition, San Francisco, Oct. 5–8, 1983.

Meador, R. A., and Cox P. T.: "Dielectric Constant Logging, A Salinity Independant Estimation of Formation Water Volume," SPE paper 5504, presented at the 50th Annual Meeting, 1975.

Rau R. N., and Wharton, R. P.: "Measurement of Core Electrical Parameters at UHF and Microwave Frequencies," SPE paper 9380, presented at the 55th Annual Technical Conference and Exhibition, Dallas, 1980.

Schlumberger: "Well Evaluation Developments Continental Europe 1982," (1982) pp. 65–71.

Sen, P. N.: "The Dielectric and Conductivity Response of Sedimentary Rocks," SPE paper 9379, presented at the 55th Annual Technical Conference and Exhibition, Dallas, 1980.

Wharton, R. P., Hazen, G. A., Rau, R. W., and Best, D. L.: "Electromagnetic Propagation Logging: Advances in Technique and Interpretation," SPE paper 9267, presented at the 55th Annual Technical Conference and Exhibition, Dallas, 1980.

CHAPTER SIXTEEN

SONIC (ACOUSTIC) LOGGING AND ELASTIC FORMATION PROPERTIES

Elastic formation properties control the transmittal of elastic waves through subsurface formations, and the science of seismic evaluation is based on the physics of rock elasticity. Sonic logging is a localized downhole branch of geophysics. A wealth of information concerning formation properties can be gathered by properly combining surface and downhole measurements. For example:

Sonic logs and check-shot surveys can be used to calibrate seismic surveys.

Sonic and density logs can be combined to provide synthetic seismic traces.

Sonic and density logs can be combined to deduce mechanical properties of the formation, and these, in turn, can be used to deduce pore pressure, rock compressibility, fracture gradients, sanding problems, etc.

Sonic and other logs can be combined to deduce porosity, lithology, and fluid saturations.

Borehole measurements can produce a vertical seismic profile (VSP), which can "see" *below* the bottom of the well.

Sonic tools may be used for cement bond logging in cased holes.

Since the elasticity of subsurface formations is basic to all of these measurements and their interpretation, an understanding of how elastic waves propagate through a medium is in order.

PROPAGATION OF ELASTIC WAVES

In a infinite medium, two types of sound waves are propagated: compressional waves and shear waves. In a finite medium (e.g., a borehole), guided waves may be propagated.

Compressional Waves

Compressional (or pressure) waves are longitudinal; that is, the direction of propagation is parallel to the direction of particle displacement (fig. 16.1). Gases and liquids, as well as solids, tend to oppose compression, and compressional waves can, therefore, be propagated through all three.

Shear Waves

Shear waves are transverse; that is, the direction of their propagation is perpendicular to the direction of particle displacement (fig. 16.2).

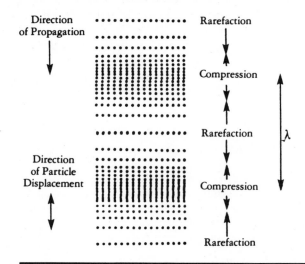

FIGURE 16.1 *Compressional Wave. Courtesy Schlumberger Well Services.*

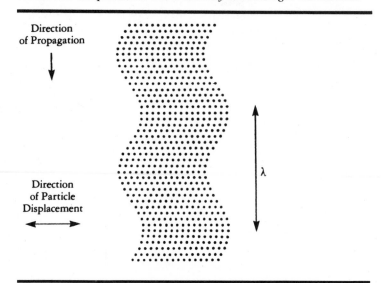

FIGURE 16.2 *Shear Wave. Courtesy Schlumberger Well Services.*

Shear waves can be propagated through solids, owing to their rigidity. On the other hand, gases (and liquids having negligible viscosity) cannot oppose shearing, and shear waves cannot be propagated through them. In practice, viscous fluids do permit some propagation of shear waves, though they become highly attenuated.

Guided Waves

Guided waves, or interface waves, include:

RAYLEIGH WAVES. Rayleigh waves occur at the mud/formation interface and are a combination of two displacements, one parallel and the other perpendicular to the interface. Their speed is slightly less than that of shear waves ($V_{Rayleigh}$ is 86% to 96% of V_{shear}). Energy leaks away from the interface as compressional waves are set up in the mud. The waves are then referred to as pseudo-Rayleigh waves.

STONELEY WAVES. Stoneley waves travel in the mud by interaction between the mud and the formation. Their amplitude decays exponentially, in both the mud and the formation, away from the borehole boundary. These low-frequency waves are called *tube waves*. Stoneley-wave velocity is lower than the mud compressional-wave velocity.

ELASTIC CONSTANTS

Proper interpretation of any measurement made using elastic-wave data requires an understanding of the elastic properties of a medium.

The properties derived from testing rock samples in the laboratory (e.g., by measuring the strain for a given applied stress) are static elastic constants. Dynamic elastic constants are determined by measuring elastic-wave velocities in the material. Sonic logging and waveform analysis provide the means for obtaining continuous velocity measurements and thus knowledge of the mechanical properties of the rock in situ.

The speed at which a wave travels through a medium may be expressed in two ways. Geophysicists think in terms of velocity— that is, distance traveled per unit of time. Subsurface formation velocities range from 6000 to 25,000 ft/s. Log analysts think in terms of time—that is, the time taken to travel one unit of distance. A convenient unit of measurement is the microsecond per foot ($\mu s/ft$), which is given the symbol Δt. With these definitions in mind, the dynamic elastic constants of a medium can be expressed as a function of bulk density, ρ_b, and travel time for compressional and shear waves, Δt_c and Δt_s, respectively, as shown in table 16.1.

Discussions later in the text will show how both compressional and shear velocities can be measured by sonic tools or deduced from other means. Practical applications of such data will be explained.

TABLE 16.1 *Dynamic Elastic Constants**

μ	Poisson's Ratio	$\dfrac{\text{lateral strain}}{\text{longitudinal strain}}$	$\dfrac{\frac{1}{2}(\Delta t_s/\Delta t_c)^2 - 1}{(\Delta t_s/\Delta t_c)^2 - 1}$
G	Shear modulus	$\dfrac{\text{applied stress}}{\text{shear strain}}$	$\dfrac{\rho_b}{\Delta t_s^2} \times a$
E	Young's modulus	$\dfrac{\text{applied stress}}{\text{normal strain}}$	$2G(1 + \mu) \times a$
K_B	Bulk modulus	$\dfrac{\text{applied stress}}{\text{volumetric strain}}$	$\rho_b \left(\dfrac{1}{\Delta t_c^2} - \dfrac{4}{3\Delta t_s^2} \right) \times a$
C_B	Bulk compressibility	$\dfrac{\text{volumetric deformation}}{\text{applied stress}}$	$\dfrac{1}{K_B}$

Source: Courtesy Schlumberger Well Services.
*Coefficient $a = 1.34 \times 10^{10}$ if ρ_b is in g/cc and Δt in μs/ft.

SONIC LOGGING TOOLS

Sonic logging tools attempt to measure the formation properties Δt_c and Δt_s by means of an apparatus suspended in the mud column. In order to be successful in this task, a number of "tricks" are resorted to: These include borehole compensation, long-spacing tools, and waveform recording.

The curves recorded on a sonic log are the interval transit time, Δt, in μs/ft (the reciprocal of speed); caliper; gamma ray and/or SP; and integrated travel time (discussed later in this chapter). A typical sonic log is illustrated in figure 16.3.

The tools available for sonic measurement include the borehole compensated (BHC) tools, slim-tool versions that can be run through tubing, and the long-spacing sonic tools.

In seismic data gathering, a disturbance is made at the surface by explosives or by use of an airgun in water. In sonic logging, an acoustic pulse, produced by alternate expansions and contractions of a transducer, is emitted by a transmitter (see fig. 16.4). This transmitter pulse generates a compressional wave through the mud. Part of the acoustic energy traverses the mud, impinges on the borehole wall at the critical angle of incidence, passes along the formation close to the borehole wall, reenters the mud, and arrives at a receiver, where it is converted into an electrical signal (fig. 16.5).

OPERATING PRINCIPLES

Sonic tools are comprised of transmitter transducers that convert electrical energy into mechanical energy and receiver transducers that do the reverse. In its simplest form, the measurement is made in an uncompensated mode (fig. 16.6). At time 0, the transmitter emits a small shock wave that travels through the mud to the borehole wall where it is refracted through the formation. Part of the energy traveling through the formation is in turn refracted back into the mud column and finds its way to the first receiver at time T_1 and to the second receiver at time T_2. The difference in the two times is

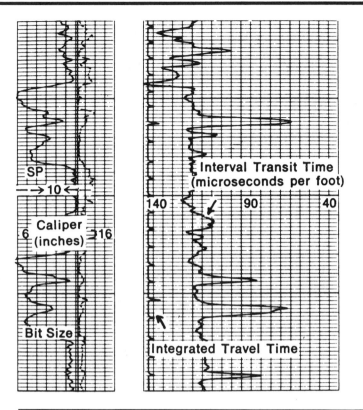

FIGURE 16.3 *Sonic Log. Courtesy Schlumberger "Services Catalog," Schlumberger Limited 1977, 1978, p. 10.*

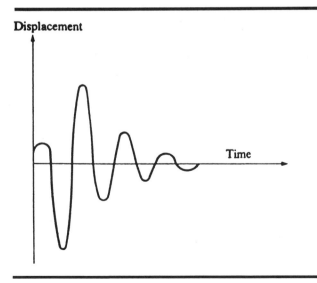

FIGURE 16.4 *Typical Transmitter Pulse. Courtesy Schlumberger Well Services.*

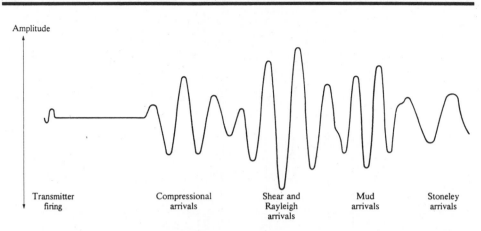

FIGURE 16.5 *Signal Generated at Receiver by Various Wave Arrivals. Courtesy Schlumberger Well Services.*

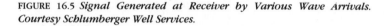
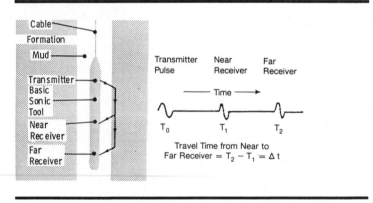

FIGURE 16.6 *Basic Sonic Device.*

referred to as Δt and represents the time it takes a compressional wave to travel through the formation a distance equal to the spacing between the two receivers. Formation travel time, Δt, is expressed in μs/ft or in μs/m.

This early form of sonic tool worked on the assumption that in mud, the travel paths to the two receivers were equal. This was true in the case of a smooth borehole of unchanging size but was not true if the borehole was of varying size or if the sonde tilted with respect to the axis of the borehole. These difficulties were overcome by the introduction of the BHC (borehole compensated) sonic tool. Figure 16.7 illustrates the principle of the BHC sonic tool. Two transmitters and four receivers are employed, and two values of Δt are measured and averaged. This system eliminated errors in Δt resulting from sonde tilt and hole-size variation. Even so, there are practical limits to the working range of the tool.

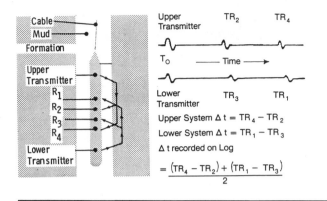

FIGURE 16.7 *Borehole Compensated Sonic Tool.*

In large boreholes, for example, the time taken for a compressional wave to travel from the transmitter to the formation, through the formation, and back through the mud to a receiver may exceed the time taken for a direct transmission from the transmitter to the receiver through the mud. The critical factors in determining when this condition will exist are the transmitter–receiver spacing, the hole size, and the travel time in the formation. The highest formation Δt that can be measured with conventional borehole compensated sonic tools with 3-ft spacing is 175 μs/ft in a 12¼-in. hole and 165μs/ft in a 14-in. hole. These limitations are not serious ones if the formation is a reservoir rock with a Δt in the normal range of 40 to 140 μs/ft. It does become a serious defect if the rock is a shale with a long transit time and the purpose of the log is to compute integrated travel time for geophysical purposes. Mathematically, the variables can be related as follows:

$$S_{\min} = (D - d) \left\{ \frac{V_{\text{form}} + V_m}{V_{\text{form}} - V_m} \right\}^{\frac{1}{2}},$$

where

$S_{\min}$ is the minimum spacing possible for recording a formation velocity V_{form},
D is the hole diameter,
d is the sonde diameter,
V_{form} is the formation velocity,
V_m is the mud velocity.

Put another way, the maximum formation Δt that can be recorded is given by

$$\Delta t_{\max} = \Delta t_m \left\{ \frac{K - 1}{K + 1} \right\},$$

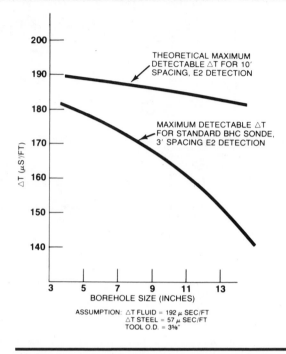

FIGURE 16.8 *Maximum Detectable Δt for 3- and 10-ft Spacings ($\Delta t_f = 192$ s/ft, tool OD = 3⅝ in.).*

where Δt_m is the mud travel time and

$$K = \left(S/(D - d)\right)^2.$$

Figure 16.8 illustrates these relationships graphically.

LONG-SPACING SONIC TOOLS

The long-spacing sonic (LSS) tool was introduced in an attempt to overcome environmental problems. For example, when a shale formation is drilled, the shales exposed to the mud frequently change their properties by absorption of water from the drilling mud. The travel time for elastic waves therefore also changes. In order to read the travel time in the undisturbed formation away from the borehole, a longer transmitter–receiver spacing is required. Typically, a long-spacing sonic tool will have a transmitter–receiver spacing of 8, 10, or 12 ft. Figure 16.9 compares a conventional borehole compensated sonic log with a long-spaced sonic log and shows how the LSS tool can be more reliable in areas with problem shales.

Lengthening the spacing on a sonic device achieves two ends:

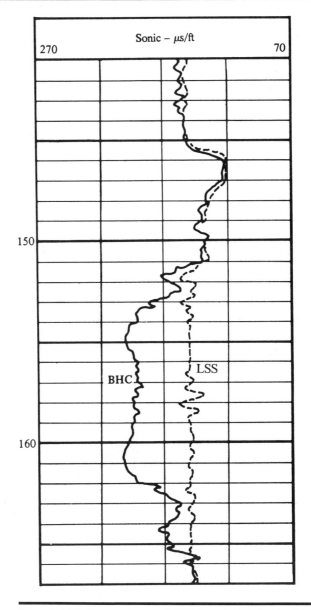

FIGURE 16.9 *Comparison of BHC and LSS Sonic Logs. Courtesy Schlumberger Well Services.*

1. A valid sonic log may be recorded in a bigger hole with a long-spacing device than with a conventionally spaced tool.
2. The zone investigated by the tool is deeper into the formation with a long-spacing device than with a conventionally spaced tool.

Deeper investigation into the formation is of great value when logging through intervals of shale that have had their properties altered by the drilling process. If Δt of the formation in the undisturbed state is less than Δt of the formation in the altered state, the shortest route for a compressional wave will be through undisturbed formation, or *deep* in the formation (see fig. 16.10).

Long-spacing sonic tools make their measurements in a *depth derived* mode. That is, the tool compensates for changes in borehole size by combining travel times measured and memorized when the tool is at one depth with travel times recorded at a shallower depth when an alternate combination of transmitters and receivers is activated. Two transmitters spaced 2 ft apart are located 8 ft below a pair of receivers that are also 2 ft apart (fig. 16.11). The first Δt reading is memorized and combined with a second Δt reading measured after the sonde has been pulled the appropriate distance further along the borehole; compensation for changes in the hole size is thus achieved.

Note (fig. 16.11) that the first Δt reading = T1R1 – T1R2 and is memorized and that the second Δt reading = T2R2 – T1R2. Then the 8–10-ft interval time DT = ½(memorized first Δt reading + second Δt reading). The 10–12-ft measurement, known as DTL, uses T2 for the memorized first Δt and R1 for the second Δt reading.

Both DT and DTL are recorded on Tracks 2 and 3 of the long-spaced sonic log (fig. 16.12). The four transit time curves can be displayed in Track 1 if desired, either in real time or in the playback phase. The log illustrated in figure 16.12 was recorded through a shaly interval, thus accounting for the slight difference between DT and DTL due to alteration of the formation. Nevertheless, the two 10-ft curves (TT1 and TT4) virtually overlap each other, even though measured with different transmitter receiver couples, thereby confirming the validity of the measurement.

CYCLE SKIPPING AND NOISE

The actual travel time measurement is determined at the first arrival peak. However, the tool's internal trigger mechanism for detecting this peak is subject to some errors. Figure 16.13 illustrates two common problems. In the first, the bias level is set too high and the travel time is triggered on a later peak, causing an erroneously long time to be measured (this is known as *cycle skipping*). In the second, the bias is set too low and the travel time is triggered on noise, causing an erroneously short travel time (this problem is known as noise). It is not always possible to distinguish the difference between cycle skipping and noise in the BHC mode because two measurements are being averaged by the tool. Figure 16.14 illustrates a bad sonic log that had to be rerun as a result of cycle skipping and noise.

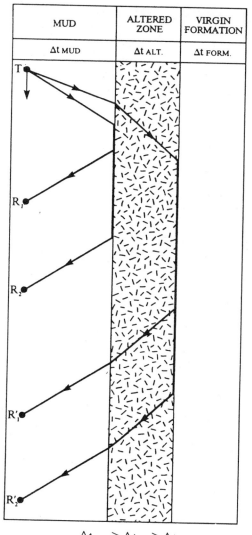

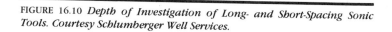

FIGURE 16.10 *Depth of Investigation of Long- and Short-Spacing Sonic Tools. Courtesy Schlumberger Well Services.*

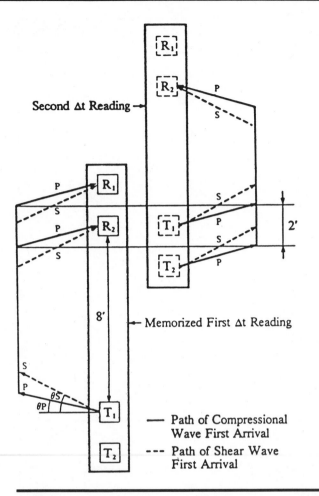

FIGURE 16.11 *Long-Spaced Sonic "Depth Derived" Principle. Courtesy Schlumberger Well Services.*

WAVEFORM RECORDING

Waveforms may be recorded with a long-spacing sonic tool. The longer spacing allows a larger time separation for the compressional and shear wave arrivals than that provided by the short-spacing tools. Various LSS transmitter–receiver combinations permit four waveforms to be recorded at 6-in. intervals. Figure 16.15 illustrates composite waveforms received at the near and far receivers when the upper transmitter is fired.

Digitization of the waveforms is normally made at a 5-μs sample interval for 512 samples—that is, 2560 μs. A delay of 200 to 500 μs is an input parameter selected by the logging engineer.

Waveform recording considerably extends the range of sonic logging applications, in both open holes and cased holes. The

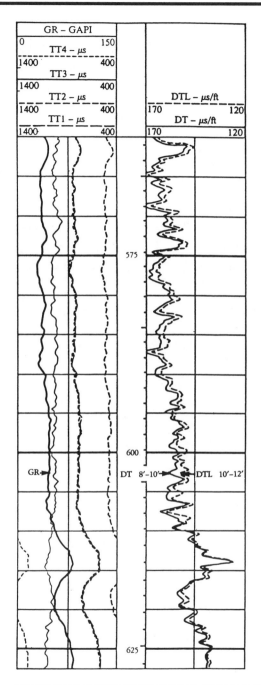

FIGURE 16.12 *LSS Log with Four Transit Time Curves Displayed in Track 1. Courtesy Schlumberger Well Services.*

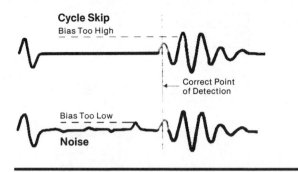

FIGURE 16.13 *Cycle Skipping and Noise.*

principal advantage of waveform recording is that the shear wave velocity of the formation is specifically detected. The objective is to distinguish between the compressional and shear wave arrivals and to measure the compressional and shear interval transit times. In cased holes, the formation arrivals are usually distinct from the casing arrivals. This permits a viable sonic measurement where, without waveform recording, previous sonic devices would have been ineffective.

The data processing methods used to extract shear wave arrival times are somewhat complex and mirror seismic processing methods—that is, multiple waveforms are stacked. It is quite common, however, to "see" shear arrivals on variable density displays of the sort shown in figure 16.16.

SONIC POROSITY (ϕ_s)

The fact that compressional waves travel faster through solid matrix material than through fluid is the basis for the method used to determine formation porosity from sonic logs. Figure 16.17 is a schematic in which the pore space and the solid matrix have been separated for purposes of illustration. If Δt_f is the time taken to travel through the pore space and Δt_{ma} is the time taken to travel through the matrix, the total travel time measured will be Δt, and the porosity will be given by

$$\Delta t \;=\; \phi \Delta t_f \;+\; (1 - \phi)\Delta t_{ma} \qquad \text{or} \qquad \phi_s \;=\; \frac{\Delta t - \Delta t_{ma}}{\Delta t_f - \Delta t_{ma}}\,.$$

This is known as the Wyllie time-average equation. To solve this graphically, use is made of figure 16.18.

QUESTION #16.1

Use figure 16.18 to find ϕ_s when Δt is 81 μs/ft and the matrix velocity is 18,000 ft/s. ($\Delta t_{ma} = 55.5$ μs/ft.)

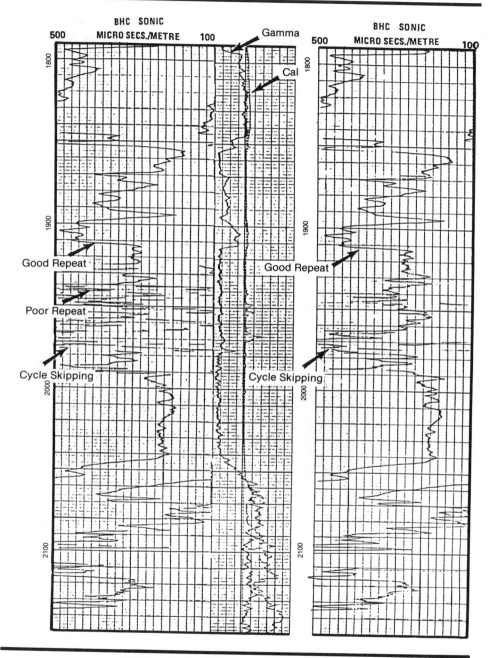

FIGURE 16.14 *Cycle Skipping.*

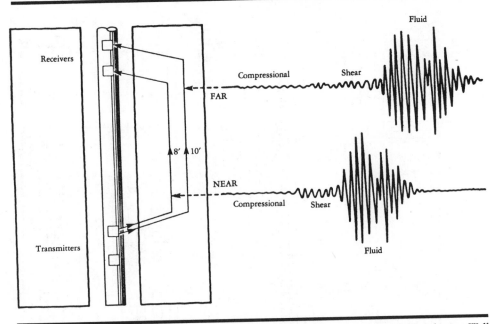

FIGURE 16.15 *Long-Spaced Sonic Waveform. Courtesy Schlumberger Well Services.*

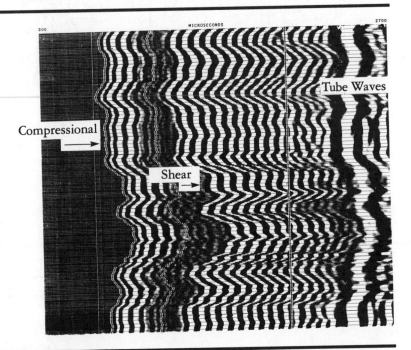

FIGURE 16.16 *Variable Density Display with Compressional and Shear Wave First-Arrival Detections Highlighted. Courtesy Schlumberger Well Services.*

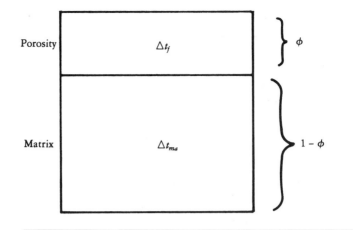

FIGURE 16.17 *Sonic Porosity Measurement.*

Matrix Travel Times

Matrix travel time will depend on the matrix itself. Table 16.2 gives a partial listing of common matrix materials. A more exhaustive list is provided in Appendix 16A.

Fluid Travel Time

Fluid travel time is a function of the temperature, pressure, and salinity of a solution (see fig. 16.19). To use the figure, enter the chart on the x-axis with the formation temperature, and move up to the appropriate salinity line. For this purpose, the fluid in the pore space is normally considered to be mud filtrate. Proceed across to the right to find the fluid velocity, at zero pressure. Now project parallel to the diagonal lines to a point directly above the estimated formation pressure. Proceed to the right and read the fluid velocity corrected for pressure, temperature, and salinity. Use of this chart will improve porosity determination by eliminating empirical changes in the matrix velocity to some incorrect value in order to force a porosity value consistent with other measurements. To convert from ft/sec to μs/ft, use may be made of the relation:

$$\Delta t \ (\mu s/ft) = 10^6/V \text{ in ft/s.}$$

QUESTION #16.2

Temperature	= 270°C.
Salinity	= 200,000 ppm.
Formation Pressure	= 6000 psi.

Find V_f/ft/s and convert this value to a Δt_f in μs/ft.

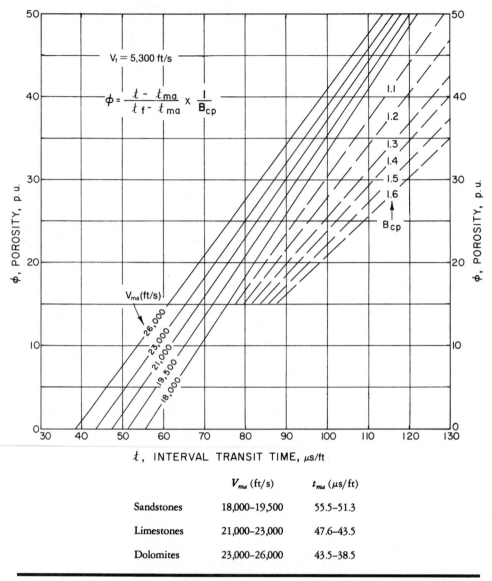

t, INTERVAL TRANSIT TIME, μs/ft

	V_{ma} (ft/s)	t_{ma} (μs/ft)
Sandstones	18,000–19,500	55.5–51.3
Limestones	21,000–23,000	47.6–43.5
Dolomites	23,000–26,000	43.5–38.5

FIGURE 16.18 *Sonic Porosity Determination. Courtesy Schlumberger Well Services.*

TABLE 16.2 Δt_{ma} for Common Matrix Materials and Fluids

Material	Travel Time Δt (μs/ft)	Velocity (ft/s)	Density g/cc
Dolomite	43.5	23,000	2.87
Limestone	47.5	21,000	2.71
Sandstone	55.6	18,000	2.65
Anhydrite	50.0	20,000	2.97
Gypsum	52.5	19,000	2.35
Salt	67.0	15,000	2.03
Water (fresh)	200	5,000	1.00
Water (100,000 ppm NaCl)	189	5,300	1.06
Water (200,000 ppm NaCl)	176	5,700	1.14
Oil	232	4,300	
Air	919	1,088	
Casing	57	17,000	

COMPACTION EFFECTS AND THE HUNT TRANSFORM

Unconsolidated formations exhibit travel times that are longer than can be accounted for by the Wyllie time-average equation. This can be handled in two ways, conventionally and by the Hunt transform. The conventional method adapts the Wyllie time-average equation by introducing a fudge factor, B_{cp}, such that

$$\phi_S = \frac{\Delta t - \Delta t_{ma}}{\Delta t_f - \Delta t_{ma}} \times \frac{1}{B_{cp}},$$

where B_{cp} is some number greater than 1. Note on figure 16.18 that lines have been drawn for B_{cp} values of 1.1 to 1.6, allowing direct determination of porosity in uncompacted formations provided B_{cp} is known. A good rule of thumb is to estimate B_{cp} from the transit time in shales adjacent the formation of interest. Then

$$B_{cp} = \Delta t_{sh}/100.$$

Thus if, in a shallow sand-shale sequence, Δt_{sh} is 130 μs/ft, a B_{cp} of 130/100, or 1.3, should be used.

The Hunt transform is based on empirical observations from sonic logs and porosity determinations from other means. Figure 16.20 shows the generalized form of the Hunt transform and plots Δt against porosity for sandstone, limestone, and dolomite. An acceptable equation relating porosity to Δt for this transform is given by

$$\phi_S = \frac{1}{\rho_{ma} - \rho_f} \left(1 - \frac{\Delta t_{ma}}{\Delta t} \right).$$

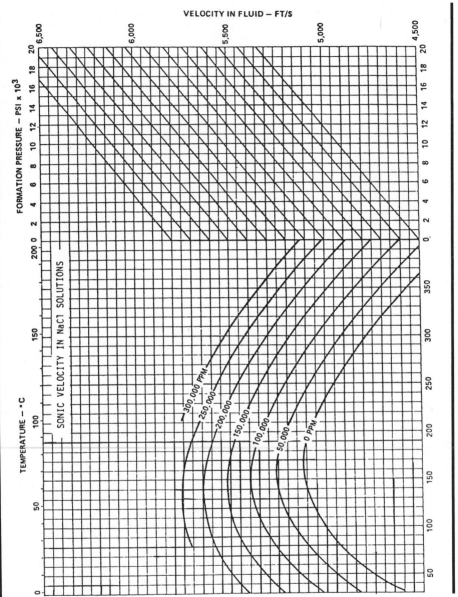

FIGURE 16.19 *Determination of V_f (Hence Δt_f). After Wyllie, by permission of Academic Press.*

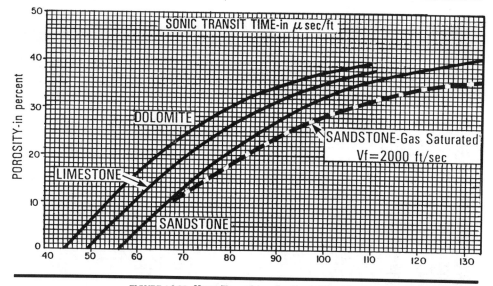

FIGURE 16.20 *Hunt Transform for Sonic Porosity. Reprinted by permission of the SPWLA from Raymer and Hunt 1980. SPWLA 21st Symposium.*

Note that Δt_f does not appear as a term in this equation. The assumption is made that the fluid is liquid (not gas) and is built in to the coefficient $1/(\rho_{ma} - \rho_f)$. In sandstones this coefficient is very close to 5/8.

Compaction effects manifest themselves on sonic logs as a decrease of Δt with depth. This is particularly evident in shales. The deeper a shale is buried, the more compact it becomes and the shorter Δt. In cases where there is no escape for the water in the shale, compaction ceases and overpressure results. In such cases, the Δt shale at that depth is anomalously high. Thus, Δt_{sh} itself becomes an indicator of formation pressure. By making readings on a sonic log in shales only and plotting these Δt_{sh} values against depth, a normal gradient may be defined. Departures from this gradient may indicate overpressure. Figure 16.21 illustrates a normal compaction gradient on a sonic log.

APPLICATIONS

R_{wa}: Sonic–Induction Log

A common combination tool is the induction–sonic tool, which provides, on one run in the hole, both a sonic log and an induction log (fig. 16.22). This combination is particularly useful in shaly sand sequences. Note that a curve called R_{wa} is displayed in Track 1 along with the SP in figure 16.22. This is apparent water resistivity derived from both the induction and sonic measurements. Δt is converted, in the surface equipment, to a value of ϕ_S, the sonic porosity. This is then combined with the value of R_t from the induction log to give

$$R_{wa}(\text{sonic}) = (\phi_S)^m \times R_t/a .$$

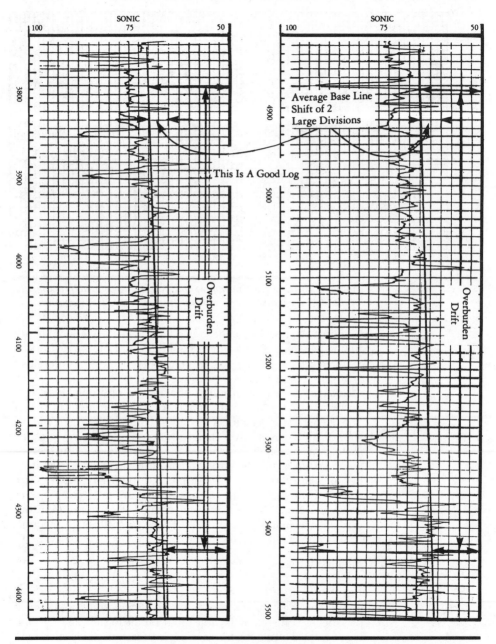

FIGURE 16.21 *Sonic Base-Line Draft Due to Overburden Pressure Effects.*

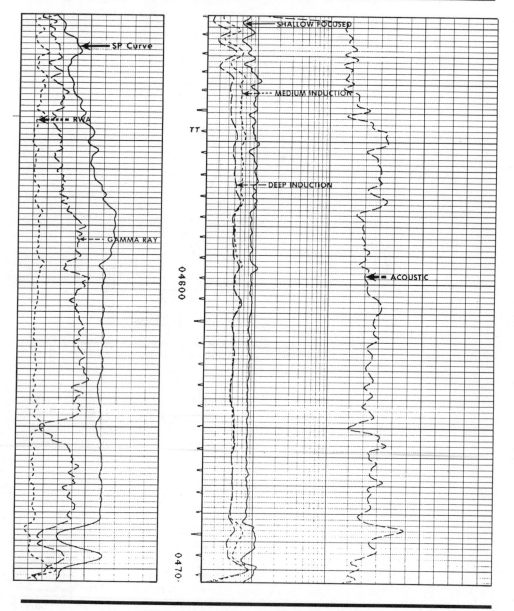

FIGURE 16.22 *Induction–Sonic Log Presentations with R_{wa} Curve.*

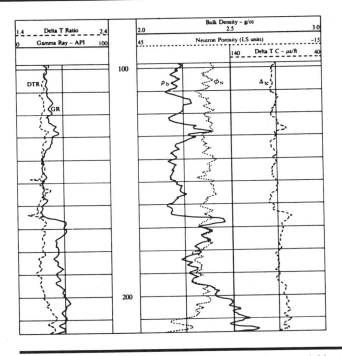

FIGURE 16.23 *Delta T Ratio in Micaceous Sands. Courtesy Schlumberger Well Services.*

This curve affords a quick and easy visual method of spotting potential productive zones. A minimum value of R_{wa} can be picked in a water-bearing zone and then any point on the log exhibiting an R_{wa} value greater than three or four times that minimum is immediately of interest.

Δt RATIO

Determination of the ratio $\Delta t_s/\Delta t_c$ provides a useful contribution to formation evaluation. The principal areas of application are lithology identification, formation fluid determination, and rock mechanical properties. The first example (figure 16.23) demonstrates an interesting property of the Δt ratio in micaceous sands. In the clean upper zone, the Δt ratio correlates very well with the gamma ray curve. The lower zone is not shaly, as the gamma ray might suggest, but contains radioactive mica and heavy components. The Δt ratio remains substantially the same in both zones, indicating a shale-free sandstone since it is unaffected by the other minerals. Thus, the Δt ratio provides, in this type of lithology, a clay indicator unaffected by mica or other nonshale radioactive components such as feldspar.

Obviously, it is preferable to measure Δt_s and then compute the $\Delta t_s/\Delta t_c$ ratio and use it as a lithology indicator. However, if the lithology is known, the known $\Delta t_s/\Delta t_c$ ratio for that lithology can be

TABLE 16.3 *Delta T Ratio for Rock Types*

Formation Lithology	Ratio $\Delta t_s/\Delta t_c$
Anhydrite	2.45
Basalt	1.55
Chalk	2.45
Chert	1.60
Clay	3.20
Claystone	1.90
Diabase	1.70
Diorite	1.75
Dolomite	1.80
Epidosite	1.70
Gabbro	1.60
Gneiss	1.80
Granite	1.70
Gypsum	2.45
Hornstone	1.85
Limestone	1.90
Limestone (silty)	2.1
Limestone (argillaceous)	2.3
Marble	1.80
Mudstone	1.85
Pyrite	1.70
Quartz	1.55
Quartzite	1.50
Salt	2.15
Sandstone	1.60
Shale	1.7–1.75
Siltstone	1.80

Source: After Mason 1984 by permission of the SPE-AIME.

used to compute Δt_s, which can then be used for calculations of the formation elastic constants, in turn useful for construing formation mechanical properties. Of help in this respect is table 16.3, which lists the $\Delta t_s/\Delta t_c$ for common rock types. These data are applied as follows: Read the conventional Δt_c from the sonic log and multiply by the ratio of $\Delta t_s/\Delta t_c$ read from table 16.3.

$$\Delta t_s = \Delta t_c \times (\Delta t_s/\Delta t_c).$$

There is evidence to suggest that grain size has an effect on the $\Delta t_s/\Delta t_c$ ratio, and that the calculated value of Δt_s should be refined using an exponent such that

$$\Delta t_s = \Delta t_c \times (\Delta t_s/\Delta t_c)^\alpha,$$

where α is a number less than 1 in fine-grained sediments and greater than 1 in coarse-grained sediments. Suggested values of α are given in table 16.4.

TABLE 16.4 *Travel Time Correction Factor α, for Grain Size*

Wentworth's Classification Grain Size (mm)		$\Delta t_s/\Delta t_c$ Correction Factor α
Silt	$\frac{1}{16}$	0.90
Very fine grained	$\frac{1}{16}$–$\frac{1}{8}$	0.95
Fine grained	$\frac{1}{8}$–$\frac{1}{4}$	1.00
Medium grained	$\frac{1}{4}$–$\frac{1}{2}$	1.05
Coarse grained	$\frac{1}{2}$–1.0	1.10
Very coarse grained	1.0–2.0	1.15
Granules	2.0–4.0	1.20
Granules	>4.0	>1.30

Source: After Mason 1984 by permission of the SPE-AIME.

Mechanical Properties of a Formation

Once the values of Δt_c and Δt_s are established, they may be combined with bulk density values to calculate the elastic constants—bulk compressibility, Poisson's ratio, Young's modulus, and shear modulus (see table 16.1). These parameters may in turn be used for calculations in areas of interest to the drilling and completion engineers, including:

Prediction of wellbore pressure at which shear failure will occur and result in sand production and need for gravel packing
Avoidance of accidental fracturing while drilling or cementing
Differential pressure requirements for hydraulic fracture initiation

Figure 16.24 is an example of a computed log showing an analysis of formation strength and the propensity to sanding problems. The same logic can be applied to such diverse fields as bit selection and design of fracture treatments.

TRANSIT TIME INTEGRATION AND SEISMIC APPLICATION

Sonic logs have important geophysical applications. To determine a time-versus-depth relation for a particular geologic column is an essential step for proper analysis of seismic data. Sonic logs can be run in such a way that the integrated travel time is displayed as a series of time ticks (see fig. 16.25).

QUESTION #16.3
On the sonic log of figure 16.25, find the integrated travel time from 8950 to 9100 ft.

Other seismic applications include the use of sonic logs, singly or in combination with density logs, to produce synthetic seismograms. The ability of an interface between two distinct geologic strata to

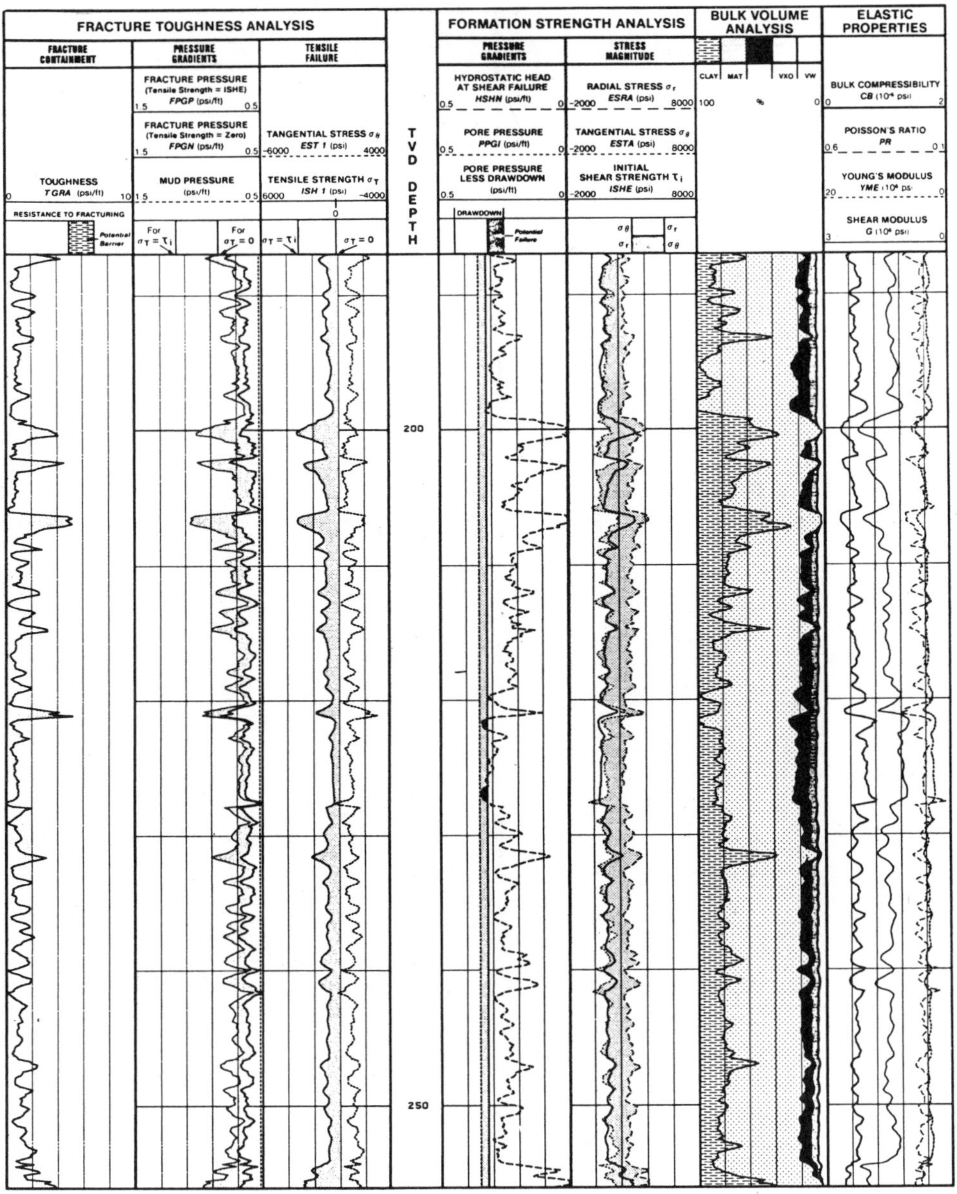

FIGURE 16.24 *Analysis of Formation Properties on a Computed Log. Courtesy Schlumberger Well Services.*

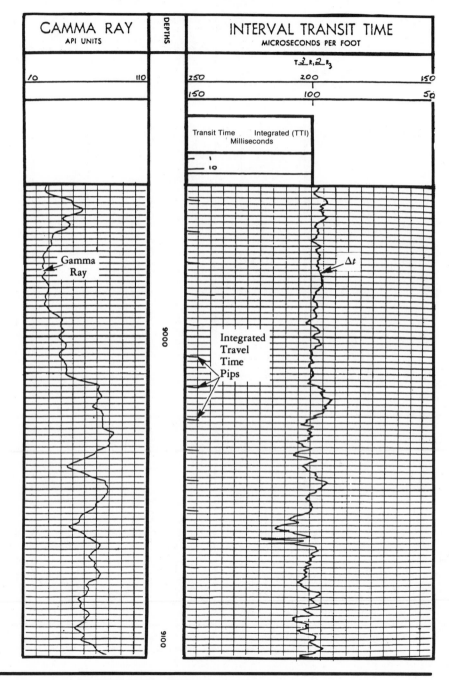

FIGURE 16.25 *Transit Time Integration.*

reflect or transmit acoustic energy is related to the relative density and travel time in each strata. A reflectivity coefficient can be calculated from the relationship

$$\frac{\rho_1 V_1 - \rho_2 V_2}{\rho_1 V_1 + \rho_2 V_2},$$

where ρ and V refer to the density and velocity, respectively, in the two strata. Thus, logs can provide artificial wiggle-trace presentations, provided the "signature" of the incident acoustic wave is known. When these are plotted on a time (rather than a depth) scale the result is a synthetic seismogram (fig. 16.26) which may be compared to an actual section for proper identification of reflector beds.

VERTICAL SEISMIC PROFILE

Another seismic application related to the sonic log is the *vertical seismic profile* (VSP). Reflections of compressional waves may be recorded by suspending a geophone in the wellbore and actuating an energy source at surface. Some of these waves will arrive at the geophone after having been reflected from beds *below* the bottom of the well. The VSP, therefore, affords a method of looking *ahead* of the drill bit. A schematic of the setup for a VSP survey is shown in figure 16.27; an example of the results is shown in figure 16.28.

CEMENT BOND LOG (CBL)

The cement bond log is also a sonic tool, similar in many respects to the BHC tool. Figure 16.29 illustrates the path of the compressional waves passing through the casing. The tool is set to work in a mode that measures the *amplitude* of the first arrival. Figure 16.30 illustrates the change in amplitude of the first-arrival peak when good bonding exists between the casing and the cement. The question of whether the cement sheath is bonded to the formation is solved by recourse to a wave-train display known as the VDL. A good bond between the cement and the formation is indicated when the later arrivals are strong, and can be visually identified by the sharpness of the VDL traces which are on the extreme right of the display in figure 16.31. More details on CBL tools and logs are given by Bateman (1984).

LOG QUALITY CONTROL

Sonic logs are subject to very easily detected errors such as the cycle skips and noise already discussed. More subtle errors can be pinned down if the log is run through marker beds such as a salt ($\Delta t = 67$ μs/ft), or anhydrite ($\Delta t = 50$ μs/ft) or into the casing where it should read 56 μs/ft (the travel time in steel).

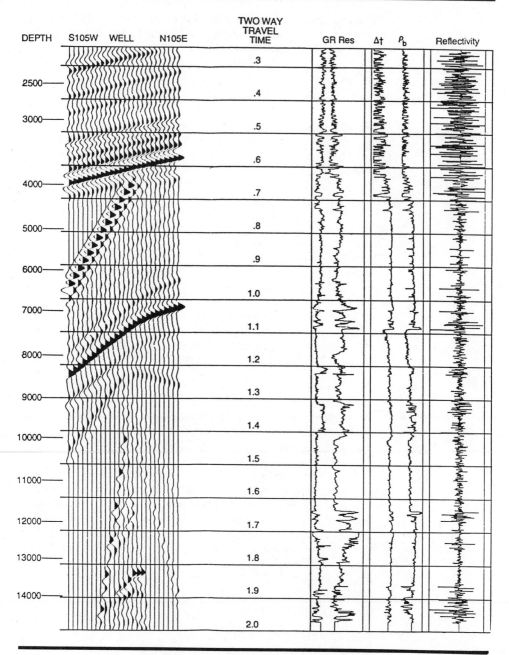

FIGURE 16.26 *Synthetic Seismogram from Density and Sonic Logs. Courtesy Schlumberger Well Services.*

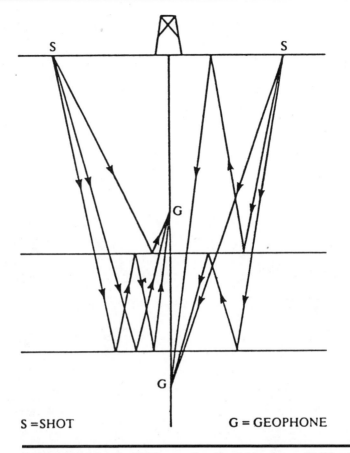

S = SHOT   G = GEOPHONE

FIGURE 16.27 *Setup for VSP Survey. Courtesy Schlumberger Well Services.*

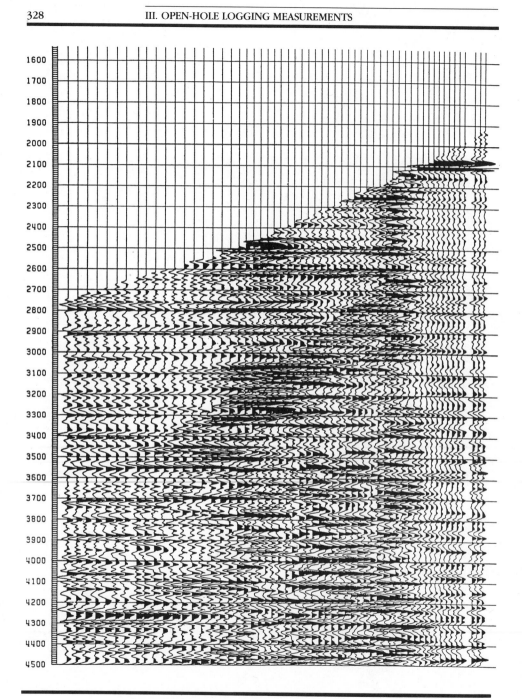

FIGURE 16.28 *Results of VSP Survey. Courtesy Schlumberger Well Services.*

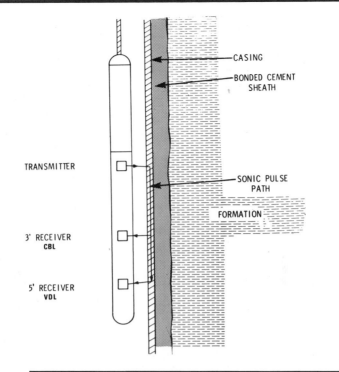

FIGURE 16.29 *CBL-VDL Tool. Courtesy Dresser Atlas, Dresser Industries, Inc.*

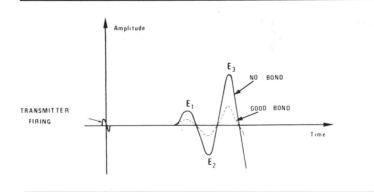

FIGURE 16.30 *Schematic of Receiver Output Signal with Bonded and Unbonded Casing. Courtesy Schlumberger Well Services.*

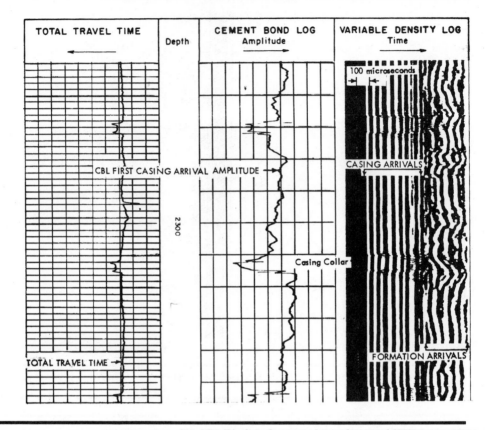

FIGURE 16.31 *CBL–VDL Display. Courtesy the Authors and SPWLA.*

APPENDIX 16A: *Compressional Wave Velocity and Interval Transit Time in Geologic Materials*

	Matrix Travel Time (μs/ft)		Matrix Velocity (ft/s)	
Material	Average Value	Range	Average Value	Range
Dunite	38.2	(34.7–41.1)	26,174	(24,305–28,807)
Gabbro	42.4	(42.4–47.6)	23,586	(20,998–23,586)
Hematite	42.9		23,295	
Dolomite	43.5	(40.0–45.0)	22,727	(22,222–25,000)
Norite	44.1	(43.5–49.0)	21,683	(20,400–22,967)
Diabase	44.6	(44.0–46.0)	22,435	(21,746–22,730)
Anorthosite	45.4		22,016	
Calcite	46.5	(45.5–47.5)	21,505	(21,053–22,000)
Aluminum	48.7		20,539	
Anhydrite	50.0		20,000	
Albitite	50.2	(49.5–50.6)	19,916	(19,752–20,212)
Granite	50.8	(46.8–53.5)	19,685	(18,691-21,367)

APPENDIX 16A : *(Continued)*

Steel	50.8		19,686	
Limestone	52.0	(47.7–53.0)	19,231	(18,750–21,000)
Langbeinite	52.0		19,231	
Iron	52.1		19,199	
Gypsum	53.0	(52.5–53.0)	19,047	(18,868–19,047)
Serpentine	53.9		18,702	
Quartzite	55.0	(52.5–57.7)	18,182	(17,390–19,030)
Quartz	55.1	(54.7–55.5)	18,149	(18,000–18,275)
Sandstone	57.0	(53.8–100.0)	17,544	(10,000–19,500)
Casing* (steel)	57.1		17,500	
Basalt	57.5		17,391	
Polyhalite	57.5		17,391	
Aluminum tube	60.9		16,400	
Trona	65.0		15.400	
Heulandite	65.1		15,355	
Halite	66.7		15,000	
Stilbite	68.0		14,699	
Sylvite	74.0		13,500	
Copper	78.7		12,700	
Carnallite	83.3		12,000	
Glacial ice	87.1		11,480	
Cement (wide variation)	95.0	(83.3–95.1)	10,526	(10,526–12,000)
Concrete (wide variation)	95.2	(83.3–125.0)	10,500	(8,000–12,000)
Shale	100.0	(60.0–170.0)	10,000	(5,882–16,667)
Anthracite coal	105.0	(90.0–120.0)	9,524	(8,333–11,111)
Bituminous coal	120.0	(100.0–140.0)	8,333	(5,906–10,000)
Sulfur	122.0		8,200	
Lead	141.1		7,087	
Lignite	160.0	(140.0–180.0)	6,250	(3,281–7,143)
Water				
200,000 ppm NaCl, 15 psi	180.5		5,540	
150,000 ppm NaCl, 15 psi	186.0		5,375	
100,000 ppm NaCl, 15 psi	192.3		5,200	
Pure	207.0		4,380	
Rubber (Neoprene)	190.5		5,248	
Kerosene, 15 psi	214.5		4,659	
Oil	238.0		4,200	
Methane, 15 psi	626.0		1,600	
Air, 15 psi	919.0		1,088	
Ethane (10°C) 0.00125 g/cc	989.6		1,010	
Carbon dioxide 0.0019776 g/cc	1176.5		850	

*Value of extensional ("first arrival") wave in thin rods.

BIBLIOGRAPHY

Asburn, B. E.: "Well Log Editing in Support of Detailed Seismic Studies," *Trans.*, SPWLA 18th Annual Logging Symposium, Houston, 1977.

Bateman, R.M.: *Cased-Hole Log Analysis,* IHRDC, Boston (1984).

Brown, H.D., Grigalva, V. E., and Raymer, L. L.: "News Developments in Sonic Wave Train Display and Analysis in Cased Holes," *Trans.,* SPWLA 11th Annual Logging Symposium, May 3–6, 1970.

Hicks, W.G., and Berry, J.E.: "Application of Continuous Velocity Logs to Determination of Fluid Saturation of Reservoir Rocks," *Geophysics* (July 1956) **21**.

Kokesh, F.P., and Blizard, R.B.: "Geometric Factors in Sonic Logging," *Geophysics* (February 1959) **24**.

Kokesh, F.P., Schwartz, R. J., Wall, W.B., and Morris, R. L.: "A New Approach to Sonic Logging and Other Acoustic Measurements," *J. Pet. Tech.* (March 1965) **17**.

Mason, K. L.: "Tricone Bit Selection Using Sonic Logs," SPE paper 13256, presented at the 59th Annual Technical Conference and Exhibition, Houston, Sept. 16–19, 1984.

Morris, R. L., Grine, D. R., and Arkfeld, T. E.: "Using Compressional and Shear Acoustic Amplitudes for the Location of Fractures," *J. Pet. Tech.* (June 1964) **16**.

Pickett, G.R.: "Acoustic Character Logs and Their Applications in Formation Evaluations," *J. Pet. Tech.* (June 1963) **15**.

Raymer, L.L., and Hunt, E.R.: "An Improved Sonic Transit Time-to-Porosity Transform," SPWLA 21st Annual Logging Symposium, Lafayette, LA, July 1980.

Thomas, D. H.: "Seismic Applications of Sonic Logs," 5th European SPWLA Logging Symposium, Paris, France, Oct. 20–21, 1977.

Tixier, M.P., Alger, R.P., and Doh, C.A.: "Sonic Logging," *J. Pet. Tech.* (May 1959) **11**.

Tixier, M.P., Alger, R.P., and Tanguy, D.R.: "New Developments in Inductions and Sonic Logging," *J. Pet. Tech.* (May 1960) **12**.

Wyllie, M.R.J., Gregory, A.R., and Gardner, G.H.F.: "Elastic Wave Velocities in Heterogeneous and Porous Media," *Geophysics* (January 1956) **21**.

Wyllie, M.R.J., Gregory, A.R., and Gardner, G.H.F.: "An Experimental Investigation of Factors Affecting Elastic Wave Velocities in Porous Media," *Geophysics* (July 1958) **23**.

Schlumberger: "Well Evaluation Developments, Continental Europe, 1982."

Schlumberger: "Seismic Applications," flyer.

Answers to Text Questions

QUESTION #16.1
$\phi_S = 19\%$

QUESTION #16.2
$V_f = 5650$ ft/s
$\Delta t_f = 177$ μs/ft

QUESTION #16.3
15 ms

CHAPTER SEVENTEEN

FORMATION DENSITY LOG

The density of a formation is one of the most important pieces of data in formation evaluation. In the majority of the wells drilled, it is used as the primary indicator of porosity. In combination with other measurements, data from the formation density log (FDL) may also be used to indicate lithology and formation-fluid type.

A conventional compensated density log is shown in figure 17.1, with the value of formation bulk density (ρ_b) in Tracks 2 and 3. The most frequently used scales are either 2.0 to 3.0 g/cc or 1.95 to 2.95 g/cc across two tracks. A correction curve, $\Delta\rho$, is sometimes displayed in Track 3 and, less frequently, in Track 2. Gamma ray and caliper curves usually appear in Track 1.

The compensated formation density (FDC) tool has an OD of 4⅝ in. and is rated 20,000 psi and 350°F. It can be used by itself or in combination with other tools, such as the compensated neutron tool (CNT). Figure 17.2 illustrates the articulated device (skid) that carries a gamma ray source and two detectors, which are referred to as the short-spacing and long-spacing detectors. The FDC tool is a pad-type tool—the backup caliper arm forces the skid against the side of the borehole.

OPERATING PRINCIPLE

Gamma rays are continuously emitted from the source. They pass through the mudcake and enter the formation, where they progressively lose energy until they are either completely absorbed by the rock matrix or return to one or the other of the two gamma ray detectors in the tool. Dense formations absorb many gamma rays; low-density formations absorb fewer gamma rays. Thus, high count rates at the detectors indicate low-density formations and low count rates at the detectors indicate high-density formations.

Gamma rays can react with matter in three distinct ways:

1. *Compton Scattering:* In this case, when a gamma ray collides with an electron orbiting some nucleus, the electron is ejected from its orbit and the incident gamma ray loses energy.
2. *Photoelectric Effect:* In this case, when a gamma ray collides with an electron it loses all its energy, and the electron is ejected from the atom.
3. *Pair Production:* In this case, a gamma ray interacts with an atom to produce an electron and a positron. These will later recombine to form another gamma ray.

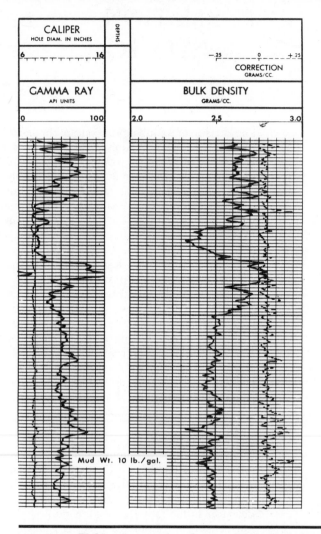

FIGURE 17.1 *FDC Log Presentation. Courtesy Schlumberger Well Services.*

If an incident beam of gamma rays strikes a target of thickness x, its intensity is reduced on passing through the target in such a way that

$$I_{out} = I_{in} \, e^{-\mu x},$$

where μ is the mass absorption coefficient (see fig. 17.3) This coefficient μ is a function of both the type of material in the target and the type of interaction that takes place. Figure 17.4 shows μ as a function of incident gamma ray energy for all three types of interaction. The conventional gamma ray source used in logging tools is made of cesium or cobalt, and the emitted gamma rays have an

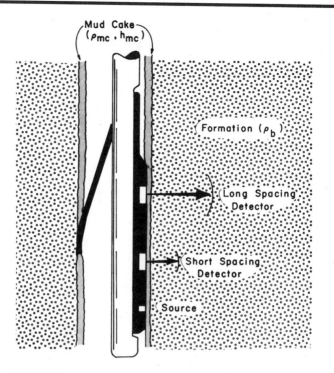

FIGURE 17.2 *Schematic of the Dual-Spacing Formation Density Logging Device (FDC). Courtesy Schlumberger Well Services.*

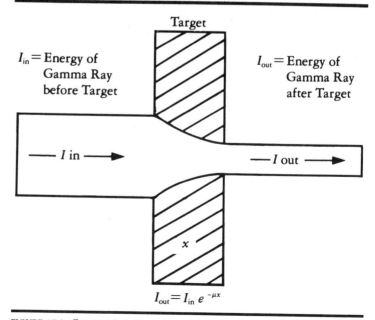

FIGURE 17.3 *Gamma Ray Mass Absorption Coefficient.*

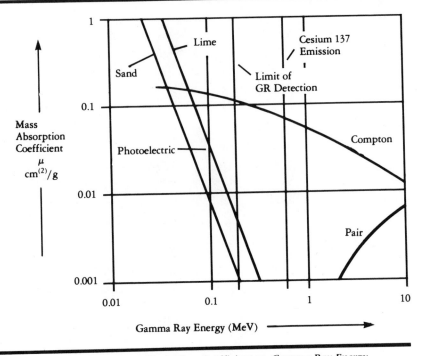

FIGURE 17.4 *Mass Absorption Coefficient vs. Gamma Ray Energy.*

energy of 0.63 MeV. Since pair production requires energies higher than 2 or 3 MeV, it is obvious from figure 17.4 that this type of interaction will not occur.

The detectors used in conventional density tools have a practical lower limit to the gamma ray energy level they can detect. This lower limit is about 0.2 MeV. Thus, the operating range is between the two vertical lines (fig. 17.4) marking the energy range between the gamma rays emitted from the source and the limit of detection by the detectors. Compton scattering, therefore, becomes the only possible form of gamma ray interaction that can be monitored by conventional density tools. The net effect of gamma ray Compton scattering and absorption is that the count rate seen at the detector is logarithmically proportional to the formation density (fig. 17.5).

$$\log(\text{count rate}) = A + B(\text{formation density}).$$

Both the near and the far spacing detectors behave in this way, so a plot of far count rates against near count rates will also produce a straight line, as shown in figure 17.6. Note that formation density increases as count rates decrease.

MUDCAKE COMPENSATION

If the mudcake has a density different from that of the formation, both the near and far count rates will change. Figure 17.7 shows

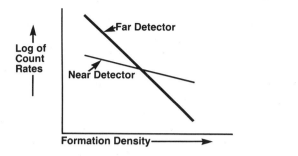

FIGURE 17.5 *Detector Count Rates vs. Formation Density Measurements.*

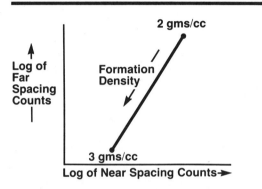

FIGURE 17.6 *Near and Far Count Rates of the Formation Density Tool.*

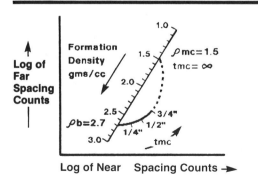

FIGURE 17.7 *Mudcake Compensation.*

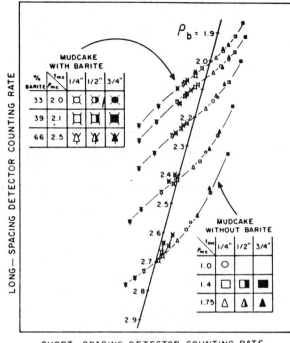

FIGURE 17.8 *Spine-and-Ribs Plot, Showing Response of Count Rates to Mudcake. Courtesy Schlumberger Well Services.*

where a plotted point would fall if a formation with a bulk density of 2.7 g/cc were to have an ever-increasing amount of mudcake of density 1.5 g/cc placed between it and the tool. In the extreme case of "infinite" mudcake thickness, both detectors will "see" only mudcake and read a value of 1.5 g/cc.

The arc describing the locus of the points is referred to as a *rib*. The zero mudcake line is referred to as a *spine*. A complete set of spine-and-ribs can be drawn for various thicknesses and densities of mudcakes, as shown in figure 17.8. Note that ribs also extend to the left of the spine for mudcakes having a density greater than the formation density; for example, in barite muds.

The surface equipment associated with the density tool computes the position of the point on the spine-and-ribs chart and then moves the point down the rib to intercept the spine. At this point, a corrected value of ρ is recorded for the log. The value of $\Delta\rho$ is calculated as the difference between ρ from the long spacing and ρ_{cor}. Thus, $\Delta\rho$ is positive in light muds and negative in heavy muds.

ELECTRON DENSITY

Since Compton scattering is an interaction between gamma rays and electrons, the density actually measured is the electron density ρ_e

TABLE 17.1 *A and Z Values for Common Elements*

Element	A	Z	$2(Z/A)$
H	1.008	1	1.9841
C	12.011	6	0.9991
O	16.000	8	1.0000
Na	22.99	11	0.9569
Mg	24.32	12	0.9868
Al	26.98	13	0.9637
Si	28.09	14	0.9968
Cl	35.46	17	0.9588
Ca	40.08	20	0.9980

Source: Schlumberger Well Services.

and not the bulk density ρ_b. Therefore, the ratio between electron density and bulk density must be investigated.

An atom of an element is distinguished by its atomic weight, A, and its atomic number, Z. For example, oxygen has an atomic weight of 16 and an atomic number of 8. Electron density is related to bulk density by the equation:

$$\rho_e = \rho_b \times 2(Z/A).$$

Thus, for oxygen, $\rho_e = \rho_b$, since $2(Z/A) = 1$. Most elements also have $2(Z/A)$ equal or close to 1, but there are important exceptions. The most important exception is hydrogen, which has a $2(Z/A)$ value of 1.9841. Table 17.1 lists the common elements along with their respective A and Z values. More values are listed in Appendix A.

The fact that hydrogen's $2(Z/A)$ value is closer to 2 than to 1 causes calibration of the density tool to be a rather special case. By definition, the tool should give correct porosities in fresh water-filled limestone. Since it measures ρ_e, the ρ_e value of both water and limestone must be known. For an element, $\rho_e = \rho_b \times 2(Z/A)$, but for a compound molecule,

$$\rho_e = \rho_b \; \frac{2Z}{\text{molecular weight}}.$$

For limestone, which has a formula of $CaCO_3$ and density of 2.71 g/cc,

$$\rho_e = \rho_b \; \frac{2(20 + 6 + 3 \times 8)}{40.08 + 12.011 + 3 \times 16}$$

$$= \rho_b \times 0.9991$$

$$= 2.71 \times 0.9991 = 2.7076.$$

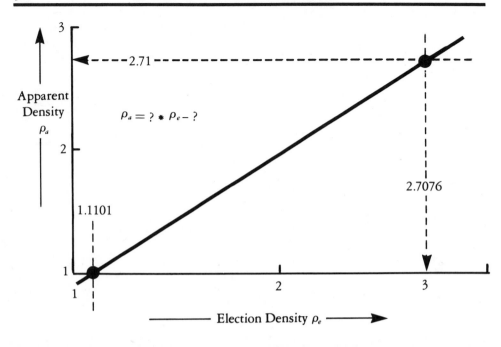

FIGURE 17.9 *Calibration of Density Tool in Fresh Water-Filled Limestone:* ρ_a vs. ρ_e.

For water, which has a formula of H_2O and a density of 1.0 g/cc,

$$\rho_e = \rho_b \frac{2(2 + 8)}{2.016 + 16}$$

$$= 1 \times 1.1101 = 1.1101.$$

Thus, the tool must be calibrated to read a ρ_e in water of 1.1101 and ρ_e in limestone of 2.7076. A linear conversion is all that is required:

$$\rho_a = A + B\rho_e,$$

where ρ_a will be the calibrated reading of the tool. By writing the two equations and solving them simultaneously, the values of A and B can be found:

$$1 = A + B \times 1.1101 \quad \text{and} \quad 2.71 = A + B \times 2.7076.$$

This can be solved graphically (fig. 17.9) or algebraically, resulting in values of $A = -0.1883$ and $B = 1.0704$. Using these values of A and B, the calibrated tool will now measure $\rho_b = \rho_a$ in water-filled limestone.

In summary, the formation density tool (FDC) reads electron density ρ_e. Since ρ_e is not exactly equal to ρ_b for all elements, a

TABLE 17.2 *Apparent Density Values for Various Minerals and Fluids*

Compound	Formula	Actual Density ρ_b	$\dfrac{2\Sigma Z}{\text{Mol. Wt.}}$	ρ_e	ρ_a (as seen by tool)
Quartz	SiO_2	2.654	0.9985	2.650	2.648
Calcite	$CaCO_3$	2.710	0.9991	2.708	2.710
Dolomite	$CaCO_3MgCO_3$	2.870	0.9977	2.863	2.876
Anhydrite	$CaSO_4$	2.960	0.9990	2.957	2.977
Sylvite	KCl	1.984	0.9657	1.916	1.863
Halite	$NaCl$	2.165	0.9581	2.074	2.032
Gypsum	$CaSO_4 2H_2O$	2.320	1.0222	2.372	2.351
Anthracite Coal		$\begin{cases} 1.400 \\ 1.800 \end{cases}$	1.030	$\begin{cases} 1.442 \\ 1.852 \end{cases}$	$\begin{cases} 1.355 \\ 1.796 \end{cases}$
Bituminous Coal		$\begin{cases} 1.200 \\ 1.500 \end{cases}$	1.060	$\begin{cases} 1.272 \\ 1.590 \end{cases}$	$\begin{cases} 1.173 \\ 1.514 \end{cases}$
Fresh Water	H_2O	1.000	1.1101	1.110	1.00
Salt Water	200,000 ppm	1.146	1.0797	1.237	1.135
"Oil"	$n(CH_2)$	0.850	1.1407	0.970	0.850
Methane	CH_4	ρ_{meth}	1.247	1.247 ρ_{meth}	1.335 ρ_{meth}—0.188
"Gas"	$C_{1.1}H_{4.2}$	ρ_g	1.238	1.238 ρ_g	1.325 ρ_g—0.188

Source: Schlumberger Well Services.

special calibration is made such that the tool reads correctly in fresh water-filled limestone. This is achieved by defining ρ_a (the apparent density) which is a linear function of ρ_e. Then $\rho_a = \rho_b$ for water and limestone.

As a result of this calibration technique, not all substances commonly found in rock formations are correctly read by the density tool. Table 17.2 gives a listing of density properties of various compounds frequently found in subsurface formations. The densities for oil and gas in table 17.2 are approximate in value.

QUESTION #17.1

For methane, $\rho_e = 1.247 \rho_b$. The value for ρ_a can be determined by

$$\rho_a = 1.0704 \rho_e - 0.1883.$$

Find ρ_a for methane $= ?$

QUESTION #17.2

If methane at a certain temperature and pressure has a density of 0.2 g/cc, what would the density tool read in pure methane?

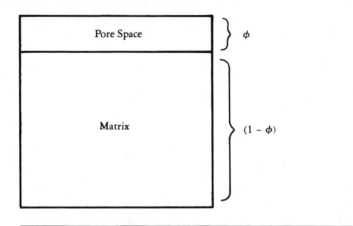

FIGURE 17.10 *Bulk-Volume Model of Porous Rock Formation.*

DENSITY POROSITY (ϕ_D)

Since the density of a mixture of components is a linear function of the densities of its individual constituents, it is a simple matter to calculate the porosity of a porous rock. For this purpose it is useful to consider the bulk-volume model of a clean formation with water-filled pore space as illustrated in figure 17.10.

Unit volume of porous rock consists of a fraction ϕ made up of water and a fraction $(1 - \phi)$ made up of solid rock matrix. Thus, the bulk density of the sample can be written as

$$\rho_b = \rho_{ma}(1 - \phi) + \rho_f \phi,$$

where ρ_{ma} refers to the matrix density and ρ_f refers to the fluid density. Simple rearrangement of the terms leads to an expression for porosity given by

$$\phi_D = \frac{\rho_{ma} - \rho_b}{\rho_{ma} - \rho_f},$$

where ϕ_D is used to denote a porosity derived from a measurement of formation density. The same concept can be illustrated graphically as in figure 17.11, where ρ_b is plotted against porosity. Note that points falling on the line connecting the matrix point (ρ_{ma}, $\phi = 0\%$) and the water point (ρ_f, $\phi = 100\%$) represent all possible cases extending from a zero-porosity rock matrix up to 100% porosity. Any intermediate value of ρ_b corresponds to some porosity ϕ.

Note that porosity, strictly speaking, is a decimal fraction and that the assumption is made that ρ_b is equal to ρ_a, the tool reading. Density porosity, given the assumptions made, can be found either using a calculator or graphically by use of figure 17.12, which

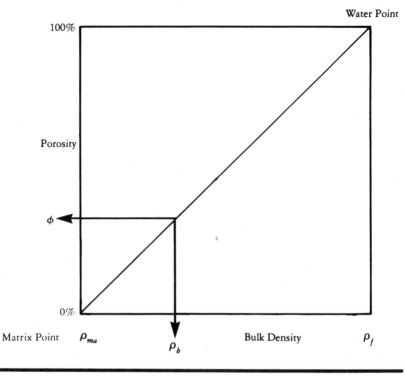

FIGURE 17.11 *Density Porosity Graph.*

generalizes the concept from figure 17.11 and builds in response lines for a number of different matrix materials and pore-filling fluids. To become familiar with this chart and its application, work the following examples:

QUESTION #17.3
In a dolomite formation,
ρ_{ma} = 2.87,
ρ_b = 2.44, and
ρ_f = 1.0 (fresh water).
Find ϕ_D = ?

QUESTION #17.4
In a sandstone formation,
ρ_{ma} = 2.65,
ρ_b = 2.40, and
ρ_f = 1.1 (salt-mud filtrate).
Find ϕ_D = ?

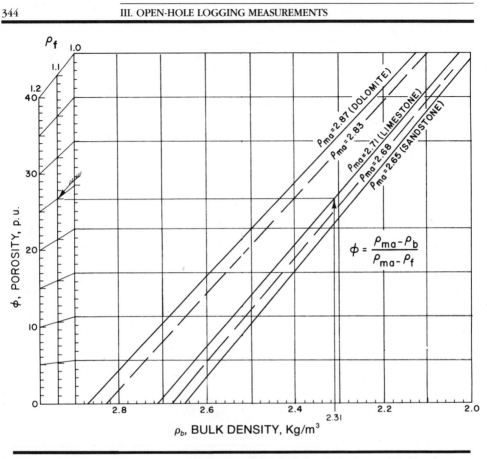

FIGURE 17.12 *Determination of Porosity from Formation Density Log. Courtesy Schlumberger Well Services.*

This derivation of porosity assumes a clean matrix of known density and water-filled pore space. These computed values of ϕ_D are incorrect when the lithology is mixed or unknown, when shale is present, when gas or light hydrocarbons are present in the flushed zone, and when pad contact with the formation is lost in washed-out holes.

Since the tool makes its reading in the zone adjacent the borehole wall, where mud filtrate has flushed away most of the original formation fluids, the choice for a value of ρ_f is dictated by the density of the mud filtrate, which in turn is a function of its salinity, pressure, and temperature. Figure 17.13 is useful for estimating ρ_f in salt water-filled formations.

QUESTION #17.5
Use figure 17.13 to find the density of sodium chloride solution at 240°F and 4000 psi if the concentration is 200,000 ppm.

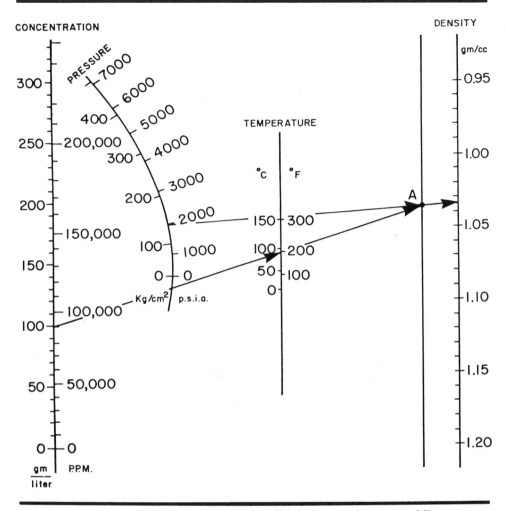

FIGURE 17.13 *Density of NaCl Solutions vs. Pressure and Temperature.*
Courtesy Schlumberger Well Services.

SHALE EFFECTS

When, in addition to matrix and fluid, a third component, shale, is
introduced, the same principle can be applied as before. Figure
17.14 illustrates a bulk-volume model of a shaly formation. The
volume fraction of shale is referred to as V_{sh}; the volume fraction of
matrix is $(1 - \phi - V_{sh})$. As before, an equation may be written for
the bulk density of such a mixture:

$$\rho_b = (1 - V_{sh} - \phi)\rho_{ma} + V_{sh}(\rho_{sh}) + \phi\rho_f.$$

Hence,

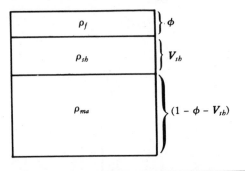

FIGURE 17.14 *Bulk-Volume Model of Shaly Formation.*

$$\phi = \frac{(\rho_{ma} - \rho_b) - V_{sh}(\rho_{ma} - \rho_{sh})}{\rho_{ma} - \rho_f}$$

or

$$\phi = \phi_D - V_{sh}(\phi_{D_{sh}}),$$

where $\phi_{D_{sh}}$ is the apparent density porosity of the shale.

Note that, since most shale densities are lower than common reservoir-rock matrix densities, ϕ_D in a shaly formation will always be greater than the true effective porosity. It is common practice among log analysts to correct density readings for shale effects by the simple expedient of estimating shale content (V_{sh}) from a gamma ray or SP log and reading the density tool response in a shale bed ($\phi_{D_{sh}}$). Although the procedure is not strictly valid, since the clay materials disseminated in a sand may not be the same as those deposited in a pure shale, it is still widely used at the quick-look level and gives satisfactory results in most cases.

In this context, it is worth remembering that shales are, in fact, porous formations albeit with effectively zero permeability and hence zero *effective* porosity. If the clay mineral in a shale has a matrix density of 3 g/cc (which is a good average), a shale with a bulk density of 2.5 g/cc has a total porosity of 25%. The same shale logged on an apparent sandstone-porosity scale would appear to have a porosity of about 9%.

GAS EFFECTS

To quantify the effect of gas on the response of the density tool, both the gas saturation and the effective density of the gas (ρ_g) must be known. Once again, a bulk-volume model is useful in quantifying the relationship between formation bulk density and the unknown quantities.

Given a clean, gas-bearing formation (see fig. 17.15), the expression for formation bulk density is

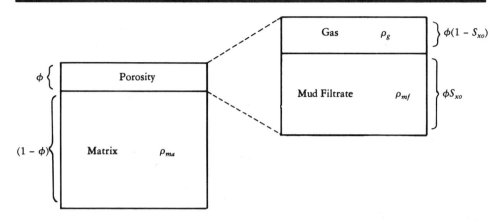

FIGURE 17.15 *Bulk-Volume Model of a Clean Gas-Bearing Formation.*

$$\rho_b = \rho_{ma}(1 - \phi) + \rho_{mf}\phi S_{xo} + \rho_g\phi(1 - S_{xo}).$$

Hence,

$$\phi = \frac{\rho_{ma} - \rho_b}{\rho_{ma} - \rho_g - S_{xo}(\rho_{mf} - \rho_g)}.$$

For a first approximation, it may be assumed that $\rho_{ma} \gg \rho_g$ and $\rho_{mf} \gg \rho_g$ and therefore

$$\phi \approx \frac{\rho_{ma} - \rho_b}{\rho_{ma} - S_{xo}}.$$

This still leaves ϕ undefined, since S_{xo} has yet to be determined. If the residual gas saturation is known or can be reasonably guessed at, ϕ can be approximated. However, a better approach is to confront the problem head on. As we have seen in chapter 14, S_{xo} can be determined from Archie's equation and a knowledge of ϕ, R_{xo}, and R_{mf}. The expression for ϕ in a gas-bearing formation also requires a value for S_{xo}. Until a value for ϕ is obtained, however, S_{xo} cannot be calculated. Fortunately, there is a way out of this seemingly endless loop. With some approximations and assumptions we can write

$$S_{xo} = \frac{(R_{mf}/R_{xo})^{1/2}}{\phi},$$

whence

$$\phi S_{xo} = (R_{mf}/R_{xo})^{1/2}.$$

If this expression for ϕS_{xo} is substituted into the equation for ϕ, we have

$$\phi = \frac{(\rho_{ma} - \rho_b) + (R_{mf}/R_{xo})^{1/2}(\rho_{mf} - \rho_g)}{\rho_{ma} - \rho_g}.$$

All the terms in this expression are either known or read from the appropriate log, with the exception of ρ_g, which may be calculated from a knowledge of gas gravity to air and reservoir temperature and pressure. An approximation for ρ_g is given by the algorithm

$$\rho_g = \frac{0.18}{(7644/\text{depth}) + 0.22}.$$

where depth is in feet and ρ_g is in g/cc. Alternatively, the graph of figure 17.16 may be used. Once a value for ρ_g has been determined, allowance must be made for the $2(Z/A)$ effects of the density tool calibration, and the actual ρ_g must be converted to an apparent density using

$$\rho_g \text{ (as seen by the density tool)} = 1.325\rho_g - 0.188.$$

QUESTION #17.6
In a gas-bearing sandstone at 10,000 ft the density log reads 1.99 g/cc, or 40% apparent sandstone porosity. Estimate the true porosity, given that R_{mf} at formation temperature is 0.2 $\Omega \cdot$m and R_{xo} is 20 $\Omega \cdot$m. Assume ρ_{mf} is 1.0 g/cc.

DEPTH OF INVESTIGATION
The depth of investigation of the density tool is quite shallow. Figure 17.17 compares the depth of investigation of the density tool with that of the MSFL and CNL tools. Most of the density-tool signal comes from a region less than 8 in. from the borehole wall. The CNL tool, by contrast, gathers most of its signal from the region within 12 in. of the borehole wall. Thus, the density tool is less affected by light hydrocarbons than is the CNL tool. In situations where deep invasion has occurred, there may be very little hydrocarbon effect on the density tool.

LITHOLOGIC DENSITY TOOL
The Litho-Density Log, run with the Litho-Density Tool* (LDT), is an improved and expanded version of the standard formation density log. In addition to the bulk density (ρ_b), the new tool also measures the photoelectric absorption index (P_e) of the formation. This new parameter enables a lithological interpretation to be made without prior knowledge of porosity.

---

*Mark of Schlumberger.

DEPTH IN FEET X 10³

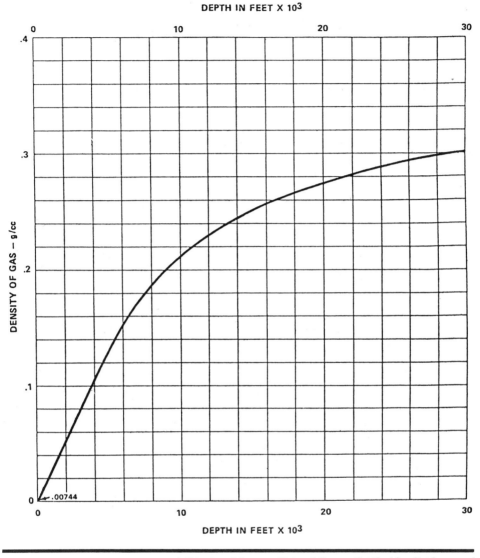

DEPTH IN FEET X 10³

FIGURE 17.16 *Density of Average Natural Gas vs. Depth. After Stephens and Spencer, Pennsylvania State University.*

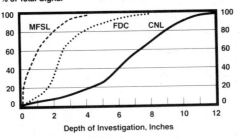

% of Total Signal

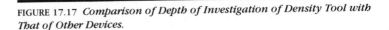

FIGURE 17.17 *Comparison of Depth of Investigation of Density Tool with That of Other Devices.*

Physical Principle

Gamma rays interact with matter in various ways depending upon their energy. However, only two reactions are of interest when dealing with relatively low energy gamma rays originating from the chemical sources presently used in logging tools. These reactions are the Compton scattering of gamma rays by electrons and the photoelectric absorption of gamma rays by electrons. Compton scattering has already been discussed in the context of the conventional density tool measurements.

The photoelectric effect occurs when a gamma ray collides with an electron and is absorbed in the process, so all of its energy is transferred to the electron. The probability of this reaction taking place depends upon the energy of the incident gamma rays and the type of atom. The photoelectric absorption index of an atom increases with increasing atomic number Z.

$$P_e = (0.1 \times Z_{\text{eff}})^{3.6}$$

The Compton effect occurs over a wide energy range, whereas the photoelectric effect occurs only when lower energy gamma rays are involved, as indicated in figure 17.18, which plots the mass absorption coefficient μ against the gamma ray energy. The figure also shows that the photoelectric effect, unlike Compton scattering, is dependent on the type of formation.

Measurement Theory

The LDT is similar to a conventional FDC device and uses a skid containing a gamma ray source and two gamma ray detectors held against the borehole wall by a spring-actuated arm (fig. 17.19). Gamma rays are emitted from the tool with an energy of 662 keV and are scattered by the formation, losing energy until they are absorbed through the photoelectric effect.

At a finite distance from the source, there is a gamma ray energy spectrum as shown in figure 17.20. This figure also shows that an increase in the formation density results in a decrease in the number

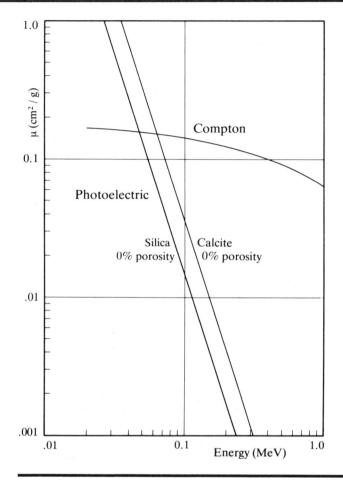

FIGURE 17.18 *Gamma Ray Mass Absorption Coefficients. Courtesy Schlumberger Well Services.*

of gamma rays detected over the whole spectrum. For formations of constant density but different photoelectric absorption coefficients (fig. 17.21), the gamma ray spectrum is only altered at lower energies.

Notice in figure 17.21 that region H only supplies information relating to the density of the formation, whereas region L provides data relating to both the electron density and the photoelectric absorption index. By comparing the counts in the energy windows H and L, the photoelectric absorption index can be measured. The gamma ray spectrum at the short-spacing detector is only analyzed for a density measurement, which is used to correct the formation density determined from the long-spacing spectrum for the effects of mudcake and rugosity.

The LDT skid and detector system produce greater counting rates than are obtained with the conventional density tools. Thus, there

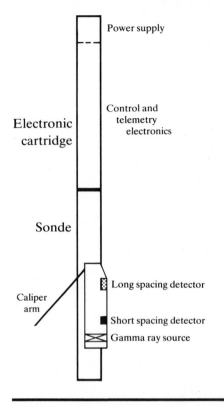

Power supply

Control and
telemetry
electronics

Electronic
cartridge

Sonde

Long spacing detector

Caliper
arm

Short spacing detector

Gamma ray source

FIGURE 17.19 *The Lithologic Density Tool. Courtesy Schlumberger Well Services.*

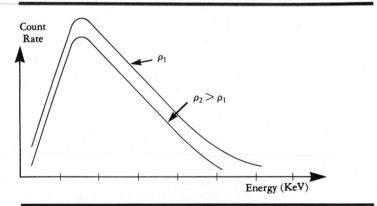

FIGURE 17.20 *Variation in Gamma Ray Spectrum for Formations of Different Densities. Courtesy Schlumberger Well Services.*

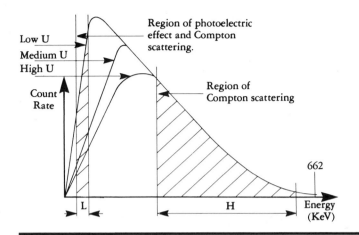

FIGURE 17.21 *Variation in Gamma Ray Spectrum for Formation of Constant Density but Different Photoelectric Capture Cross Sections. Courtesy Schlumberger Well Services.*

are lower statistical variations and better repeatability of the measurements. The geometry of the skid is such that the density reading has a sharper vertical resolution than that of the conventional density measurement. The P_e measurement also has a high resolution, which is useful in identifying fractures and laminar formations.

Interpretation of the P_e Curve

The photoelectric absorption coefficient is virtually independent of porosity, there being only a slight decrease in the coefficient as the porosity increases. Similarly, the fluid content of the formation has little effect. Simple lithologies, such as pure sandstone, anhydrite, etc., can be read directly from logs using the P_e curve (PEF) alone (see fig. 17.22). Look for the following readings in the most commonly occurring reservoir rocks and evaporites.

Material	P_e
Sand	1.81
Limestone	5.08
Dolomite	3.14
Salt	4.65
Anhydrite	5.05

Although there is a degree of variation in the log readings, due to impurities, four main lithologies can be identified in this example: sandstone up to 265 m, anhydrite from 255 to 210 m, dolomite from 210 to 185 m, and halite above 185 m. In some ambiguous cases, the density or neutron porosity readings must also be referred to. A more exhaustive treatment of interpretive technique for the P_e measurement is given in chapter 23.

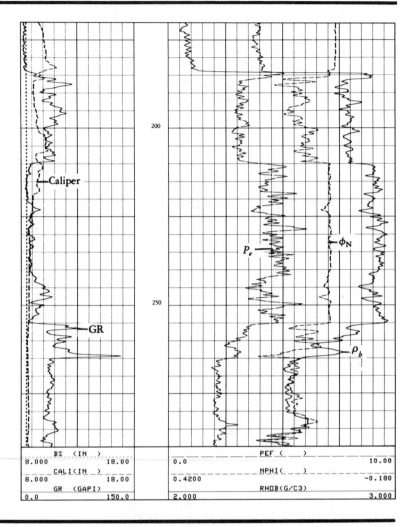

FIGURE 17.22 *Identification of Simple Lithologies with the Photoelectric Absorption Coefficient. Courtesy Schlumberger Well Services.*

DENSITY LOG QUALITY CONTROL

Practical calibration of the density tools is accomplished by a series of standards. The primary standard (fig. 17.23) is made by using laboratory formations. Since these cannot be easily transported, a set of secondary standards is available at logging service company bases in the form of aluminum and sulfur blocks of accurately known density and geometry. These blocks, weighing 400 lb, are not easily transportable, either, so a field calibrator containing two small gamma ray sources is used to reproduce the same count rates as those found in the aluminum block. The sulfur block is used as a check for the mudcake compensation.

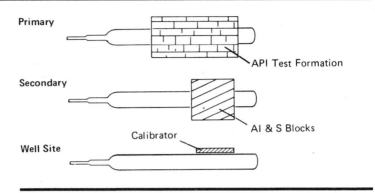

FIGURE 17.23 *FDC Calibration.*

Wellsite calibration should be performed before and after each log is run. Shop calibration should be run at least every 60 days and a copy of it attached to the main log. It is important to note that the field calibrator, the skid with the detectors, and the source form a matched set. If any item of the three does not match the serial numbers on the master calibration, the log should be rejected.

Natural benchmarks for checking the validity of a density log are salt, which has a ρ_a of 2.032 g/cc, and anhydrite, which has a ρ_a of 2.977 g/cc. These minerals may not appear in the wellbore being logged, and even if they do, they may not be 100% pure. So even these benchmarks should be used with caution. In general, density logs are either well calibrated (and therefore correct) or very noticeably bad.

Apart from the natural benchmarks, the next-best quality check is a review of the $\Delta\rho$ curve. If the short-spacing detector fails, then the whole compensation mechanism is thrown out of kilter. If $\Delta\rho$ is generally within the limits of ±0.05 g/cc, the log may be assumed to be correct. However, if $\Delta\rho$ is negative in light muds, something is wrong. Likewise, positive values for $\Delta\rho$ in heavy (barite) muds is a danger signal.

BIBLIOGRAPHY

Ellis, D., Flaum, C., Roulet, C., Marienbach, E., and Seeman, B.: "Litho-Density Tool Calibration," SPE paper 12048, presented at the Annual Technical Conference and Exhibition, San Francisco, Oct. 5–8, 1983.

Garner, J. S., and Dumanoir, J. L.: "Litho-Density Log Interpretation," *Trans.,* Paper N, SPWLA 21st Annual Logging Symposium, Lafayette, LA, July 1980.

Schlumberger: "Production Log Interpretation" (1973).

Schlumberger: "Log Interpretation Principles" (1972) ch. 8.

Schlumberger: "Litho-Density Tool Interpretation," M 086108 (1982).

Schlumberger: "Well Evaluation Developments, Continental Europe" (1982).

Schlumberger: "Log Interpretation Charts" (1984).

Sherman, H., and Locke, S.: "Effects of Porosity on Depth of Investigation of Neutron and Density Sondes," paper SPE 5510, presented at the Annual Meeting, Dallas, Sept. 28–Oct. 1, 1975.

Tittman, J., and Wahl, J. S.: "The Physical Foundations of Formation Density Logging (Gamma-Gamma)," *Geophysics* (April 1965).

Tixier, M. P., Raymer, L., and Hoyle, W.R.: "Formation Density Log Applications in Liquid-Filled Holes," *J. Pet. Tech.* (March 1962) **14**.

Wahl, J. S., Tittman, J., and Johnstone, C. W.: "The Dual Spacing Formation Density Log," *J. Pet. Tech.* (December 1964) **17**.

Watson, C. C.: "Numerical Simulation of the Litho-Density Tool Lithology Response," paper SPE 12051, presented at the Annual Technical Conference and Exhibition, San Francisco, Oct. 5–8, 1983.

Answers to Text Questions

QUESTION #17.1

$\rho_a = 1.335\,\rho_{\text{meth}} - 0.188$

QUESTION #17.2

$\rho_{\text{meth}} = 0.079$ g/cc

QUESTION #17.3

$\phi_D = 23\%$

QUESTION #17.4

$\phi_D = 16\%$

QUESTION #17.5

$\rho_{mf} = 1.11$ g/cc

QUESTION #17.6

ρ_g at 10,000 ft (from fig. 17.16) $= 0.21$ g/cc

ρ_g as seen by density tool $= 1.325 \times 0.21 - 0.188 = 0.090$

$\phi = 0.293$ or 29.3%

CHAPTER EIGHTEEN

NEUTRON LOGS

The neutron is a fundamental particle found in the nucleus of all atoms except hydrogen, which contains only a proton. The neutron has approximately the same mass as the proton but carries no electrical charge. These two properties, smallness of size and electrical neutrality, make it an ideal projectile for penetrating matter. Neutrons pass through brick walls and steel plates with the greatest of ease. They can pass through steel casing and also penetrate rocks. It was logical, therefore, that they should find a place in the arsenal of logging tools. Over the years, a number of logging tools have appeared that rely on one or another of many possible ways the neutron can interact with matter. In order to understand these different tools fully, a review of some aspects of nuclear physics is required.

Two categories of neutron sources are found in the logging industry: chemical sources and pulsed sources. Chemical sources are composed of two elements, which are in intimate contact with each other and react together to emit neutrons continuously. These sources are normally either plutonium-beryllium or americium-beryllium. Chemical sources need to be heavily shielded when not in use. Pulsed sources, on the other hand, are relatively harmless when not in use. They incorporate an ion accelerator and a target and can be activated by simply switching on the accelerator. Presently, the pulsed neutron sources are used for pulsed neutron logging and for some new, esoteric tools that measure inelastic neutron collisions (carbon/oxygen-type logs).

Near the chemical sources, neutrons are found with substantially all of their initial energy of several MeV; these are called *fast neutrons*. These neutrons interact with other atoms in several ways (which will be discussed later), and they lose energy with each collision. Eventually, the neutrons reach an intermediate level of energy where they have an energy of only a few eV; these neutrons are called *epithermal neutrons*. After yet more interactions, a neutron may be slowed down to a point where it has the same energy as the surrounding matter; this energy level is a direct function of the absolute temperature. Such neutrons are called *thermal neutrons*. They have energies in the range of 0.025 eV. It is at this stage that the neutron is ripe for capture. Neutron capture is accomplished when a nucleus "swallows" a neutron, gets "indigestion," and spits out a gamma ray. The whole sequence is summarized in table 18.1.

When a neutron slows to the thermal stage, it is captured by a nucleus that then emits a gamma ray. This gamma ray is called a

TABLE 18.1 *Stages in the Life of a Neutron*

Type	Energy	Encountered
1. Fast	Several MeV	Near source
2. Epithermal	A few eV	After collisions
3. Thermal	0.025 eV	At capture

capture gamma ray. The measurements of conventional neutron logging tools, with the exception of pulsed neutron tools, are all predicated on the spatial distribution of the neutrons or capture gamma rays they produce. Pulsed neutron logs monitor the distribution as a function of time.

The first neutron tools, known as gamma ray neutron tools (GNT), consisted of a chemical source and a single detector of neutron capture gamma rays. This tool was a qualitative indicator of porosity but was badly affected by hole size and the salinity of the borehole fluid and the formation water.

In an attempt to overcome these inherent problems, the sidewall neutron porosity (SNP) tool was introduced. It relied on a single detector of epithermal neutrons. This tool overcame the salinity problems in general but had its own unique problem in that mudcake could affect its readings, and estimation of the magnitude of the error was not always easy.

The compensated neutron log (CNL) was then introduced, with two detectors of thermal neutrons. It solved most of the defects of the previous tools, yet it too encountered problems—with formations containing thermal-neutron absorbers. The present state of the art is a CNL type of tool with dual detectors of epithermal neutrons; this tool may solve the problem of thermal-neutron absorbers. Figure 18.1 illustrates a generalized neutron logging tool.

The ability of a nucleus to slow down or capture a neutron is governed by a property known as its *cross section*. In general, the slowing down cross section for any given nucleus may be quite different from its capture cross section; i.e., although some elements may be good at stopping neutrons they may not be particularly hungry to swallow them once stopped. To add to the complexities of the process of slowing down and capture, it is found that the cross sections are also a function of the energy of the neutron involved in the process. Thus, an analysis of how fast neutrons are scattered in a subsurface formation and eventually captured is a complex task. Table 18.2 lists the main elements found in the logging environment together with their slowing-down and capture cross sections. Also listed are their natural abundance and the number of collisions required to slow a fast neutron down to the thermal energy level of 0.025 eV.

Two elements, hydrogen and chlorine, dominate the behavior of all neutron tools. Hydrogen, the element with a proton for a nucleus, provides the best material for slowing down a neutron. Simple mechanics reveals that when two balls collide, the maximum

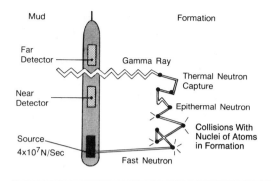

FIGURE 18.1 *Generalized Neutron Logging Tools.*

TABLE 18.2 *Slowing Down and Capture Cross Sections for 2 MeV Neutrons*

| Element Symbol | Abundance ppm | Cross Section | | Collisions to 0.025 eV |
		Capture	Slowing	
H	1,400	0.30	20.0	18
Be	—	0.01	6.1	87
B	—	700.00	3.0	105
C	320	0.00	4.8	115
N	—	1.88	10.0	130
0	466,000	0.00	4.1	150
Na	28,300	0.51	3.5	215
Mg	20,900	0.40	3.6	227
Al	81,000	0.23	1.5	251
Si	277,000	0.13	1.7	261
S	520	0.53	1.5	297
Cl	314	31.60	10.0	329
K	25,900	2.20	1.5	362
Ca	36,300	0.43	9.5	371
Fe	50,000	2.50	11.0	514
Cd	—	2500.00	5.3	1028

energy loss occurs when the two balls are of equal mass. The neutron and the proton, being of equal mass, account for the hydrogen nucleus' prodigious powers to slow down neutrons. (A good place to hide in case of nuclear attack is, therefore, under water, provided the problem of drowning has been solved first.) The other dominant element affecting neutron tools is chlorine, which is a voracious devourer of thermal neutrons, absorbing them a hundred times faster than most other elements. After accounting for the relative abundance of all the elements and their slowing-down cross sections and capture cross sections, it transpires that a neutron only has to collide with a hydrogen nucleus an average of 18 times to reach thermal energy. However, once it reaches thermal energy it is most likely to be absorbed by a chlorine nucleus. This explains why the original gamma ray neutron tools were so

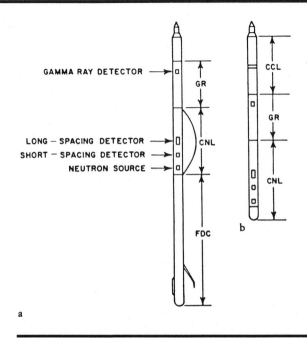

FIGURE 18.2 *Stacked Tool Arrangements for* (a) *Running a Gamma Ray, Compensated Neutron, and Compensated Density Log and* (b) *Running a Casing Collar Locator, Gamma Ray, and Compensated Neutron Log. Reprinted by permission of the SPWLA from Truman et al. 1972. SPWLA 13th Symposium.*

dependent on fluid salinity. A few parts per million of sodium chloride in the mud or the formation water could alter their response dramatically. It also explains why the SNP was such an improvement. The SNP was completely blind to capture gamma rays; it only detected epithermal neutrons. The CNL tools, theoretically, are just as blind to salinity effects, since they too ignore the capture gamma rays from chlorine. But inspection of table 18.2 shows that small additions of boron or cadmium in the formation can seriously affect the distribution of thermal neutrons.

THE COMPENSATED NEUTRON TOOL

The compensated neutron tool (fig. 18.2) measures a *neutron porosity index*, which can be related to the porosity of the formation only if the lithology and formation fluid content are known. The tool consists of a chemical neutron source and two detectors of thermal neutrons (fig. 18.3). It is run eccentered, with the source and detectors forced against the borehole wall by means of a bow spring.

Conventional compensated neutron (CNL) tools have a diameter of 3⅝ in. and are rated at 20,000 psi and 400°F. They can be run equally well in open or cased, liquid-filled holes. In an empty hole (gas filled), they do not work. It is normal practice to run these tools

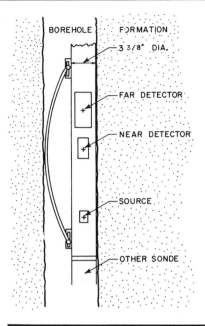

FIGURE 18.3 *Detail of a Compensated Neutron Tool. Reprinted by permission of the SPE-AIME from Alger et al. 1971, fig. 2.* © *1971 SPE-AIME.*

in combination with the density and gamma ray tools. If run alone, no caliper is recorded. Figure 18.4 shows a typical neutron–density log. The neutron porosity index is the dashed curve recorded in Tracks 2 and 3.

Broadly speaking, the compensated neutron log can be classified as an indicator of three things: porosity, lithology, and formation fluid type.

CNL OPERATING PRINCIPLE

To understand properly the operation of the CNL logging tool, the distribution of the thermal neutrons as they move away from the source must be investigated. The *thermal neutron flux* is defined as the number of thermal neutrons crossing unit area in unit time. This flux is controlled by the hydrogen content of the formation. Hydrogen is found in the water molecules filling the pore space (assuming the formation to be water bearing). Thus, the hydrogen content of the formation is a direct indication of the porosity of the formation.

Figure 18.5 shows a plot of the thermal neutron flux as a function of the distance from the source for three different values of porosity. Note that the lines cross over each other at some distance from the source. At points closer to the source than the crossover region, high thermal neutron flux means high porosity, but at points further from

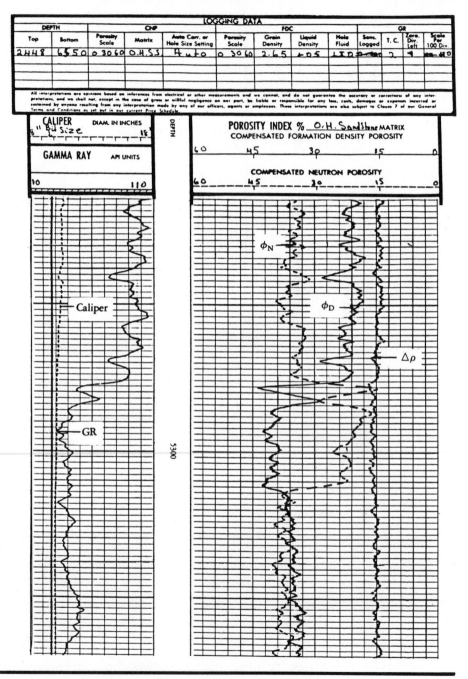

FIGURE 18.4 *Example of a Compensated Neutron Log (CNL).*

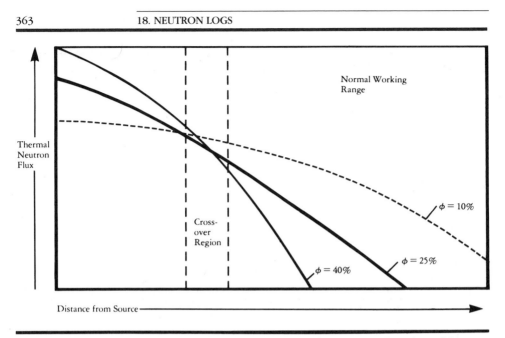

Thermal
Neutron
Flux

Normal Working
Range

$\phi = 10\%$

Cross-
over
Region

$\phi = 40\%$

$\phi = 25\%$

Distance from Source

FIGURE 18.5 *Thermal Neutron Distribution as a Function of Distance from the Source.*

the source than the crossover region, the reverse is true—that is, high thermal neutron flux indicates low porosity.

The absolute neutron count rate is a poor indication of porosity, since a number of factors affect it. The actual count rate seen at any detector spacing from the source is a function not only of porosity, but also of such environmental factors as the hole size, the mud weight, and the casing size and weight. Therefore, the CNL reading must be normalized to correct for unknown environmental effects. This is done by taking two readings of thermal neutron flux at different spacings and using them to define the slope of the response line. This slope is relatively unaltered by environmental effects, although the position of the response line on the graph may vary substantially in the y direction (see fig. 18.6). The primary measurement of the CNL tool is thus a ratio of two count rates. A high ratio is indicative of high porosity. The conversion of the ratio to a porosity value is based on laboratory experiments conducted with rock samples of known porosity. Figure 18.7 shows the results of such experiments.

To record porosity directly, it is necessary to convert the near-count/far-count ratio into a porosity (ϕ) value. We see from figure 18.7, for example, that a ratio of 2.0 means less than 10% porosity in dolomite or more than 20% in sandstone. The surface controls of the CNL tool allow the operator to select the matrix for which a porosity is required. If the operator chooses to run the CNL on a limestone setting, the conversion of ratio to porosity will follow the middle of the three response lines in figure 18.7. If, sub-

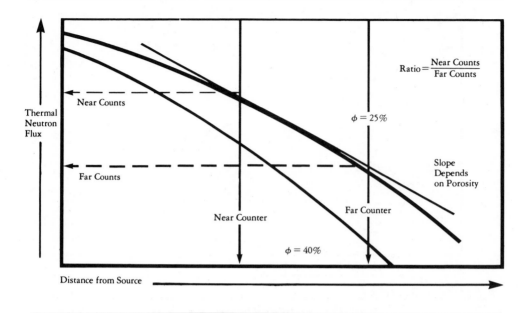

FIGURE 18.6 *CNL Borehole Compensation.*

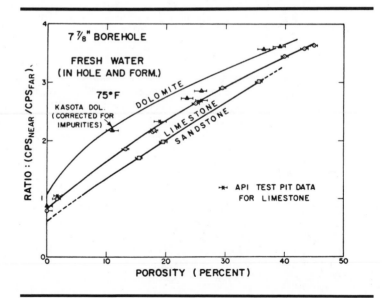

FIGURE 18.7 *Tool Response for Sandstone, Limestone, and Dolomite Laboratory Formations. Reprinted by permission of the SPE-AIME from Alger et al. 1971, fig. 4. © 1971 SPE-AIME.*

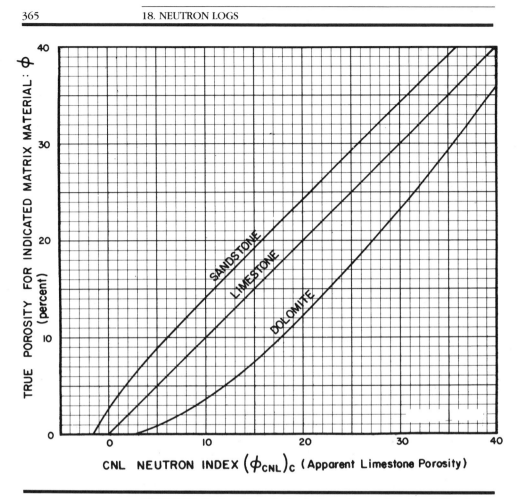

FIGURE 18.8 *Neutron Porosity Equivalence Curves. Courtesy Schlumberger Well Services.*

sequently, it is found that the actual matrix is not limestone, then the apparent limestone porosity must be converted to some other matrix porosity. This is accomplished by use of figure 18.8, which shows that the relationship between apparent limestone porosity and the porosity values for dolomite and sandstone is fairly uniform (with the exception of the very high and the very low porosity values). In the middle of the range of apparent limestone porosity values, certain approximate rules of thumb can be used. For example, if ϕ_{NLM} is the CNL porosity recorded with a limestone-matrix setting, and ϕ_T is the true porosity, table 18.3 can be used to convert the apparent porosity to dolomite and sandstone porosity values. The opposite can also be addressed; i.e., if the log is recorded on a matrix setting other than limestone, how can the recorded values be converted back to limestone? Table 18.4 provides the means to solve this problem. The following questions will acquaint the reader with these conversions.

TABLE 18.3 *Conversion to True Porosity*

True Porosity in	Correction		
Limestone	ϕ_T	$=$	ϕ_{NLM}
Sandstone	ϕ_T	$=$	ϕ_{NLM} + 4.0 pu
Dolomite	ϕ_T	$=$	ϕ_{NLM} − 7.5 pu

TABLE 18.4 *Conversion to Equivalent Limestone Response*

Panel Setting	Correction		
Limestone	ϕ_{NLM}	$=$	ϕ_{log}
Sandstone	ϕ_{NLM}	$=$	ϕ_{log} − 4.0 pu
Dolomite	ϕ_{NLM}	$=$	ϕ_{log} + 7.5 pu

QUESTION #18.1
The CNL is recorded on a limestone-matrix setting and gives a value of ϕ_{NLM} = 15%. However, the lithology is actually sandstone. What is ϕ_T, the true porosity?

QUESTION #18.2
The CNL is recorded on a sandstone-matrix setting and gives a value of ϕ_{NSD} = 22.5%. What would be the equivalent neutron porosity index (ϕ_{NLM})?

CNL ENVIRONMENTAL CORRECTIONS
In general, CNL corrections for environmental factors are small. Of all corrections, the temperature and borehole pressure corrections are the largest.

Figure 18.9 shows Schlumberger's open-hole correction chart. It is important to note on this chart that corrections depend on whether the log was run before or after January 1976 (at which point, tool modifications were introduced). The corrections are cumulative. The easiest way to become familiar with the chart is to run through a typical situation.

QUESTION #18.3
The log was run after January 1976. The automatic hole-size correction was applied. Mudcake is ½ in. thick. Mud weight is 10 lb/gal. Both borehole and formation salinities are 150,000 ppm. Tool standoff is zero, pressure is 10,000 psi, and the temperature is 250°F. If the log porosity reads 30%, what is the corrected porosity reading?

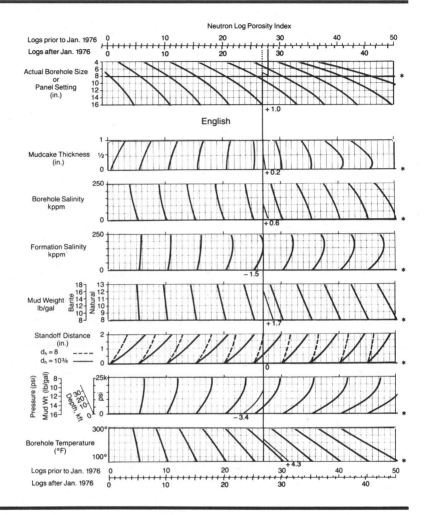

FIGURE 18.9 *Dual Spacing Neutron Log (CNL) Correction Nomograph for Open Hole. Courtesy Schlumberger Well Services.*

Cased-hole corrections are found using figure 18.10. Again, a practical exercise will serve to introduce the chart.

QUESTION #18.4

The open-hole diameter was 12¼ in. Casing size is 9⅝ in. OD, with a weight of 40 lb/ft. The cement thickness is 1.3 in. Other items are as for the open-hole example above. If the log porosity reads 20%, what is the corrected porosity reading?

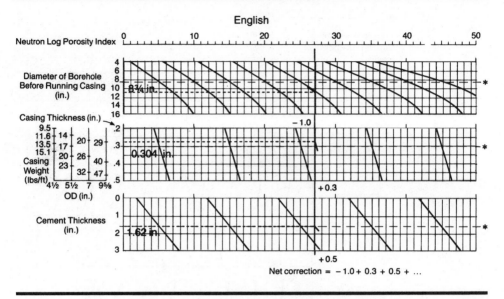

Net correction = − 1.0 + 0.3 + 0.5 + ...

FIGURE 18.10 *Dual Spacing Neutron Log (CNL) Correction Nomograph for Cased Hole. Courtesy Schlumberger Well Services.*

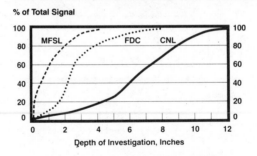

FIGURE 18.11 *Depth of Investigation of FDC, CNL, MSFL Tools (φ = 35%).*

DEPTH OF INVESTIGATION

The depth of investigation of the CNL tool depends on both the porosity and the salinity of the mud filtrate (invading water) and formation water. It can be expressed in terms of a geometric factor J:

$$\phi_{CNL} = J(\phi_{CNL})_{invaded} + (1 - J)(\phi_{CNL})_{uninvaded}.$$

J is a function of the depth of invasion measured from the borehole wall. Figure 18.11 shows a typical set of J curves for the MSFL, the FDC, and the CNL tools. Note that the CNL tool reads much deeper into the formation than either of the other two tools. If the radial distance within which 90% of the measured signal received is measured, the FDC might be found to read 5 in. into the formation and the CNL to be reading 9 in. into the formation. This is a generalization, however; the figures vary substantially depending on the formation fluid and the porosity.

MATRIX SETTINGS AND LITHOLOGY EFFECTS

When choosing a CNL logging scale and matrix setting, it is good practice to remain consistent with the standard operating procedure for the particular lithology expected in the well. This should be:

a. *Sand/Shale Sequences:* Run the log with a sandstone-matrix setting and with a porosity scale of 60% to 0%, left to right across Tracks 2 and 3. If a density log is also recorded, scale it as an apparent sandstone porosity on the same scale (see fig. 18.12).

b. *Carbonates/Evaporites/Generally Unknown Lithology:* Run the log with the CNL on a limestone-matrix setting with a porosity scale of 45 to –15%, left to right across Tracks 2 and 3. If a density log is also recorded, a scale of 1.95 to 2.95 g/cc is required (see fig. 18.13).

The advantage of the compatible sandstone scales is (1) that gas can be easily spotted (neutron curve reads less than the density curve) and (2) shales can be distinguished from sands (neutron reads more than density). The advantage of the compatible limestone scales is that the curves coincide in limestone, and the neutron curve reads less than the density curve in sands and more than the density in dolomite and shales. These scaling schemes are covered more fully in the sections dealing with porosity and lithology and in chapter 22.

GAS EFFECTS

Since the neutron tool is sensitive to hydrogen, it will respond with an abnormally low reading where the concentration of hydrogen is low—for example, in gas-bearing formations.

QUESTION #18.5
On the log shown in figure 18.12, find a gas-bearing interval and give its top and base.

Making corrections for gas effects on neutron logs requires additional data, such as a density log. The mechanics of making the corrections are covered in chapter 22.

In a manner that is entirely analogous to the approach taken with the density tool, the response of the neutron tool in gas-bearing formations may be written as:

$$\phi_N = \phi_{N_{ma}}(1 - \phi) + \phi_{N_{mf}}\phi S_{xo} + \phi_{N_{hy}}\phi(1 - S_{xo}),$$

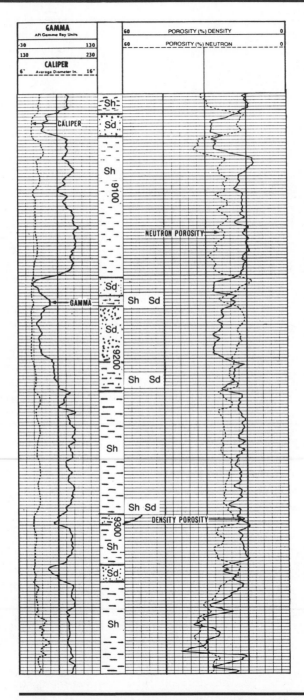

FIGURE 18.12 *Example of Neutron–Density Presentation in Sand/Shale Sequences. Courtesy WELEX, a Halliburton Company.*

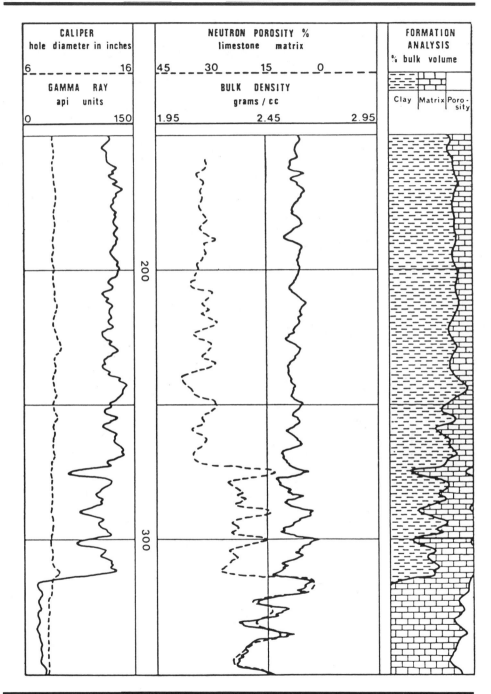

FIGURE 18.13 *Example of CNL–FDC Presentation in Complex Lithology.*
Reprinted by permission of the Indonesian Petroleum Association from
Goetz et al. 1977.

where

$\phi_{N_{ma}}$ is the response to matrix (usually considered equal to zero),

$\phi_{N_{mf}}$ is the response to mud filtrate (usually considered equal to one), and

$\phi_{N_{hy}}$ is the response to hydrocarbon.

Rearrangement of the terms gives

$$\phi = \frac{\phi_N - \phi_{N_{ma}}}{\phi_{N_{hy}} - \phi_{N_{ma}} + S_{xo}(\phi_{N_{mf}} - \phi_{N_{hy}})} .$$

If it is assumed that $\phi_{N_{ma}}$ and $\phi_{N_{hy}}$ are very small and $\phi_{N_{mf}} = 1$, then

$$\phi = \phi_N / S_{xo}$$

Since S_{xo} is always less than 1, it follows that ϕ_N in hydrocarbon-bearing formations is always less than ϕ.

SHALE EFFECTS

In general, response of the neutron log may be written in the form

$$\phi_N = \phi_T + V_{sh}\phi_{N_{sh}},$$

where:

ϕ_N is the log reading
ϕ_T is the true porosity
V_{sh} is the bulk volume of shale
$\phi_{N_{sh}}$ is the response of the neutron tool in pure shale.

Typical values of $\phi_{N_{sh}}$ for the CNL lie between 20 and 45%. The value of V_{sh} can be estimated from the gamma ray–SP–; or neutron–density combinations. A corrected porosity may thus be found using

$$\phi_T = \phi_N - V_{sh}\phi_{N_{sh}}.$$

OTHER NEUTRON LOGGING TOOLS

Not included in this discussion are a number of other neutron logging tools, including the pulsed neutron log, the inelastic gamma (carbon/oxygen) log, and the Chlorine log.* These tools find their applications mostly in cased holes, and the reader is referred to other sources (see, e.g., Bateman 1984) for a more detailed discussion.

---

*The Chlorine log is offered by N. L. McCullough.

Likewise, the discussion of old (GNT) neutron logs and their interpretation is left for another occasion. Dual-spacing epithermal neutron logs are, at the time of writing, in various stages of design and testing and no doubt more will be heard about these devices in the future.

CNL CALIBRATION AND QUALITY CONTROL

The CNL is calibrated in a large water-filled tank. This shop calibration is carried out at 60-day intervals. Practical quality control can be monitored by the usual criteria for all curves plus two natural benchmarks—salt ($\phi_{NLM} = -1.5\%$) and anhydrite ($\phi_{NLM} = 0.0\%$). Apparent neutron and density porosities should agree in clean, water-bearing zones where lithology is known.

BIBLIOGRAPHY

Alger, R. P., Locke, S., Nagel, W. A., and Sherman, H.: "The Dual Spacing Neutron Log—CNL," SPE paper 3565, presented at the 46th Annual Meeting, New Orleans, 1971.

Allen, L. S., Tittle, C. W., Mills, W. R., and Caldwell, R. L.: "Dual-Spaced Neutron Logging for Porosity," *Geophysics* (1967), **32**, 60–68.

Bateman, R. M.: "Cased-Hole Log Analysis and Reservoir Performance Monitoring" IHRDC, Boston, 1984.

Goetz, J. F., Prins, W. J., and Logar, J. F.: "Reservoir Delineation by Wireline Techniques," 6th Annual Convention of the Indonesian Petroleum Association, Jakarta, May 1977.

Poupon, A., Clavier, C., Dumanoir, J., Gaymard, R., and Misk, A.: "Log Analysis of Sand-Shale Sequences—A Systematic Approach," *J. Pet. Tech.* (July 1970) **23**, 867–881.

Schlumberger: "Log Interpretation Charts" (1972).

Schlumberger: "Log Interpretation Charts" (1984).

Segesman, F., and Liu, O.: "The Excavation Effect," SPWLA, *Trans.*, 12th Annual Logging Symposium, Dallas, May 1971.

Sherman, H., and Locke, S.: "Effect of Porosity on Depth of Investigation of Neutron and Density Sondes," SPE paper 5510, presented at the 50th Annual Meeting, Dallas, 1975.

Tittle, C. W.: "Theory of Neutron Logging," *Geophysics* (1961) **26**, 27–29.

Tittman, J., Sherman, H., Nagel, W. A., and Alger, R. P.: "The Sidewall Epithermal Neutron Porosity Log," *Trans.,* AIME (1966) **237**, 1351–1362.

Tixier, M. P., Martin, M., and Tittman, J.: "Fundamentals of Logging," Lecture 6, Petroleum Engineering Conference, University of Kansas, Lawrence, KA, 1956.

Truman, R. B., Alger, R. P., Connell, J. G., and Smith, R.L.: "Progress Report on Interpretation of the Dual-Spacing Neutron Log (CNL) in the U.S.," SPWLA *Trans.*, 13th Annual Logging Symposium, Tulsa, 1972.

Wahl, J. S., Nelligan, W. B., and Frentrop, A. H.: "The Thermal Neutron Decay Time Log," *Soc. Pet. Eng. J.* (December 1970) 365–379.

Wahl, J. S., Tittman, J., and Johnstone, C. W.: "The Dual Spacing Formation Density Log," *Trans.,* AIME (1964) **231**, 1411–1416.

Weinberg, A. M., and Wigner, E. P.: *The Physical Theory of Neutron Chain Reactors,* University of Chicago Press, Chicago (1958).

Welex: "Open-Hole Services," G-6003.

Answers to Text Questions

QUESTION #18.1
$\phi_T = 19.2\%$

QUESTION #18.2
$\phi_{NLM} = 18\%$

QUESTION #18.3
$\phi_{cor} = 34.5\%$

QUESTION #18.4
$\phi_{cor} = 30\%$

QUESTION #18.5
9052 and 9066 ft   or
9151 and 9161 ft

CHAPTER NINETEEN

DIPMETER SURVEYS

The dipmeter survey records data regarding the way in which subsurface layers of rock have been deposited and subsequently moved. The raw data consist of *orientation information,* which indicates where the downhole tool is located with respect to vertical and geographic coordinates, and *correlation information*, which can be used to determine the attitude of bedding planes with respect to the tool. The field log does not indicate formation dip. Computer processing of the raw data is required before any geological information can be extracted. The two important computer-processed parameters are the *bed-dip magnitude* and the *dip azimuth*. By studying the way in which these two parameters vary with depth, a great deal of valuable information may be obtained. It is worthwhile to review the definitions of dip and dip azimuth:

The *dip angle* is the angle formed between a normal to a bedding plane and vertical—a horizontal bed has a dip angle of 0° and a vertical bed has a dip of 90° (see fig. 19.1).

The *dip azimuth* is the angle formed between geographic north and the direction of greatest slope on a bedding plane. Dip azimuth is conventionally measured clockwise from north, so that a plane dipping to east has a dip azimuth of 90°, and one to west 270° (fig. 19.2).

APPLICATIONS
Dipmeter surveys have a variety of applications. At the lowest level, the raw data may be used:

To compute a deviation survey and true vertical depth
To compute the integrated hole volume
As an aide to fracture detection

At a higher level, computed dipmeter results may be used to determine:

Gross geologic structural features crossed by the wellbore
Sedimentary details within sand bodies
The depositional environment
True stratigraphic thickness and true vertical thickness

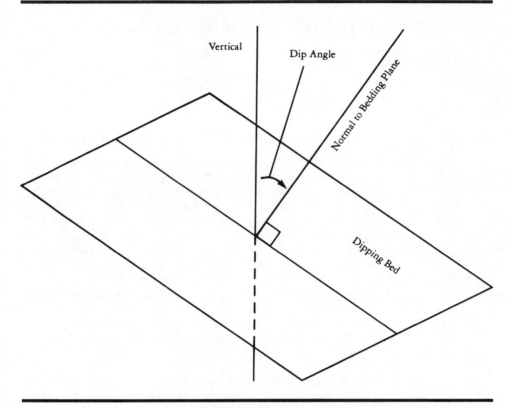

FIGURE 19.1 *Definition of Dip.*

At the highest level, computed dipmeter results from many wells may be combined to produce:

Structural cross sections
Trend surface maps

Thus, the most important applications of the dipmeter survey are in exploration drilling to help identify local structure and stratigraphy and in development drilling to help map the productive horizons and indicate directions to follow for further field development.

TOOLS AVAILABLE
A number of dipmeter tools are available. Historically, 3-arm dipmeter tools were used for many years; but these have now been entirely superseded by 4-arm and 6-arm tools. Figure 19.3 illustrates a commonly used 4-arm dipmeter tool. All currently used dipmeter tools have the following common characteristics.

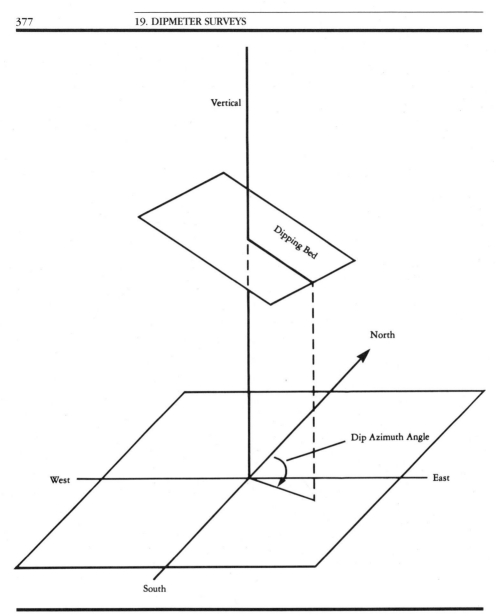

FIGURE 19.2 *Definition of Dip Azimuth.*

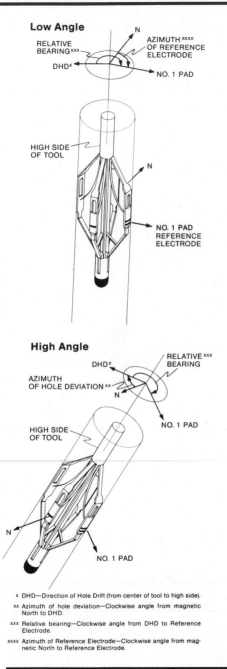

Low Angle

High Angle

x DHD—Direction of Hole Drift (from center of tool to high side).

xx Azimuth of hole deviation—Clockwise angle from magnetic North to DHD.

xxx Relative bearing—Clockwise angle from DHD to Reference Electrode.

xxxx Azimuth of Reference Electrode—Clockwise angle from magnetic North to Reference Electrode.

FIGURE 19.3 *4-Arm Dipmeter Tool. Courtesy Schlumberger Well Services.*

An orientation section that measures:
a. tool deviation from vertical
b. tool azimuth with respect to north
c. the orientation of the reference electrode pad to either north or the low side of the hole

A caliper section, which measures two or more hole diameters

A microelectrode array that records the resistivity of the formation in the very localized area where the pads contact the formation (correlation traces)

A gross-correlation device that will supply a moderately deep resistivity curve or a gamma ray or SP curve.

Orientation, until recently, was measured with a pendulum to indicate deviation from vertical and a magnetic compass to indicate tool rotation relative to magnetic north. Recently introduced tools use fluxgate magnetometers, gyroscopes, and/or accelerometers in order to deduce the tool position and orientation.

The microresistivity pads carry small button electrodes for water-based muds and knife-edge electrodes for oil-based muds, although the latter are not always very effective.

FIELD PRESENTATION

In the field the norm is to supply a 5-in. print of the orientation curves, the correlation traces, and the caliper curves. All data are recorded on magnetic tape. If the service company is authorized to compute the raw data, the field engineer will retain the digital data tape. Otherwise, it should be delivered along with the field prints to the operator's representative.

On rare occasions it may be desirable to compute dip results from the film rather than from the digital data tape, in which case a film on a very expanded scale (60 in. = 100 ft) will be required. Figure 19.4 illustrates a 5-in. dipmeter presentation; figure 19.5 illustrates the far more detailed 60-in. presentation.

DIPMETER COMPUTATION

The basic principle of dipmeter computation may be grasped by consideration of the case of a thin bed crossing a borehole. If the dip of the bed is greater than zero then at its intersection with the borehole it will appear shallower in the well on the up-dip side of the wellbore than on the down-dip side. If the bed in question has a resistivity different from the resistivity of the surrounding beds then a resistive (or conductive) anomaly will exist that appears higher on one side of the borehole than the other. The computation of dip and dip azimuth thus reduces to a problem of trigonometry. Any plane can be uniquely defined by three points in space. A 4-arm dipmeter provides four points. If the bedding planes are uniformly thick and plane at the intersection with the wellbore, only three of the dipmeter's available four points are necessary to compute a dip.

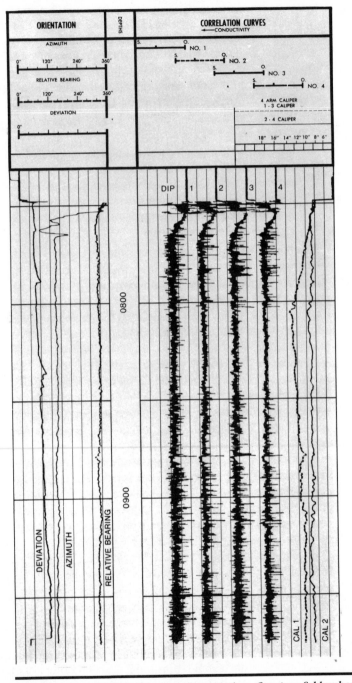

FIGURE 19.4 *5-in.-100 ft Dipmeter Presentation. Courtesy Schlumberger Well Services.*

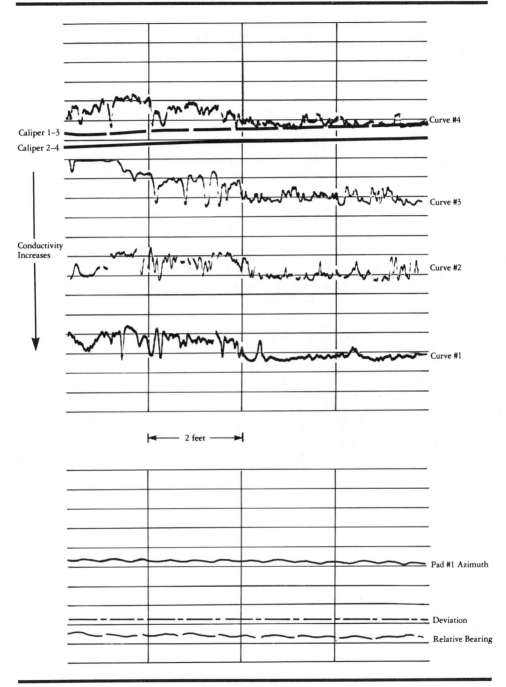

FIGURE 19.5 *60-in.-to-100 ft Dipmeter Presentation.*

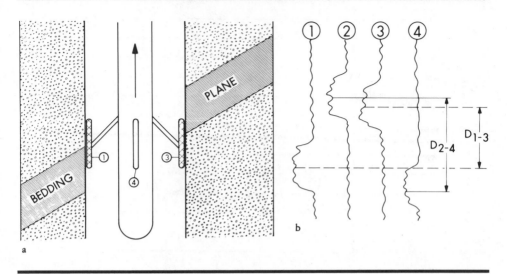

FIGURE 19.6 (a) *Cross Section of 4-Arm Dipmeter Tool and the Formation Around It.* (b) *Schematic of Dipmeter Log.*

However, it is frequently found that one or another of the correlation traces is substandard as a result of hole conditions or recording technique, and the four traces allow a margin of safety. Figure 19.6(a) shows a cross section of a borehole with a 4-arm dipmeter tool. Figure 19.6(b) is a schematic of the correlation curves that might be recorded by a 4-arm dipmeter. Displacement of an anomaly shown on one correlation curve to another correlation curve is the key to computing the formation dip. Figure 19.7 illustrates the displacements to be expected when a dipping plane cuts across a borehole.

Dip computation starts with the correlation of one trace to another in order to discover the relevant displacement. Correlation can be made optically using the 60-in.-per-100-ft record and a special apparatus known as an optical correlator, or by computer. Optical correlation is rarely used today since it requires a skilled specialist, is time consuming, and has no way of making allowance for tool acceleration and deceleration. Computer-based correlation can be made with a variety of techniques, such as pattern recognition, Fourier analysis, and conventional correlograms. The most commonly used technique builds correlograms: A short length of one correlation curve is compared with an equal length of another correlation curve. Starting some distance uphole and sliding down to some distance downhole, the two segments are cross multiplied. This correlation function reaches a maximum at the point where the curves coincide. Three parameters are used to control the process of correlation: the correlation interval, the search angle, and the step distance (see fig. 19.8).

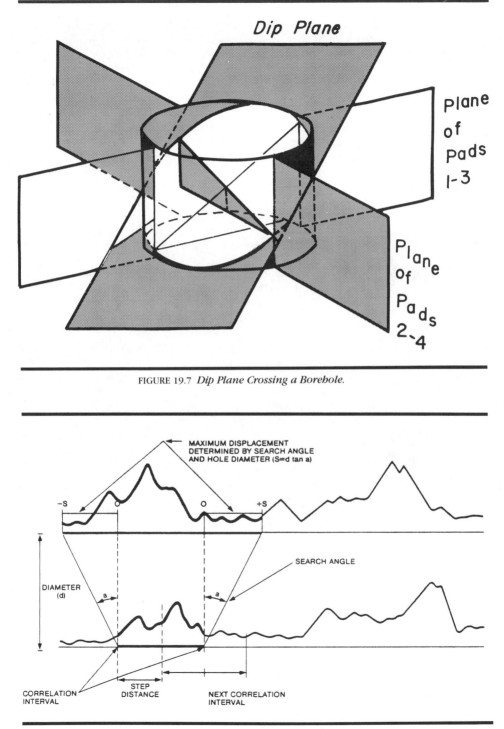

FIGURE 19.7 *Dip Plane Crossing a Borehole.*

FIGURE 19.8 *Correlation Interval, Search Angle, and Step Distance.*

The correlation interval is chosen to be from a fraction of a foot up to several feet, depending on the information sought. For fine stratigraphy, and if the quality of the raw data merits, a correlation interval of 3 in. to 2 ft may be used. For standard work, 2 to 6 ft may be used. For structural information, 6 to 18 ft may be used.

The search angle defines how far up and down the hole to seek a correlation, and, depending on the hole size, this in turn reflects the analyst's guess of the highest dip to be expected. Computer time becomes of interest at this point. To search for a 30° dip in a 12-in hole requires correlation to start 12 × tan 30° uphole and stop at 12 × tan 30° downhole (a search over 13.86 in.). A search for a 60° dip requires starting at 12 × tan 60° uphole and ending at 12 × tan 60° downhole (a search over 41.57 in.). So, though the dip searched for is doubled, the time taken has tripled! Some programs are structured to make an initial search in the 30 to 35° range and then search further only if a satisfactory correlation has not been found.

The step distance defines the depth increment to be used for the next round of correlations. It is usually set to half the correlation interval. Thus, a dipmeter computation at 4 ft × 2 ft × 35° means that a correlation interval of 4 ft was used with a step of 2 ft and a search angle of 35°.

Since only three points are required to define a plane, a 4-arm-dipmeter survey forms an overdetermined system. Any three curves of the four can provide a dip. Three items may be selected from a choice of four in twelve ways. Potentially, therefore, many dip computations may be made at the same depth. It is found in practice that these do not all agree. For the same reason that four-legged stools tend to wobble on an uneven floor but three-legged stools do not, a number of dip values are possible simply due to nature not providing us with bedding planes that are perfect planes at the scale of one borehole diameter. Add to this the effects of borehole rugosity, floating pads, and so on, and the result is a scatter of possible dips. The choice of the correct dip then becomes an exercise in common sense. In general, this exercise has come to be known as *clustering*. Simply stated: If at any level in the well the majority of the possible dips agree with each other *and* agree with the majority of the dips at adjacent levels in the well, then those are the most probable dips. The criteria for judging the worth of any type of dipmeter computation is of course its ability to reflect the known geologic facts.

COMPUTED LOG PRESENTATIONS

There are a number of ways of presenting the computed answers of a dipmeter survey. These include:

Tadpole or arrow plots
SODA plots
Listings

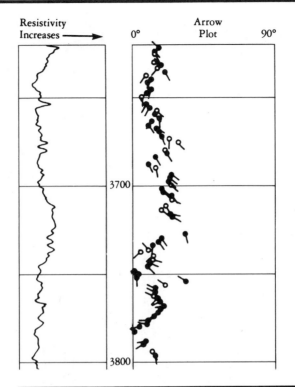

FIGURE 19.9 *Tadpole Plot. Courtesy Schlumberger Well Services.*

Azimuth frequency plots
Histograms
Polar plots
Stick plots
Stratigraphic plots

A typical *tadpole plot* is shown in figure 19.9. The dip magnitude is read from the position of the base of the tadpole on the plot. The dip azimuth is read by observing the direction in which the tail of the tadpole points. The azimuth convention is to measure angles clockwise from north. Thus, a north dip points uphole, an east dip to the right, a south dip downhole, and a west dip to the left.

SODA plots separate dip and azimuth as distinct points on separate tracks of the answer plot. A typical *listing* is shown in figure 19.10. In addition to the dip and dip azimuth, these listings may include further details such as dip quality, hole volume, and so on. *Azimuth frequency* diagrams (or rose plots) present statistical information regarding some depth interval in the well, usually 100 or 500 ft. Within that interval, a polar plot is built by plotting the number of dips with a dip azimuth of a particular direction in a circular histogram (see fig. 9.11). These diagrams are most useful for making a quick scan of the geologic column for trends in dip direction with

DEPTH	DIP ANGLE	DIP DIR	GRADE		HOLE ANGLE	HOLE DIR		HOLE DIAMETER 1-3	2-4	ECCENTRICITY SIZE	ECCENTRICITY DIR	INTEGRATED VOLUME
14003.0	1.7	206	1	•	2.4	102	•	10.8	10.5	0.23	332	297.08
14005.0	2.3	202	1	•	2.4	101	•	10.7	10.4	0.24	325	296.64
14007.0	5.2	302	1	•	2.4	102	•	10.6	10.3	0.24	316	295.43
14009.0	5.5	315	1	•	2.4	102	•	10.5	10.3	0.19	309	294.22
14011.7	4.9	310	1	•	2.4	102	•	10.4	10.3	0.14	301	293.06
14013.7	5.1	296	1	•	2.4	103	•	10.3	10.2	0.14	295	291.90
14015.7	2.3	313	1	•	2.2	104	•	10.1	10.1	0.00	•••	290.75
14017.7	2.4	313	1	•	2.2	105	•	9.9	10.0	0.14	13	289.64
14019.7	3.7	273	1	•	2.2	105	•	9.9	9.9	0.00	•••	288.56
14021.7	3.0	270	1	•	2.2	105	•	9.8	9.8	0.00	•••	287.49
14023.7	2.7	296	1	•	2.2	107	•	9.7	9.8	0.14	355	286.44
14025.7	3.3	299	1	•	2.2	107	•	9.7	9.7	0.00	•••	285.40
14027.7	1.6	303	1	•	2.2	109	•	9.6	9.7	0.14	344	284.38
14029.7	2.1	291	1	•	2.2	110	•	9.5	9.7	0.20	339	283.37
14031.7	3.2	283	1	•	2.2	109	•	9.4	9.6	0.20	331	282.36
14033.7	2.8	291	1	•	2.2	110	•	9.4	9.5	0.14	325	281.38
14035.7	1.9	330	1	•	2.2	110	•	9.4	9.4	0.00	•••	280.40
14037.7	2.0	328	1	•	2.2	110	•	9.4	9.5	0.14	312	279.44
14039.7	1.5	251	2	•	2.2	111	•	9.5	9.6	0.14	308	278.47
14041.7	3.0	275	1	•	2.2	111	•	9.5	9.6	0.14	304	277.47
14043.7	3.1	253	1	•	2.2	113	•	9.5	9.6	0.14	301	276.48
14045.7	2.4	247	1	•	2.2	114	•	9.4	9.5	0.14	298	275.48
14047.7	1.6	89	2	•	2.2	115	•	9.4	9.5	0.14	295	274.51
14049.7	2.8	20	1	•	2.2	115	•	9.5	9.5	0.00	•••	273.54
14051.7	2.9	24	1	•	2.2	114	•	9.6	9.6	0.00	•••	272.55
14053.7	0.8	221	1	•	2.2	114	•	9.6	9.7	0.14	285	271.55
14055.7	1.9	243	1	•	2.2	115	•	9.7	9.7	0.00	•••	270.53
14057.7	2.0	278	3	•	2.2	116	•	9.7	9.7	0.00	•••	269.51
14059.7	3.2	231	1	•	2.2	116	•	9.7	9.7	0.00	•••	268.48
14061.7	1.3	267	1	•	2.2	115	•	9.7	9.7	0.00	•••	267.45
14063.7	1.9	27	1	•	2.2	115	•	9.7	9.7	0.00	•••	266.43
14065.7	2.6	14	1	•	2.2	115	•	9.7	9.7	0.00	•••	265.41
14067.7	9.7	240	1	•	2.2	115	•	9.7	9.7	0.00	•••	264.30
14069.7	9.9	246	1	•	2.2	115	•	9.8	9.8	0.00	•••	263.35
14071.7	7.0	234	1	•	2.2	114	•	9.8	9.8	0.00	•••	262.31
14073.7	3.5	224	1	•	2.2	115	•	9.8	9.8	0.00	•••	261.26
14075.7	3.3	281	1	•	2.2	113	•	9.8	9.7	0.14	142	260.21
14077.7	2.7	284	1	•	2.2	112	•	9.7	9.7	0.00	•••	259.17
14079.7	0.3	297	1	•	2.2	111	•	9.6	9.7	0.14	219	258.15
14081.7	1.5	95	1	•	2.2	110	•	9.5	9.7	0.20	213	257.13
14083.7	3.6	111	2	•	2.2	108	•	9.5	9.6	0.14	206	256.13
14085.7	0.8	314	2	•	2.2	108	•	9.5	9.7	0.20	201	255.14
14087.7	2.4	243	1	•	2.2	109	•	9.5	9.7	0.20	196	254.13
14089.7	3.4	116	1	•	2.2	108	•	9.5	9.7	0.20	191	253.13
14091.7	2.0	45	1	•	2.2	106	•	9.5	9.7	0.20	185	252.12
14093.7	0.5	350	1	•	2.2	105	•	9.5	9.7	0.20	180	251.12
14095.7	0.8	357	1	•	2.2	104	•	9.5	9.7	0.20	173	250.11
14097.7	4.4	357	1	•	2.2	102	•	9.5	9.8	0.25	167	249.11
14099.7	3.1	107	2	•	2.2	99	•	9.5	9.8	0.25	160	248.09
14101.7	2.3	166	2	•	2.2	99	•	9.6	9.9	0.24	155	247.07
14103.7	4.4	107	1	•	2.2	99	•	9.6	9.9	0.24	150	246.04
14105.7	3.0	99	1	•	2.2	99	•	9.5	9.9	0.28	146	245.01
14107.7	1.5	246	1	•	2.2	98	•	9.5	9.9	0.28	142	243.98
14109.7	0.0	260	1	•	2.2	98	•	9.4	9.9	0.31	139	242.95
14111.7	2.4	293	1	•	2.2	98	•	9.4	10.0	0.34	136	241.94
14113.7	2.9	315	1	•	2.2	98	•	9.5	10.0	0.31	132	240.91
14115.7	2.5	322	1	•	2.2	99	•	9.6	10.1	0.31	128	239.88
14117.7	1.2	256	2	•	2.2	100	•	9.6	10.1	0.31	125	238.82
14119.7	1.8	264	1	•	2.2	101	•	9.7	10.2	0.31	123	237.76
14121.7	0.9	296	1	•	2.2	101	•	9.8	10.2	0.28	121	236.69
14123.7	2.4	266	1	•	2.2	101	•	9.8	10.2	0.28	119	235.60
14125.7	0.9	304	1	•	2.2	102	•	9.9	10.2	0.24	117	234.51
14127.7	0.7	344	1	•	2.2	102	•	10.3	10.5	0.19	112	233.41
14129.7	2.7	275	1	•	2.2	101	•	10.4	10.7	0.24	106	232.23
14131.7	1.6	270	1	•	2.2	101	•	10.5	10.7	0.19	101	231.01
14133.7	1.0	284	1	•	2.2	101	•	10.5	10.6	0.14	98	229.79
14135.7	1.0	207	1	•	2.2	102	•	10.3	10.4	0.14	97	228.57
14137.7	1.5	272	1	•	2.2	102	•	10.3	10.4	0.14	96	227.40
14141.7	2.1	264	1	•	2.2	101	•	10.4	10.6	0.19	95	225.06

FIGURE 19.10 *Dipmeter Listing. Courtesy Vizilog Inc.*

depth. Conventional *histograms* (fig. 19.12) of both dip and dip azimuth can also be presented.

Polar plots can be built in two ways. One way, the rose plot, has just been described. The other way is to scale the plot with zero dip at the outside and 90° at the middle. In this way, the azimuth of the *lowest* dip becomes more apparent. This type of plot (see fig. 19.13) is popular for picking structural dip. *Stick plots* (fig. 19.14) show a series of short lines inclined to the horizontal. Each line represents the dip angle as projected in some line of cross section. It is normal to distort the horizontal and vertical scales on these plots to fit the

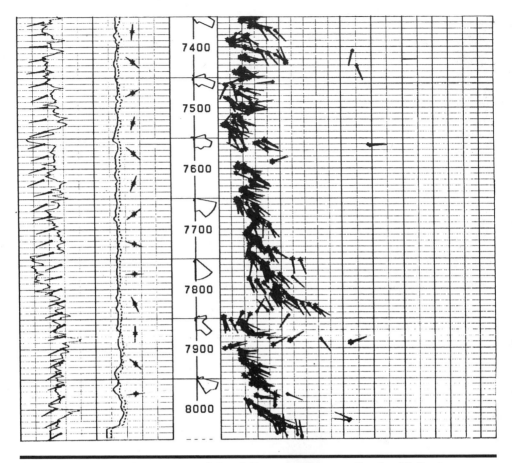

FIGURE 19.11 *Azimuth Frequency Diagrams. Courtesy Vizilog Inc.*

geologist's mapping requirements. Stick plots are normally produced and used in multiwell projects to draw cross sections. They are particularly helpful where the interwell correlation is not immediately obvious from the conventional logs alone. Note that, on a stick plot, a flat dip is flat whatever the azimuth of the section, but a highly dipping bed will appear (a) as an inclined line in the section viewed 90° from the dip azimuth but (b) flat if viewed in the direction of the dip azimuth.

Stratigraphic plots (fig. 19.15) attempt to give a visual representation of the stratigraphy of the beds. Each dip may be represented by the trace of the bedding plane on the borehole wall. If the trace could be "unwrapped" and laid on a flat surface, a sine wave would be visible, its amplitude a reflection of the dip magnitude and its low point an indication of the dip azimuth.

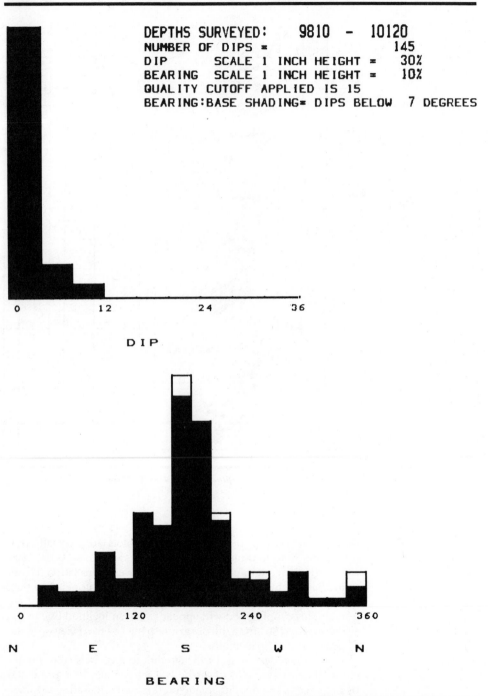

FIGURE 19.12 *Histogram of Dip and Dip Azimuth. Courtesy Vizilog Inc.*

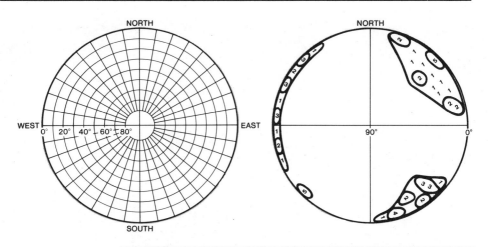

FIGURE 19.13 *Polar Plot. Courtesy Schlumberger Well Services.*

INTERPRETATION

Dipmeter plots may be interpreted by observing the variation of dip and dip azimuth with depth in conjunction with the open-hole logs. To aid in this type of analysis, it is helpful to highlight certain types of patterns. Conventionally, a group of dips of more-or-less constant azimuth showing an increase in dip magnitude with depth is colored red. A group of dips of more-or-less constant azimuth showing a decrease in dip magnitude with depth is colored blue. A group of dips of constant low-dip magnitude and more-or-less constant azimuth is colored green. Figure 19.16 illustrates these patterns.

Broadly speaking, dip interpretation may be split into two parts, structural and sedimentary. Gross structural characteristics, such as unconformities, folds, anticlines, and synclines, produce patterns that vary gradually over hundreds of feet. Sedimentary characteristics, such as cross-bedding, appear only within sedimentary beds and are localized to a few feet to tens of feet. Some of these patterns and their associated geologic features are described.

Folded Structure

Figure 19.17 shows a folded structure. Note that, in the shallow part of the well, dips are moderate and to the north. In the deeper section, the well has crossed the axial plane of the fold and dips are more pronounced and to the south. Where the well crosses the axial plane, dips are flat; it is here that a hydrocarbon trap exists. From the dips on the flanks, it is possible to compute both the tilt of the axial plane and the plunge of the fold.

Unconformity

Figure 19.18 illustrates an unconformity. A series of sediments in the deeper part of the well were originally deposited flat. Thereafter,

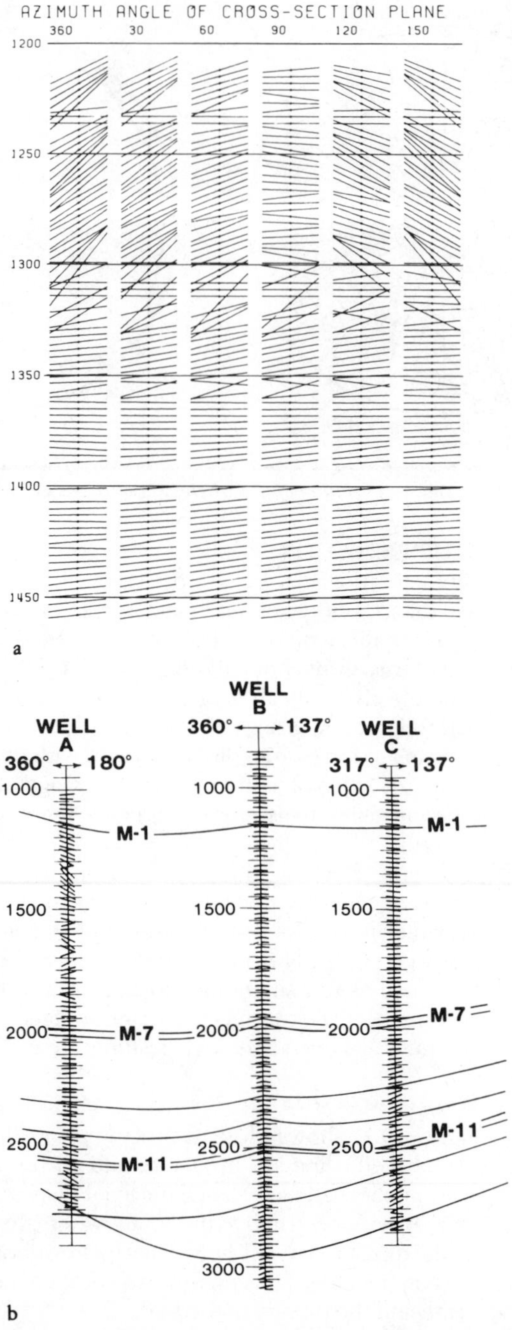

a

b

FIGURE 19.14 *Stick Plots in* (a) *Single Well (Six Different Sections) and* (b) *Three Wells (Helps Interwell Correlation). Courtesy Schlumberger Well Services.*

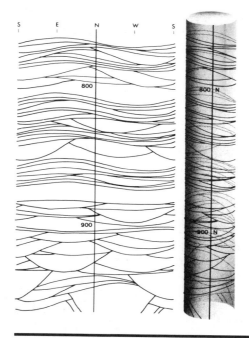

FIGURE 19.15 *Stratigraphic Plot. Courtesy Schlumberger Well Services.*

these sediments were tilted and then eroded and a new set of beds was deposited. At the interface between the old and new sediments, an abrupt change of dip takes place.

Faults

Faults may be picked from dip patterns by observing the drag patterns, if present, on either side of a fault. A normal fault with drag is diagrammed in-figure 19.19. Above the intersection of the well-bore with the fault, a red pattern will develop (dip increasing with depth). Below the intersection of the wellbore with the fault, a blue pattern will develop (decreasing dip with depth). At the intersection of the wellbore with the fault plane, the dip of the fault plane itself may be seen occasionally. Note that the fault dips down in the direction of the azimuth of the drag pattern; it strikes perpendicular to that direction.

Current Bedding

Among the sedimentary details that may be inferred from a dipmeter plot is the direction of transportation of sediments by streams. Figure 19.20 shows the sort of pattern to be expected in such a case. Here, blue patterns develop, with the dip azimuth in the patterns pointing downstream. Depending where the well is drilled, it may be of interest to move upstream toward the source or downstream to finer sediments or broader deposits.

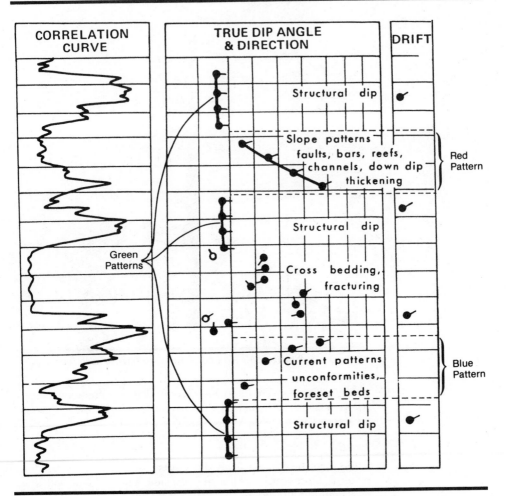

FIGURE 19.16 *Common Dip Patterns and Coloring Code.*

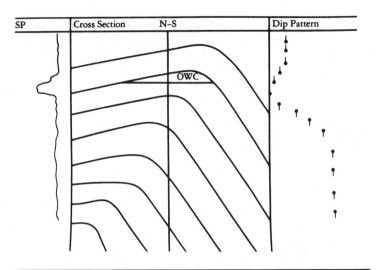

FIGURE 19.17 *Folded Structure.*

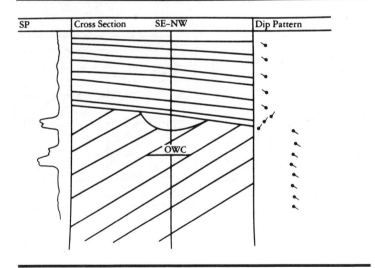

FIGURE 19.18 *Unconformity.*

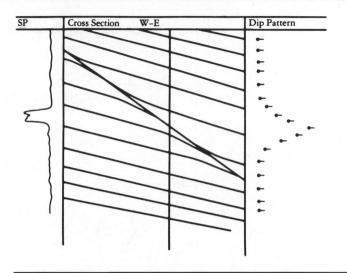

FIGURE 19.19 *Normal Fault with Drag.*

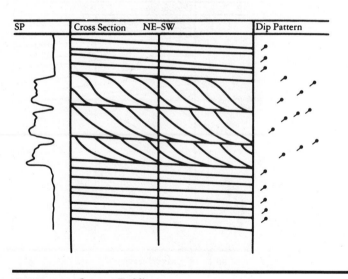

FIGURE 19.20 *Current Bedding.*

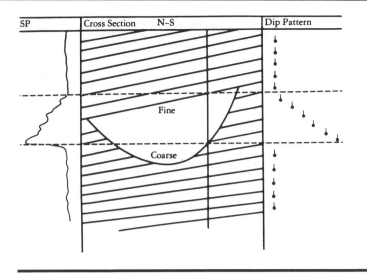

FIGURE 19.21 *Channel Cut and Fill.*

Channel Cut and Fill

A common type of deposit results when a channel is cut and then refilled with reservoir sand. A red pattern will develop together with a characteristic SP shape, broadening to the base. In drilling such plays, it is useful to know in which direction the channel extends and in which direction it thickens. Referring to figure 19.21, note that the well was drilled off the axis of the channel. Had it been drilled to the north, a thicker section of sand would have been found. To move to the center of a channel, offset the well in the same direction that the red-pattern tadpoles point. To follow the channel, move at right angles to the red-pattern dip azimuth, in this case either east or west.

Buried Bar with Shale Drape

Another common feature is a buried bar over which subsequent shale deposits have been draped. Here, dips within the sand body decrease with depth (blue); but above the sand body, dips in the shale increase with depth (red). The SP will usually show a characteristic pattern, broad at the top. To follow the bar, wells should be offset at right angles to the dip azimuth seen within the bar. To drill a thicker section, a well should be offset in the opposite direction to the dip seen in the bar.

FRACTURE FINDING

Another application of the dipmeter survey is in the detection of fractures. There are many methods available for fracture detection, and none, on its own, is a completely reliable diagnostic. The dipmeter should be viewed as just one of many methods for fracture finding and should be used to complement other methods.

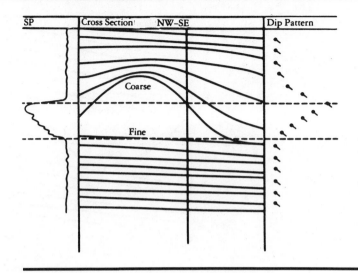

FIGURE 19.22 *Buried Bar with Shale Drape.*

The theory is very simple. A fracture will be invaded with mud filtrate and will, therefore, offer a less resistive path to electric current. If one of the dipmeter pads happens to lie in front of a fracture, it will record a low resistivity value. Another pad at the same depth may not be in front of a fracture and will record a higher resistivity. Comparison of adjacent pad traces should reveal the presence of a fracture when the two resistivity values are different. Various ways are available to display the curves in order to highlight such differences. Figure 19.23 shows one such presentation. Note that since the orientation of the dipmeter tool is known, the orientation of the fracture can also be deduced.

LOG QUALITY CONTROL
Good dip information requires good raw data. To ensure good raw data, the following guidelines are suggested.

1. Recondition hole prior to running the dipmeter.
2. Use a swivel-head adapter to reduce tool rotation while logging.
3. Log at 1800 to 2400 ft/hour to reduce tool jerking. Slow down even more if the tension on the line is erratic.
4. Reject sections of log where the tool rotates once in less than 60 ft of hole.
5. Make repeat sections and/or overlaps of 100 to 200 ft every time logging is stopped for film or tape changes.
6. Inspect the raw log for dead correlation curves, insensitive curves, stuck calipers, and so on. At last resort, three good correlation curves are sufficient, but four are much better.

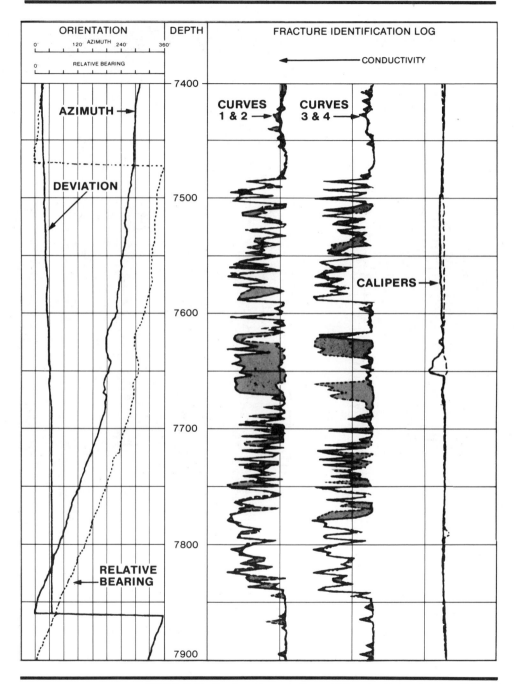

FIGURE 19.23 *Fracture Presentation of Dipmeter Data. Courtesy Schlumberger Well Services.*

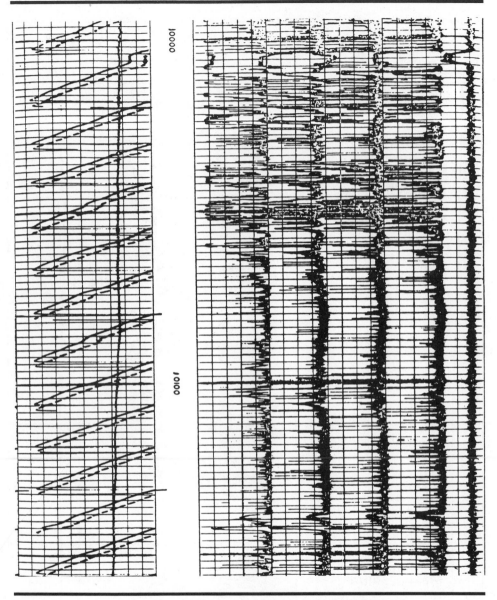

FIGURE 19.24 *Dipmeter Field Print—Excessive Tool Rotation.*

7. Carefully inspect the orientation curves for nonsense readings such as a hole deviation less than zero.
8. On a computed log, check the dip against field controls for consistency. Many dipmeter surveys have been off by 90 or 180° because of incorrect pad wiring or erroneous computation.

Figure 19.24 shows a field print of a dipmeter survey. From the orientation curves, it is evident that the tool was rotating too frequently. Such a log should be rerun.

APPENDIX 19A: DIPMETER INTERPRETATION RULES

METHOD OF PLOTTING

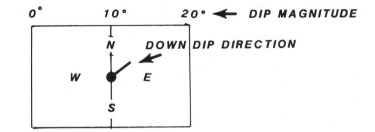

EXAMPLE 10° N45°E

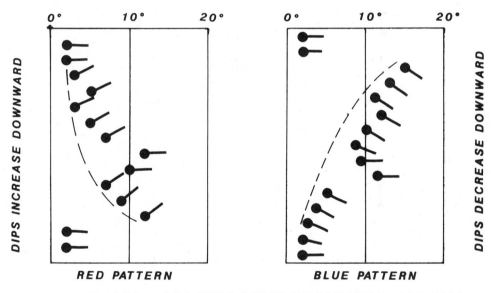

RED PATTERN   BLUE PATTERN

THESE PATTERNS, ALONG WITH OTHER INFORMATION, ARE USED FOR BOTH FAULT AND STRATIGRAPHIC INTERPRETATIONS

Appendix 19A has been reproduced with kind permission of J. A. Gilreath, Schlumberger Offshore Services, New Orleans, Louisiana.

MISSING and REPEAT SECTIONS

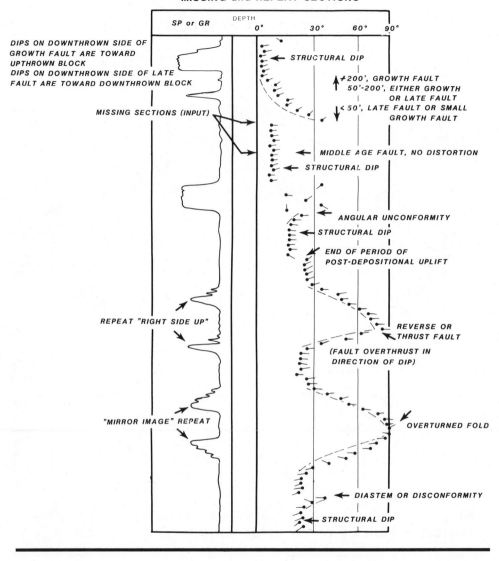

STRATIGRAPHIC INTERPRETATION
CONTINENTAL ENVIRONMENT

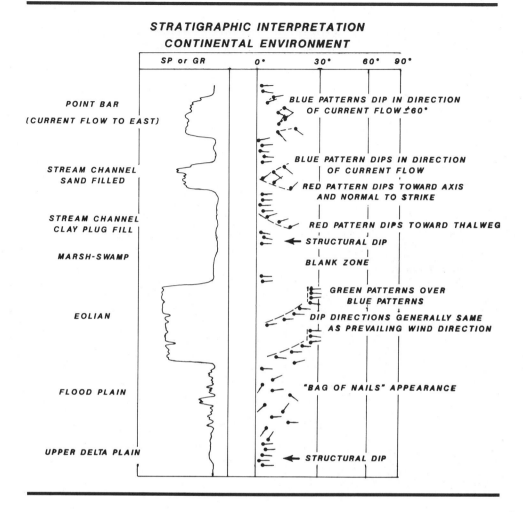

STRATIGRAPHIC INTERPRETATION
CONTINENTAL SHELF DELTA DOMINATED

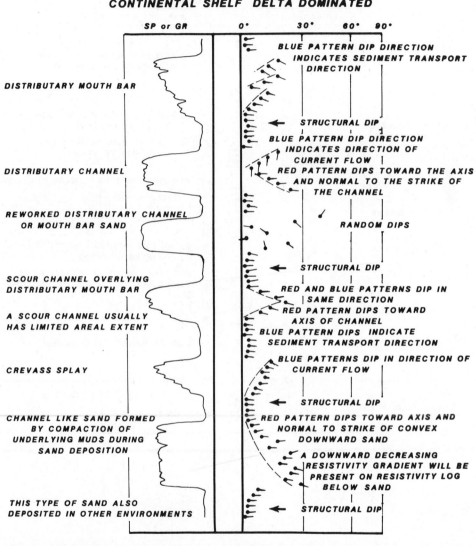

SP or GR   0°   30°   60°   90°

DISTRIBUTARY MOUTH BAR

BLUE PATTERN DIP DIRECTION
INDICATES SEDIMENT TRANSPORT
DIRECTION

← STRUCTURAL DIP

DISTRIBUTARY CHANNEL

BLUE PATTERN DIP DIRECTION
INDICATES DIRECTION OF
CURRENT FLOW
RED PATTERN DIPS TOWARD THE AXIS
AND NORMAL TO THE STRIKE OF
THE CHANNEL

REWORKED DISTRIBUTARY CHANNEL
OR MOUTH BAR SAND

RANDOM DIPS

← STRUCTURAL DIP

SCOUR CHANNEL OVERLYING
DISTRIBUTARY MOUTH BAR

RED AND BLUE PATTERNS DIP IN
SAME DIRECTION
RED PATTERN DIPS TOWARD
AXIS OF CHANNEL

A SCOUR CHANNEL USUALLY
HAS LIMITED AREAL EXTENT

BLUE PATTERN DIPS INDICATE
SEDIMENT TRANSPORT DIRECTION

BLUE PATTERNS DIP IN DIRECTION OF
CURRENT FLOW

CREVASS SPLAY

← STRUCTURAL DIP

CHANNEL LIKE SAND FORMED
BY COMPACTION OF
UNDERLYING MUDS DURING
SAND DEPOSITION

RED PATTERN DIPS TOWARD AXIS AND
NORMAL TO STRIKE OF CONVEX
DOWNWARD SAND

A DOWNWARD DECREASING
RESISTIVITY GRADIENT WILL BE
PRESENT ON RESISTIVITY LOG
BELOW SAND

THIS TYPE OF SAND ALSO
DEPOSITED IN OTHER ENVIRONMENTS

← STRUCTURAL DIP

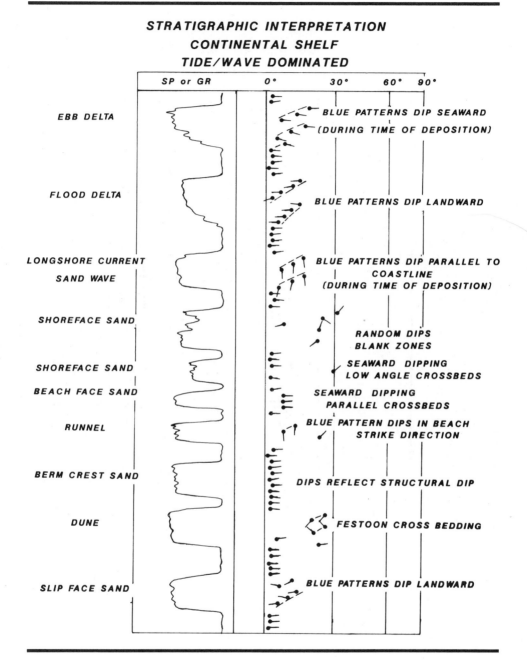

STRATIGRAPHIC INTERPRETATION
CONTINENTAL SHELF
TIDE/WAVE DOMINATED

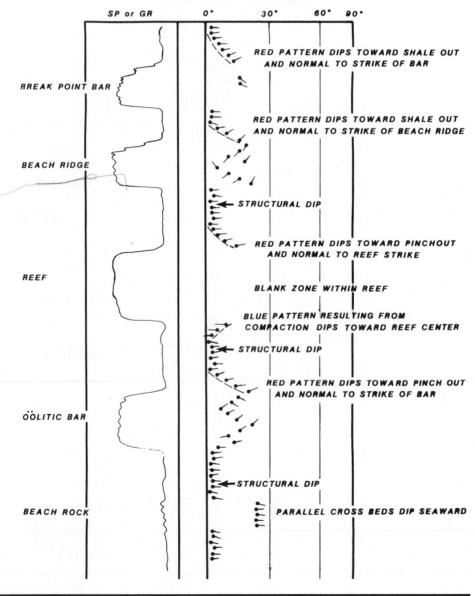

STRATIGRAPHIC INTERPRETATION

CONTINENTAL SHELF

TIDE/WAVE DOMINATED

CONTINENTAL SLOPE AND ABYSSAL ENVIRONMENTS

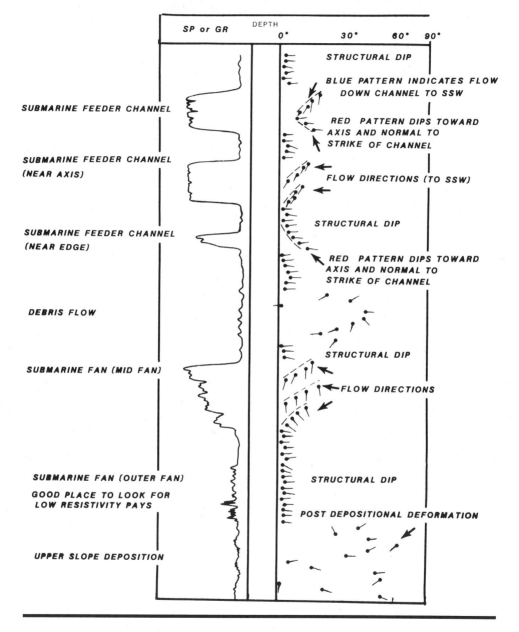

BIBLIOGRAPHY

Fitzgerald, D. D.: "Obtaining Valid Dipmeter Surveys in Deviated Wells," *World Oil* (November 1980).

Gilreath, J. A., and Stephens, Ray W.: "Interpretation of Log Response in a Deltaic Environment," paper presented at the American Association of Petroleum Geologists Marine Geology Workshop, Dallas, April 1975.

Schlumberger: "Dipmeter Interpretation—Volume I: Fundamentals" (1981).

Schlumberger: "Dipmeter Interpretation—Volume I: Fundamentals" (1983).

PART IV

ANALYSIS OF LOGS AND CORES

CHAPTER TWENTY

INTERPRETATION TECHNIQUES

GENERAL PHILOSOPHY

Interpretation of well logs can be made on a variety of levels, depending on the requirements of the user. These distinct levels can be classified as:

1. One point from one zone in one well: This method might be used, for example, to pick a top, or the location of some marker bed as a guide to further drilling, choosing a casing point.
2. A few zones in one well: This method might be used for completion and testing decisions.
3. An entire well: This method might be used for reserve studies, correlation of the well to nearby wells, and completion and testing decisions.
4. Many wells in a field: This method might be used in some field study as a starting point for mapping porosity, oil-in-place, and so on, over an entire field.

According to the requirements, the analyst may use nothing more than a quick-look evaluation or consume hours of computer time in the analysis of the log data. Consciously or not, all analysts use some kind of physical model as a basis for analysis. It is worthwhile to reflect on these models before discussing the techniques of mathematical interpretation.

Suppose a device existed that read "porosity." Suppose further that this device could be made as small or as large as desired (fig. 20.1). Now assume that it is 100 times smaller than a grain of sand. In a porous formation, this tiny tool will either find itself in a pore space and read 100% porosity, or it will find itself in a sand grain and read 0% porosity—that is, it only has two possible states. If the tool was placed at a large number of random points in the formation, a statistical distribution of 100% and 0% porosity readings would be developed that would give a description of the porosity of the formation.

Now assume the device is made very large, for example, a cubic mile in size. This large tool measures a porosity of 20%. This is not a very useful piece of information since there is no knowledge of the homogeneity of the cubic mile being measured. In fact, half of the measured volume could have a porosity of 40% and half could have a porosity of 0% and the difference would never be known. In reality, porosity tools investigate a formation volume of a few liters so that variations at the granular level are not measurable. In that

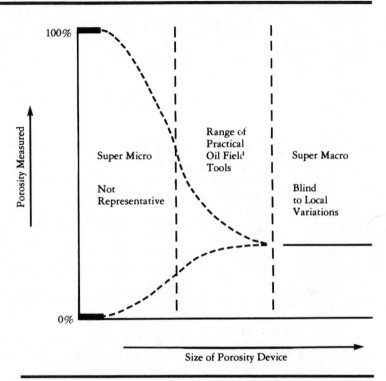

FIGURE 20.1 *Effect of Tool Size on Parameter That Is Measured.*

case, there is no guarantee that the porosity 10 ft from the borehole is the same as the porosity near the wellbore. The lesson is that any petrophysical measurement is a function of the way it is measured. Rather like Heisenberg's uncertainty principle, a very accurate value of porosity can be known (small device), but with no guarantee that the measurement is representative; or a macro-reading of the value of porosity can be made, but with no certainty as to what volume of rock it represents.

QUESTION #20.1
If a field of 10 acres extends to a depth of 10,000 ft and 20 wells are drilled with a hole size of 12 in., what proportion of the total reservoir volume has actually been sampled?

A similar problem is that of determining the length of the coastline of an island. Figure 20.2 shows the mythical isle of Fubar. A first attempt at defining the length of its coastline is made using the map, and a figure of 1000 km is obtained. Later, a more detailed map, on a much larger scale, is found, and more details are found as the coastline bends into estuaries and juts back out at peninsulas. This

FIGURE 20.2 *Coastline of Fubar.*

map when measured gives a coastline of 1500 km. Next, a hardy hiker sets out with a measuring wheel and actually walks all around Fubar at the high-tide level on the beach, dodging round rocks and into crevices and caves and, somewhat exhausted, comes up with a figure of 3000 km. Not to be outdone, a mad marine biologist takes a ten-year leave of absence from Fubar U. and remeasures the coastline taking into account the extra distance involved in passing the line over each barnacle on each rock on the beach and comes up with a figure of 5000 km. What, therefore, is the *correct* figure for the length of the coastline of Fubar?

The moral of the story is that with any problem of measurement, the answer is a function of how the measurement is made. Never was this more true than in well logging. Each practical logging tool is a trade off between local precision, which can only be extrapolated field-wide with great danger, and global averaging, which is blind to local inhomogeneities. Where different tools investigate similar rock volumes, their measurements may be combined with impunity. Where one tool reads a local property and another a global one, their combination can produce interpretive results that are either meaningless or flat wrong.

The analyst should bear all of this in mind before launching into the purely mathematical aspect of log analysis. In figure 20.3, an attempt has been made to place these concepts in perspective by cross plotting vertical resolution against radial investigation for a number of commonly used logging tools.

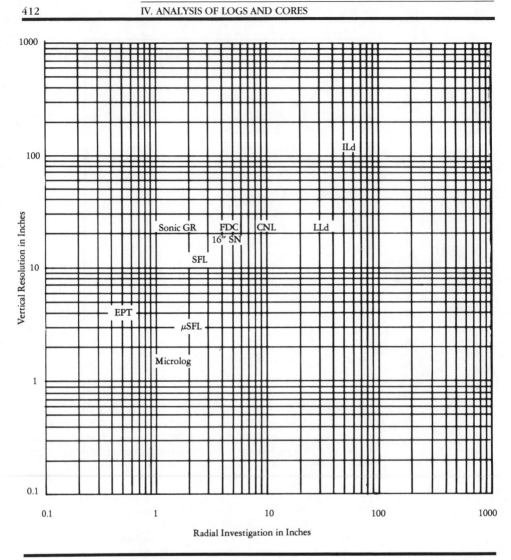

FIGURE 20.3 *Vertical and Radial Investigation Characteristics of Common Logging Tools.*

MODELS

In order to apply quantitative analysis to well logs, some physical model is required to relate log response to the mineral and fluid content of the formation and their distribution therein. Common models used include:

Clean formations
Shaly formations
Multimineral mixed lithologies

Clean Formations

Clean formations conform to Archie's model where porosity may be deduced from a single porosity device (e.g., the density tool) and all formation conductivity is due to the connate water in pore space. This water saturation can be determined from the standard relationship

$$S_w^n = F \times R_w/R_t.$$

Sand/Shale Sequences

Sand/shale sequences present no particular problem provided that the thickness of each sand or shale layer is large compared with the vertical resolution of the logging tool. The shale sections can then be distinguished from the clean sand sections, which are handled in the conventional manner. Problems do arise when the sand and shale laminae are very thin or when the shale particles are distributed randomly among the sand grains. In the case of very thin strata, the logging tools will provide averages that are not representative of either the sand or the shale strata.

Treatment of shaly sand models requires considerable care so that the log readings are matched to the type, amount, and distribution of the clays present in the formation. Their specific treatment is reviewed in chapter 27.

Multimineral Mixed Lithologies

Where the effects of clay materials do not pose serious modeling problems, another problem with mixed minerals may exist. Carbonate and evaporite sequences may be present in the column logged. This case is most easily handled by logical application of the response equations of the individual tools—the method is discussed in chapter 28.

STRUCTURED APPROACH

A frequently asked question is "Where do I begin?" This section will suggest a structured approach to log analysis. Dividing the task into logical units, or sequential steps, will clarify the process. These steps are:

1. Data gathering
2. Quality checks
3. Reconnoitering
4. Picking the model
5. Determining computation parameters
6. Calculations
7. Reporting conclusions

Data Gathering

Find *all* the logs. You may think you will need only an induction-sonic log, but get the rest of the logs too. Maybe the dipmeter will pinpoint a fault. Get the formation water analysis, core analysis, and drillstem test results; you need all the data you can get. The search for data may reveal that the file is full of previous log analysis results and the job you had planned to do is unnecessary!

Quality Checks

Read the headings on the log. The log headings contain all sorts of useful information on mud type, hole size, operating difficulties, bad tools, and so on. Look for bad logs, base-line shifts, noise, and so on.

Reconnoitering

Start with a log print on a scale of 1 or 2 in. per 100 ft (1/500 or 1/1000 for metric). Visually examine the log from top to bottom, looking for trends and/or formations of interest. Use a colored marker pen to quickly delineate shales, evaporites, and porous and permeable zones. Make a few quick calculations for lithology. If necessary, determine if the formation of interest is a shaly sand or a carbonate.

Picking the Model

Having made the reconnoitering pass to locate formations of interest, a model should be picked so you will know what analysis techniques to use. For a shaly sand, try to determine if it is a laminated shale or a dispersed shale. For a carbonate reservoir, try to determine which minerals are likely to be present. Geological reports, sidewall cores, or bit-cuttings analysis carry a wealth of information about lithology.

Determining Parameters

You will require a variety of parameters: This assortment depends on the model used, and on the complexity of the case. As a bare minimum, you will need to know the value of R_w, the F-to-ϕ relationship, and what to use for determination of porosity and R_t. The more sophisticated models require tool-response endpoints for various lithologies, shale responses to porosity tools, shale resistivities, and so on. Computer-generated crossplots can help with this parameter selection.

Calculations

Calculation of saturation and porosity values is the least of your worries. This can be accomplished by the use of nomograms or by means of a pocket calculator. Programmable calculators save time. Interactive and batch computer programs are also available. Service companies offer computer log analysis.

Log Used	Objective	Parameters Required
SP, GR	(Find reservoir rocks)	
Neutron, Density Sonic, P_e	ϕ (Find porosity and lithology)	Neutron matrix setting $\rho_{ma}, \rho_f, \rho_b$ $\Delta t_{ma}, \Delta t_f, \Delta t$
Laterolog Induction	S_w (Find water saturation)	R_t, R_w, ϕ a, m, n
SFL or MSFL	S_{xo} (Find S_{xo})	R_{xo}, R_{mf}, ϕ a, m, n
Neutron, Density	ρ_{hy} (Determine hydrocarbon type)	ϕ_N, ϕ_D, S_{xo}
	h (Find pay thickness)	S_w maximum$\Big\}$ pay cut-off ϕ minimum
	OIP (Find oil in place)	ϕ, S_w, h
	N (Find reserves)	OIP, area, recovery factor, β_o or β_g

FIGURE 20.4 *Log Analysis Outline.*

Reporting Conclusions

An analysis should end with a definitive conclusion. It could be simply that sand A is indicated to be productive and should be tested, or it could be in the form of a computer log indicating the locations of oil, gas, lithology, net pay, and so on. Whatever form the report takes, do document two things: (1) the model used and (2) the parameters (R_w, ρ_{ma}, etc.) used, so that another analyst can make a judgment in an informed manner.

Figure 20.4 offers some flow-charted steps that the reader may find useful for performing step-by-step desk top analysis. A summary of the most useful log analysis equations is in the Appendix.

BIBLIOGRAPHY

Goetz, J. F., Prins, W. J., and Logar, J. F.: "Reservoir Delineation by Wireline Techniques," *The Log Analyst* (September–October 1977) **18.**

Raymer, L. L., and Burgess, K. A.: "The Role of Well Logs in Reservoir Modeling," SPE paper 9342, presented at the 55th Annual Technical Conference and Exhibition, Dallas, Sept. 21–24, 1980.

Timur, A.: "Open Hole Well Logging," SPE paper 10037, presented at the International Petroleum Exhibition and Technical Symposium of the SPE held in Beijing, China, 18–26 March, 1982.

Answers to Text Question

QUESTION #20.1
1 acre-ft = 43,560 cu ft
$\qquad$ 10 acre $\times$ 10,000 ft = 3.365×10^9 cu ft
$\qquad$ 20 wells $\times$ 10,000 ft $\times$ $\pi(0.5)^2$ = 1.571×10^5 cu ft
Proportion = 1:27,730

CHAPTER TWENTY-ONE

QUICK-LOOK INTERPRETATION, CROSSPLOTS, AND OVERLAYS

Quick-look log analysis refers to a number of techniques for plotting log data in a reasonably effortless and simple way that reveals either the formation content or the formation lithology. These methods are widely used by log analysts for wellsite evaluations. Their great appeal lies in their simplicity and subtlety. Most quick-look methods can be applied without any special equipment and produce quite acceptable results. Broadly speaking, there are three branches to quick-look analysis:

Compatible overlays of curves
Crossplots of selected curve readings
Simple algorithms for calculators

In general, *compatible overlays* manage to eliminate some unknown quality by taking a ratio, while revealing some other quality that is of interest—water saturation, for example. *Crossplots* are indispensable for computer–generated analysis but also offer a quick and convenient means of determining end points such as R_w. *Algorithms* offer a quick and simple means to calculate items of interest, such as porosity and water saturation. They are widely incorporated into log analysis routines on both programmable calculators and computers.

COMPATIBLY SCALED OVERLAYS

Theory

Compatibly scaled overlays compare two or more log curves. In practice, one curve will be overlaid on the other and a tracing made so that the end result is a composite with two curves. In general, the relative deflection between the two curves is indicative of some formation property of interest. It is necessary for all overlays that the curves being compared are compatibly scaled (i.e., both must be in the same system of units). Apples must be compared with apples and not oranges. Many of the quick-look overlays in use today are quite subtle in that they eliminate some quantity that is either unknown or of no interest. The most commonly used overlays are:

SP with R_{xo}/R_t
R_0 with R_t } for hydrocarbon
EPT porosity with porosity   detection

Neutron porosity with density porosity $\left.\vphantom{\begin{array}{c}a\\b\end{array}}\right\}$ for lithology, porosity,
Density porosity with sonic porosity and hydrocarbon
typing

SP with R_{xo}/R_t

This overlay is by no means universally applicable since it requires conditions that result in healthy SP developments. Where those conditions are met (i.e., in wells drilled with fresh mud having salty connate water), it is an elegant way to detect hydrocarbons without the need to know porosity. Thus, it is a natural candidate for wells where only an induction–SP log is available. It is usually produced at the time the log is run since it requires some manipulation of the raw data in the service company's surface equipment. Inputs required are an SP, a deep resistivity measurement (usually a deep induction), and a shallow resistivity measurement (usually a short–normal, SFL, or similar). The theory behind the method depends only on Archie's equation and the SP relationship, which may be written as

$$SP = -K \log (R_{mf}/R_w).$$

Note that for the purposes of this overlay R_{mf} and R_w are used rather than their "equivalent" values R_{mfe} and R_{we}.

Archie's equation, if written for both the invaded and undisturbed zones, allows a ratio to be taken that eliminates F, the porosity-dependent formation factor:

$$S_w^n = F \times R_w/R_t, \text{ and}$$

$$S_{xo}^n = F \times R_{mf}/R_{xo}, \text{ so}$$

$$\left(\frac{S_w}{S_{xo}}\right)^n = \frac{R_w}{R_{mf}} \times \frac{R_{xo}}{R_t} .$$

If it is assumed that S_{xo} is related to S_w, for example, by the old rule of thumb

$$S_{xo} = S_w^{1/5},$$

then the quantity $(S_w/S_{xo})^n$ can be replaced by $S_w^{8/5}$, if n is assumed to be equal to 2. The term R_{mf}/R_w can then be replaced by $(R_{xo}/R_t) S_w^{5/8}$ and the SP equation rewritten as

$$-SP = K \log [(R_{xo}/R_t)S_w^{5/8}]$$

or

$$-SP = K [\log (R_{xo}/R_t) + \log S_w^{5/8}].$$

In a water-bearing zone with $S_w = 1$ the term $K \log S_w^{5/8}$ is equal to 0 and thus the term $K \log (R_{xo}/R_t)$ is numerically equal to $-SP$. In an oil-bearing zone with S_w less than 1, however, the term $\log S_w^{5/8}$ will be less than 1 and hence the term $K \log (R_{xo}/R_t)$ will be numerically less than $-SP$. Provided that there is no substantial SP reduction resulting from the presence of hydrocarbons—which is usually the case in all but very shallowly invaded formations (see chapter 9)—a comparison of the actual SP with the quantity $K \log (R_{xo}/R_t)$ will have the following characteristics: In wet zones, the two curves will track. In hydrocarbon-bearing zones, the $K \log (R_{xo}/R_t)$ curve will separate from the SP curve (see fig. 21.1). Note that in the lower sand, which is wet, the SP and the massaged R_{xo}/R_t ratio closely coincide and in the upper sand, which is hydrocarbon-bearing, the two curves separate. In shales, since the R_{xo}/R_t ratio is close to 1, the massaged R_{xo}/R_t is effectively zero. In practice, some experimentation is usually required in order to obtain a valid overlay with the two traces aligned in both shales and wet zones. This requires the correct K value to be used for the formation temperature in question and correct offsetting of either the SP baseline or the massaged R_{xo}/R_t curve.

By way of summary, use the SP–R_{xo}/R_t overlay method by all means, if you have sand/shale sequences and good SP development (high R_{mf}/R_w ratio), and you do not envisage recording a porosity log. This method cannot be used if oil-based mud is employed. A variation of the method, though not widely employed, uses the SP to generate a pseudo-R_{xo}/R_t ratio, which may then be compared with the actual R_{xo}/R_t ratio. A refinement to the method allows a more accurate computation of the real R_{xo}/R_t ratio by taking into account invasion effects. In deeply invaded formations, for example, the ratio of R_{SFL} to R_{ILd} will not be as large as the real R_{xo}/R_t ratio (see chapter 22). This refinement requires that a solution to the tornado chart be carried in the surface-equipment software package.

R_0 with R_t: The F Overlay

Another popular and extremely effective overlay is the F overlay, which effectively compares R_0 with R_t. If Archie's equation is written as

$$S_w^n = F \times R_w/R_t$$

and logarithms are taken of both sides of the equation, we have

$$n \log S_w = \log F + \log R_w - \log R_t,$$

F is available from porosity measurements, for example, a density log. Bulk density may be converted to an F by employing the standard relationships:

$$\phi = (\rho_{ma} - \rho_b)/(\rho_{ma} - \rho_f)$$

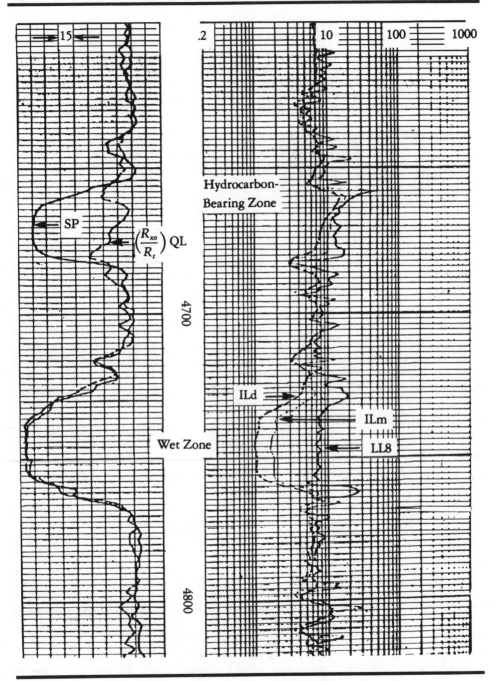

FIGURE 21.1 *SP with R_{xo} / R_t Overlay.*

and

$$F = a/\phi^m.$$

In clean formations of known lithology, this conversion is a simple matter for the service company's surface equipment. It merely requires values of ρ_{ma}, ρ_f, a, and m to be dialed in; F then appears as an output. In practice, this F curve is recorded on a logarithmic scale as a separate log. The analyst then uses the logarithmic F curve in conjunction with a logarithmic R_t curve (a deep induction or deep laterolog) in the manner of a slide rule to normalize the two curves so that one will overlay the other in clean, wet zones. By so doing, the log F curve has in fact been shifted by an amount equal to log R_w. Since the product of F and R_w is R_0, the net effect is an overlay that compares R_0 to R_t (see fig. 21.2). Wherever the two curves separate, with R_t greater than R_0, the implication is that S_w is less than 1, and, therefore, hydrocarbons are to be expected.

Note that in the water-bearing section (fig. 21.2), the normalized F curve (i.e., R_0) coincides with the R_t curve and that in the hydrocarbon-bearing section, the R_t curve separates from the R_0 curve. This separation can be quantified in S_w units by using an appropriate logarithmic scaler (see fig. 1A in the Appendix for a full-scale cut-out version of this scaler), as illustrated (not to scale) in figure 21.3. This scaler is used as follows: Place the scaler across the overlay in such a way that its long axis is parallel to the depth lines on the log. Place the $S_w = 100$ mark on the normalized F curve. Now read off the S_w value at the R_t curve.

Another useful feature of the F overlay is that R_w need not be known. The act of normalizing the F curve to the deep resistivity trace in a wet zone effectively computes R_w for you. If the F curve is on top of the resistivity log when the normalization is made, then locate the 100 line of the F scale at the top of the F log and read the resistivity value on the resistivity scale that lies immediately beneath it. This value will be numerically equal to 100 times R_w. For example, after normalizing in a wet zone, $F = 100$ lies over the R_t scale line corresponding to 20 $\Omega \cdot$m. This means that R_w is 0.2 $\Omega \cdot$m.

The Logarithmic Movable-Oil Plot

The logarithmic movable-oil plot is an extension of the F overlay. This plot requires two overlays, one to indicate S_w and a second to indicate S_{xo}. In practice, both overlays are made on a single log and the result is a movable-oil plot (MOP) on a logarithmic scale.

The production of a MOP proceeds in two stages. First, the F curve is normalized to R_t in a wet zone and the resulting R_0 curve is traced onto the resistivity log (or, if preferred, the R_t curve may be traced onto the normalized F log). Second, the F curve is normalized to an R_{xo} trace (such as an SFL, or MSFL) in a wet zone and the resulting R_{xoo} curve ($R_{xoo} = R_{xo}$ in a rock 100% saturated with a fluid of resistivity R_{mf}) is traced onto the resistivity log (or, as above, the R_{xo} curve may be traced onto the normalized

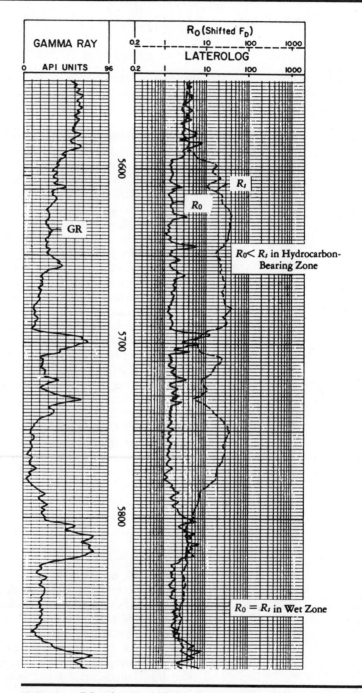

FIGURE 21.2 *F Overlay (R_0 with R_t). Courtesy Schlumberger Well Services.*

LOGARITHMIC SCALE INTERPRETATION

TO READ Sw:
PLACE 100 ON THIS SCALE ON TOP OF
THE NORMALIZED F CURVE.
READ Sw WHERE THE Rt CURVE INTER-
SECTS THE SCALE.

TO READ φ FROM F CURVE:
PLACE 10 ON THIS SCALE ON TOP OF THE
F = 100 LINE ON THE F LOG.
READ φ WHERE THE F CURVE INTER-
SECTS THIS SCALE.

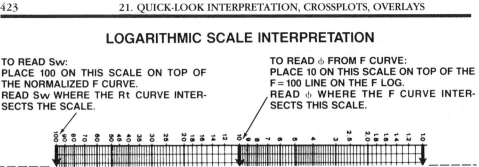

FIGURE 21.3 *F Overlay Scaler for* S_w *(Not to Scale)*.

F log). The result of such tracings will be as shown in figure 21.4. Note that the area between the R_0 and R_t curves is representative of the hydrocarbon-filled pore space. The MOP adds a further dimension by subdividing that space into two parts, one containing residual oil and the other movable oil. This presentation is thus helpful in assessing the most productive intervals (i.e., the ones with high movable oil).

Consider the bulk volume model shown in figure 21.5. Note that, in the undisturbed zone, the saturating fluids are connate water and oil and that the water saturation is S_w. In the invaded zone, however, the saturating fluids are mud filtrate and residual oil and the water saturation is S_{xo}. If S_{xo} is high, residual-oil saturation (ROS) is low, and a large fraction of the oil-in-place has been moved by the miniwaterflood effected by the drilling process. If, however, S_{xo} is close to S_w, very little of the oil-in-place has been moved; in such cases, a formation may be less productive. Thus, on the MOP, the area between the R_0 curve and the R_{xoo} curve is a measure of the residual-oil saturation, and the area between the R_{xoo} curve and the R_t curve is a measure of the movable-oil saturation.

In practice, once a tracing of this sort has been prepared it is common to color in the appropriate areas with highlighters of different colors—e.g., red for residual oil, orange for movable oil, and blue for water. Note that zero porosity point never appears on such a logarithmic plot, since *F* or R_t corresponding to zero porosity is infinity. Some arbitrary line is, therefore, chosen as the right-hand limit for the purpose of coloring in water with a blue highlighter.

The Neutron-Density Overlay

The neutron-density overlay is probably the most commonly used presentation. The two logs are nearly always run together and, together, supply an enormous wealth of information, even with a cursory inspection.

Two distinct types of neutron-density overlay will be considered: the compatible sandstone-scaled overlay and the compatible limestone-scaled overlay. The former is needed in sand/shale sequences, and the latter in carbonate/evaporite sequences. The

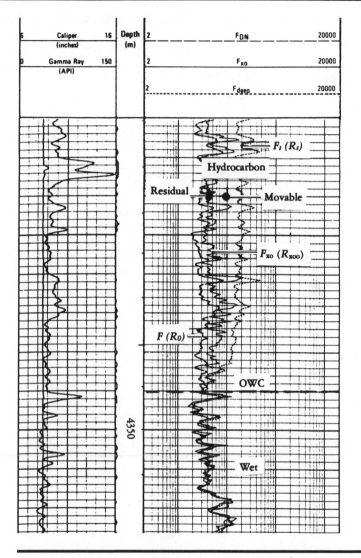

FIGURE 21.4 *Logarithmic Movable-Oil Plot. Courtesy Schlumberger Well Services.*

compatibly scaled-sandstone presentation requires that the neutron log (and here the dual-spacing thermal neutron log is the norm) be recorded on a sandstone-matrix setting and displayed on a 0 to 60% scale. The bulk density recording, ρ_b, should be converted to an apparent sandstone-porosity curve by choice of an appropriate value for ρ_{ma} (2.65 or 2.68 g/cc, for example), and displayed on the same scale as the neutron log. The choice of a scale, be it 0 to 60% or 0 to 50%, is immaterial provided both curves are on the same scale and the shale readings do not exceed the high end of the scale (i.e., a

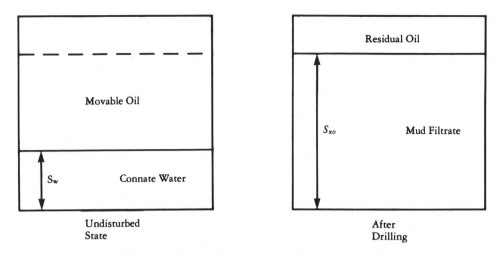

FIGURE 21.5 *Bulk-Volume Model for Movable and Residual Oil.*

presentation on a 0-to-30% scale is inappropriate if ϕ_N in the shales is 45% and, therefore, off scale).

In the neutron-density overlay of figure 21.6, note that, in shales, ϕ_N is much larger than ϕ_D (dotted curve to the left of the solid curve) and that, in water-bearing intervals, ϕ_N equals ϕ_D where the formation is clean. In gas-bearing zones, note that ϕ_N reads a lower apparent porosity than does ϕ_D and that the traces tend to be mirror images of each other. Even in oil-bearing sections, it is common to see ϕ_N reading slightly less than ϕ_D. Figure 21.6 demonstrates that on one plot it is possible to distinguish quickly porous and permeable sections from shales and, within the potential reservoir rocks so defined, distinguish gas, oil, and water.

A slightly different presentation is used in carbonate reservoirs, since the purpose is slightly different. Here, the norm is not sandstone or shale but limestone and possibly evaporites. Porosities in general are lower in carbonate reservoirs than in sandstone reservoirs, and dolomite may be present as well. The compatible limestone-scaled overlay therefore calls for the neutron log on a limestone matrix setting and the density log on an apparent limestone porosity basis. Since dolomite and anhydrite may compute apparent porosities less than zero on a limestone basis (remember ρ_{ma} for dolomite is higher than ρ_{ma} for limestone), an appropriate scale for this type of overlay is 45 to –15% (left to right). Although, in some locations, 30 to –10% is favored, the 45 to –15% scale has an added advantage: All is not lost if the bulk density curve has not been converted to an apparent porosity curve. By appropriate shifting, the neutron porosity and the bulk density curves may be overlayed so that 2.95 g/cc coincides with –15% on the neutron log and 1.95 g/cc coincides with 45% on the neutron log. The reader

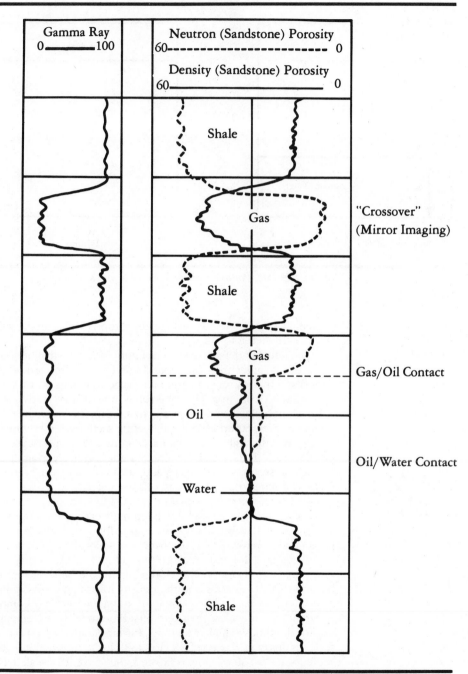

FIGURE 21.6 *Neutron–Density Overlay—Sandstone Presentation.*

may satisfy himself that 1.95 g/cc translates to approximately 45% porosity on a limestone scale, and 2.95 g/cc to –15%.

The neutron-density limestone-scaled overlay in figure 21.7 is constructed for a formation porosity of 15%. Note that in limestone the two traces coincide; in dolomite the apparent neutron porosity is higher than the apparent density porosity; and in sandstone the reverse is true. This sandstone crossover is due entirely to the matrix effects on the two porosity devices and is the result of the limestone porosity scaling used. It should not be confused with crossover produced by gas. Note that in the sandstone intervals both curves have a similar character although separated from one another. The gas effect of mirror imaging is not evident. If gas were present, it would manifest itself by an *additional* separation, with neutron porosity even lower than density porosity.

Shale and the evaporites salt and anhydrite are also shown in figure 21.7. Such a presentation is extremely valuable as a quick-look guide to rock types in the column drilled and is a very good starting point for any other, more detailed, analysis.

Density–Sonic Overlay

Although not as widely used as a neutron–density overlay, the density-sonic overlay is particularly useful for detection of secondary porosity (vugs and fractures). Figure 21.8 shows a formation with both matrix (intergranular) porosity and fracture porosity. Provided the matrix is known, the density porosity will be equal to the total porosity:

$$\phi_D \;=\; \phi_T \;=\; \phi_{matrix} \;+\; \phi_{secondary}.$$

The sonic tool, however, responding as it does to the first compressional wave arrival (see chapter 16) will respond only to the matrix porosity. Compressional waves traveling through a vertical, fluid-filled fracture, for example, will travel more slowly than those traveling through the matrix system. Since the sonic tool triggers on the first (faster traveling) arrival, the later arrivals (having passed through the fracture system) will be ignored. Thus

$$\phi_S \;=\; \phi_{ma}.$$

An overlay of ϕ_S and ϕ_D will therefore have the interesting property that in fractured or vuggy intervals the two curves will separate (see fig. 21.9). A number of scaling options are available for density-sonic overlays. Compatible limestone scaling of 45 to –15% (left to right) is generally adequate. If bulk density and Δt are to be overlayed, compatible scaling is 1.95 to 2.95 g/cc (left to right) for ρ_b and 108 to 28 μs/ft (left to right) for Δt (equivalent to 40 μs/ft per track). In metric units, a scaling of 366 to 86 μs/m is adequate (equivalent to 70 μs/m per track).

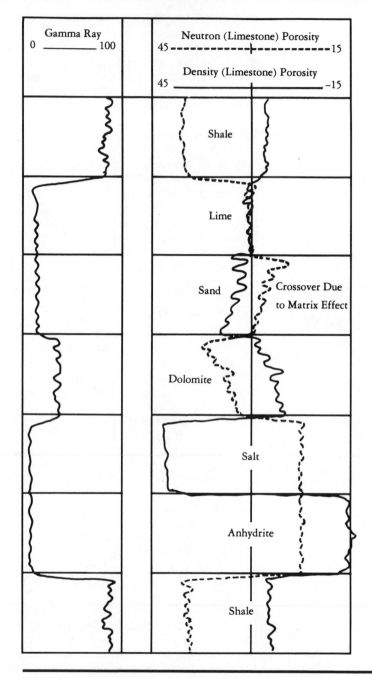

FIGURE 21.7 *Neutron–Density Overlay—Limestone Presentation.*

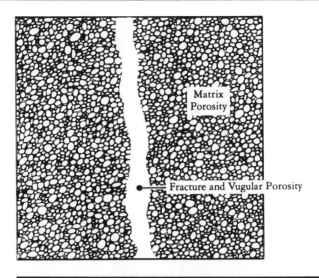

FIGURE 21.8 *Matrix and Fracture Porosity.*

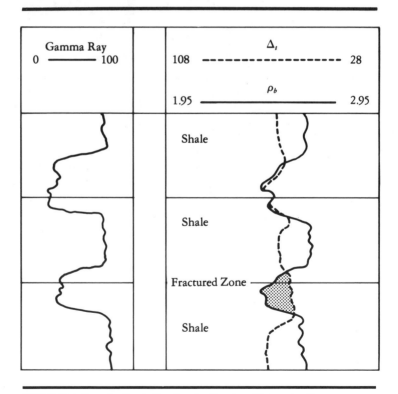

FIGURE 21.9 *Density–Sonic Overlay for Fracture Determination.*

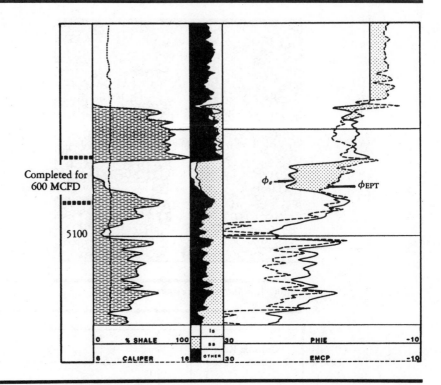

FIGURE 21.10 *EPT–Porosity Overlay for Hydrocarbon Detection. Reprinted by permission of the SPE-AIME from Eck and Powell 1983, fig. 13.* © *1983 SPE-AIME.*

EPT–Porosity Overlay

The compatible porosity overlay of an EPT-derived porosity with another porosity curve is useful for quick-look hydrocarbon detection. The theory behind the overlay is simple. An EPT-derived porosity is effectively the water-filled pore space whereas a density porosity, for example, is effectively the total porosity. If the formation is wet, the traces will match. In hydrocarbon-bearing zones, separation is probable (see fig. 21.10). The usual cautions are appropriate—that is, such an overlay requires clean formations and the correct choice of matrix parameters in computing the two porosities. Chapter 15 deals with the derivation of EPT porosity in more detail.

CROSSPLOTS

Theory

Crossplots of log data are indispensable tools for the analyst. Although widely used in conjunction with computer processing of log data, crossplots may, of course, be prepared by hand using only

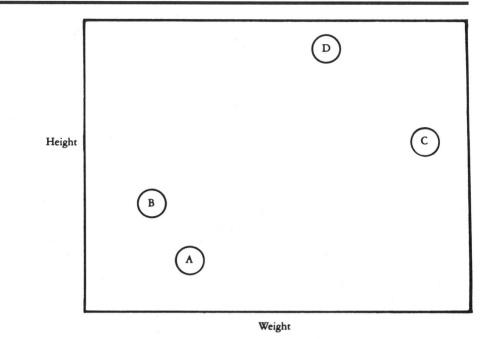

FIGURE 21.11 *Crossplot of Height vs. Weight for Animals.*

limited amounts of data "eye-ball" filtered by the analyst. The end result of using a crossplot will be the picking of some parameter (e.g., R_w) needed for quantitative log analysis. Crossplots come in a variety of flavors, and the purpose of this section will be to give the reader an overall view of crossplotting techniques. More detailed work with the individual crossplots will follow in the appropriate chapters on lithology, porosity, R_w, and saturation.

Crossplots as Maps

A crossplot may be thought of as a map. It is a two-dimensional representation of the variation of data with respect to two or more properties. For example, any animal can be classified by its weight and its height. A crossplot of animal weight versus animal height is, therefore, a valid method of "mapping" animal species. In the hypothetical plot of figure 21.11, for example, point A might represent a short fat man and B a tall skinny one. C might represent an elephant and D a giraffe. In typing a population in this manner the crossplot gives a visual meaning to what would otherwise be a listing of heights and weights, which is not as easy to assimilate.

Trend Analysis and Groupings

Our plot of height against weight, if made more complete by adding tens or hundreds of different animal species, would start to show groups of points typical of a type of animal. In logging terms,

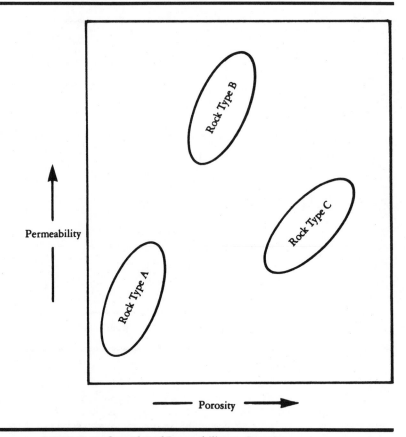

FIGURE 21.12 *Crossplot of Permeability vs. Porosity.*

crossplots can likewise be used to discern trends or groups. Suppose a well is extensively cored and core analysis made in order to find porosity and permeability throughout the section. A plot of permeability against porosity might look like the plot shown in figure 21.12. Note that three distinct rock types are revealed by the crossplot. Rock type A is of low porosity and permeability (perhaps a carbonate); rock type B is of moderate porosity and high permeability (perhaps a sandstone); and rock type C is of high porosity and moderate permeability. Crossplots provide a visualization of trends and groupings of data points that helps in understanding the nature of the population being plotted.

Extrapolations

Extrapolation is a useful feature of the crossplot. In the year 1766, Titius Wittenberg noticed in his study of the planets that a relatively simple rule of thumb could be applied to relate the orbital radius of a planet to its number (counting Mercury as 1, Venus as 2, Earth as 3, etc.). This later became known as Bode's law, which states that the

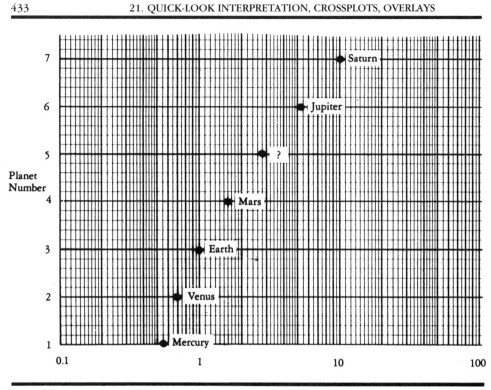

FIGURE 21.13 *Crossplot for Finding Planets (Orbital Radius Predicted by Bode's Law).*

radius of a planetary orbit, in units of the Earth's radius of orbit, is given by

$$\text{orbital radius of planet } n = 0.4 + 0.3 \times 2^{(n-2)}.$$

Figure 21.13 plots planet number against the Bode-predicted orbital radius. At the time this predictive plot was first made, no planet was known to exist between Mars and Jupiter. Observations at the expected orbital radius led to the discovery of the asteroids. Later discoveries of Neptune and Pluto, albeit with some serendipity, have also been ascribed to the same technique. The extrapolation of patterns and trends on crossplots is an extremely valuable tool for predicting such key log analysis parameters as R_w.

Frequency Plots

Computers are particularly adept at manipulating log data to produce frequency plots, which are crossplots on which the number of occurrences of a particular pair of data points is printed at a particular map location. Figure 21.14 shows a neutron–density frequency-crossplot for data from a section logged through a sand/shale sequence. For the purpose of the plot, each axis is divided

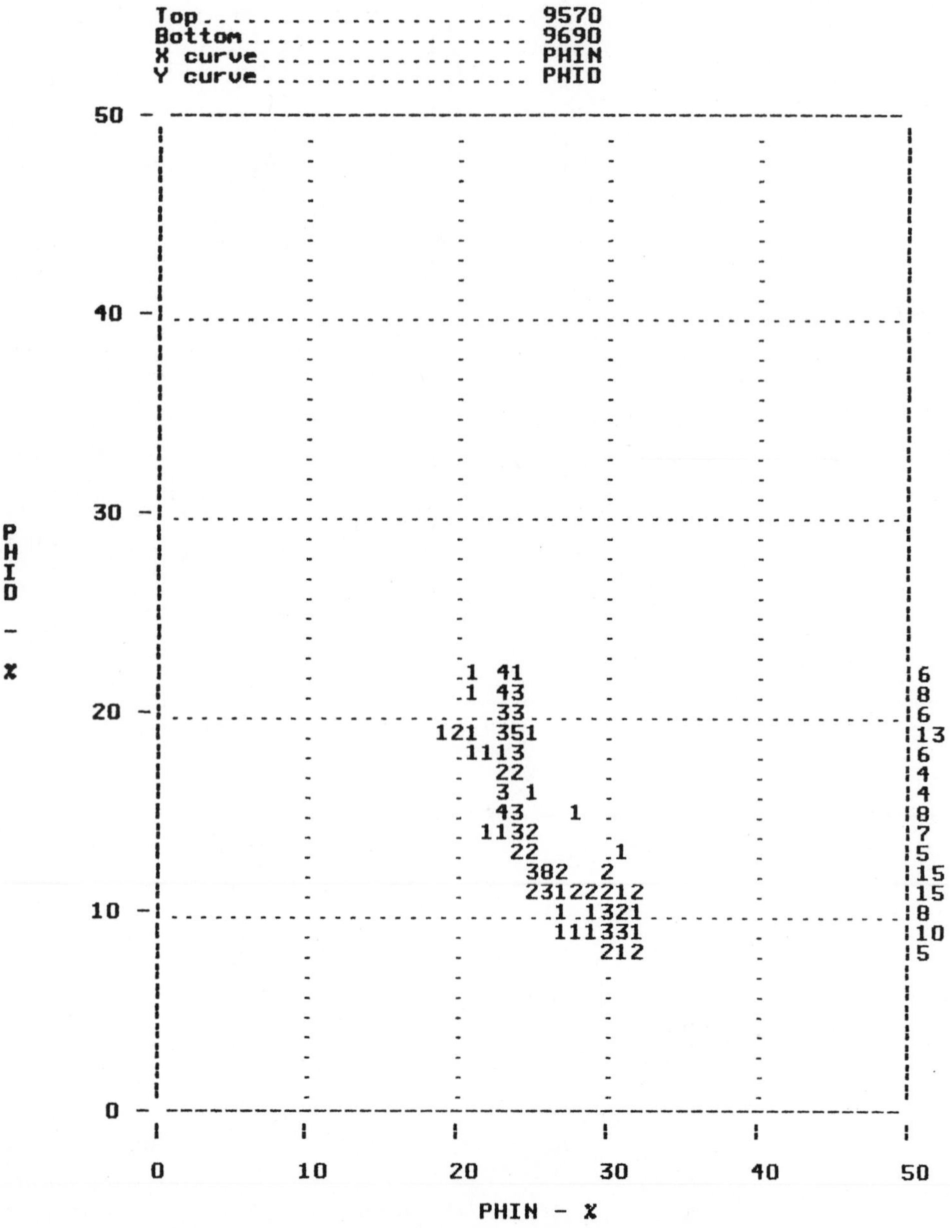

FIGURE 21.14 *Neutron–Density Frequency Crossplot.*

into one-porosity-unit cells. There are thus $50 \times 50 = 2500$ cells on the plot. Each data point on the log is inspected and its ϕ_N and ϕ_D coordinates are used to place a point on the plot. Where many points fall in the same cell their total number is accumulated and this number (or frequency) is the item printed on the final plot at that particular cell address. Note that where no data are found the plot remains blank and where the majority of the points lie there is a good visual image of the distribution of the formation properties in the logged section. Such plots are used extensively for normalization of log data (i.e., recalibration shifts), shale picks, and so on.

Z Plots

The Z plot is a companion plot to the conventional frequency crossplot (or X-Y plot) described above. The purpose of the Z plot is to map the distribution of three variables instead of two. Given the limitations of a two-dimensional sheet of paper, this task is accomplished by printing the scaled average of the third variable (z) at any given x-y cell address.

For example, on figure 21.14 there were 8 occurrences of the condition ϕ_N (PHIN—the x axis) = 26% and ϕ_D (PHID—the y axis) = 12%. The same cell on the Z plot of Figure 21.15 shows a scaled average of a third parameter, the gamma ray. At the chosen address the scaled average of the 8 values of gamma ray occurring at $\phi_N = 26\%$ and $\phi_D = 12\%$ is seen to be 8. On this particular plot the scaling of the z-axis was accomplished by applying the formula:

$$z = \frac{\text{gamma ray} - (z\text{-min})}{(z\text{-max}) - (z\text{-min})}$$

and z-min and z-max were chosen so that the least radioactive levels "scored" 0 and the most radioactive levels 10.

Reviewing the Z plot of figure 21.15, it is evident that the shaliest points lie to the southeastern corner of the plot (high z values of 8, 9, and *), and the least radioactive points lie to the northwestern part of the plot (low z values of 1, 2, and 3). Z plots are useful for data analysis where three variables are of interest.

HISTOGRAMS

Histograms of raw log data and of computed log analysis results are also useful tools for the analyst. Figure 21.16 illustrates a histogram. The y axis of such a plot is scaled either in actual number of occurrences or in percent of the total number of data points analyzed. The x axis is scaled over the useful range of the data analyzed. Among the better known uses for histograms are:

Picking minima and maxima
Rescaling logs
Checks on the validity of computed results

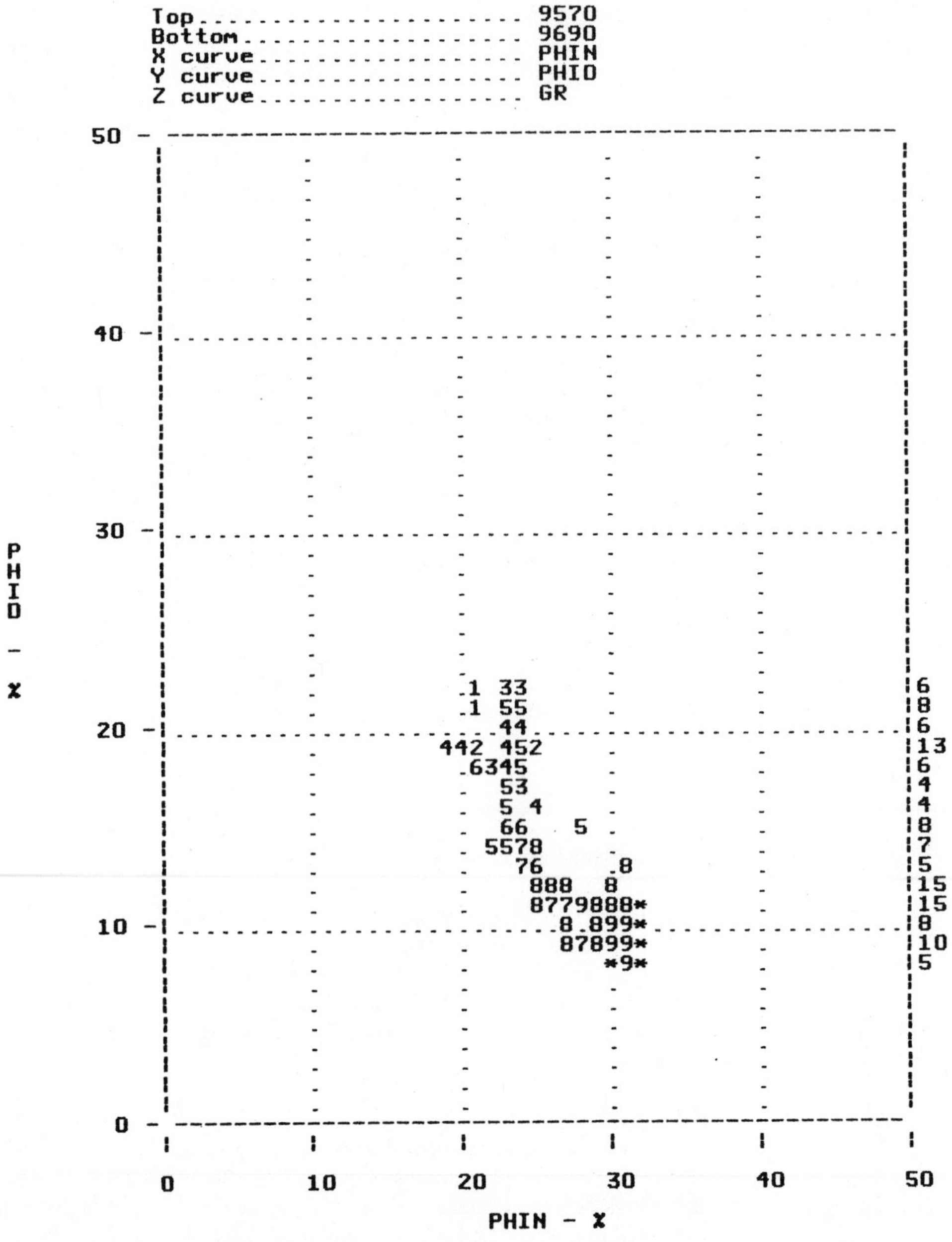
Top 9570
Bottom 9690
X curve .... PHIN
Y curve .... PHID
Z curve .... GR
PHID - %
PHIN - %
0
10
20
30
40
50

FIGURE 21.15 *Neutron–Density Crossplot with Gamma Ray in Z Axis.*

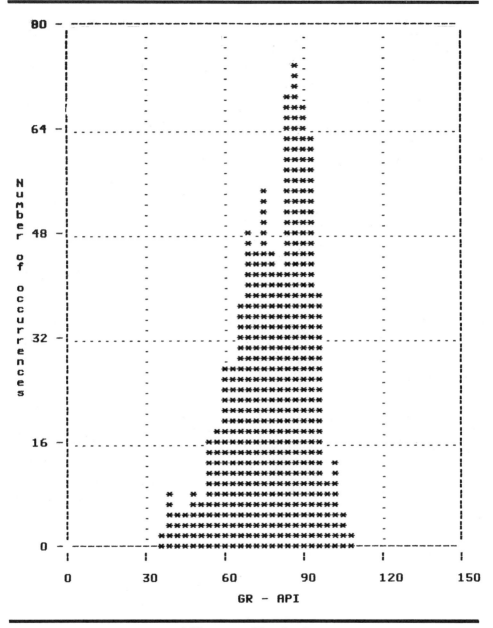

FIGURE 21.16 *Histogram of Raw Log Data.*

Illustrations of particularly useful histograms will be given in the appropriate chapters when discussing particular aspects of log analysis.

QUICK-LOOK ALGORITHMS

With the advent of the handheld calculator, especially the programmable type, log analysts have tended to condense their wisdom into algebraic manipulations to ease their daily toil. Today, much of what used to be done by reference to charts is now done using the pocket computer. This has advantages and disadvantages. On the plus side, obviously, more data points can be analyzed in a shorter time and with greater precision. On the minus side, however, the analyst is no longer free to shift points on a chart to get an immediate feel for the sensitivity of a computed answer to a change in the raw-data values used.

Algorithms for solving many of the common manipulations that log analysts use are given in the appropriate chapters as the particular analysis method is discussed. As a convenience these algorithms are also included in the Appendix.

A word of caution is in order. Although the calculator/computer is a wickedly clever device for the crunching of numbers and relieves the analyst of that chore, it also places a greater burden on the analyst—that of ensuring (a) that the model fits the circumstances and (b) that legitimate data are submitted to be processed. A more precise answer should not be confused with a more accurate one.

BIBLIOGRAPHY

Eck, M.E., and Powell, D.E.: "Application of Electromagnetic Propagation Logging in the Permian Basin of West Texas," SPE paper 12183, presented at the 58th Annual Technical Conference and Exhibition, San Francisco, Oct. 5–8, 1983.

Schlumberger: "Log Interpretation Volume II—Applications" (1974).

Schlumberger: "Data Processing Services Catalogue," Client Document M 081020 (June 1981).

CHAPTER TWENTY-TWO

POROSITY

Devices that measure porosity are sensitive to both rock matrix and the fluid filling the pore space. Thus, the measurements of "porosity tools" reflect not only porosity, but also the type of rock, the clay content, and the fluid type. As yet, no tool has been invented that reads just porosity. This characteristic of the conventional devices used to measure porosity is a blessing in disguise, since it allows the analyst to derive porosity and more from a variety of different porosity tools. The term *porosity tools* includes compensated neutron, compensated formation density (with or without photoelectric-factor measurement), and sonic tools.

Clean (shale/clay-free) water-bearing formations of known lithology represent the simplest environments for porosity determination, since the effects of mixed lithology, clay, and hydrocarbons do not confuse the issue. Before discussing techniques for their analysis, a review of the scales used on porosity logs and the procedures and techniques to be used in reading them will be presented.

Three things must be determined before reading a porosity log:

1. the type of curve recorded (e.g., bulk density, ρ_b, or apparent porosity, ϕ_D)
2. the scale (e.g., 45 to 15% or 60 to 0% on a neutron log)
3. the actual lithology of the formation and the nature of the fluid occupying the pore space.

Once all three are known, true porosity can be determined in those intervals where the particular porosity device under consideration can reasonably be expected to work—for example, with a pad-contact device such as the density tool, no readings should be attempted in washed-out zones.

Porosity is the volume fraction of pore space in the rock and is expressed as a fraction of the bulk volume of the rock. The normal convention in reservoir engineering is to express porosities in percentage units—for example, a porosity of 0.3 is referred to as 30% porosity. Another term is also frequently used—namely, the *porosity unit,* or pu. By using units rather than percentage, a lot of confusion is avoided, as for example in comparing a 20-pu sandstone with a 25-pu sandstone. The latter is 5 pu higher than the former. This avoids the confusion caused by saying one is 5% better than the other or 25% better than the other, depending on which way the difference is expressed.

Reading Porosity from the Density Log

Normal scaling for the density log is from 2.00 to 3.00 g/cc from left to right across two film tracks, or 20 divisions. Each division thus represents 0.05 g/cc. On this basis, it is a simple matter to find the bulk density, ρ_b, in g/cc. The range of bulk densities encountered in sandstone, for example, is from 1.00 g/cc (water) in 100% porosity to 2.65 g/cc at zero porosity. A spread of 1.65 g/cc therefore represents a spread of 100 pu and each of the 20 log-scaling divisions represents 0.05 g/cc. The number of pu in a scale division can be determined as follows:

$$1.65 \text{ g/cc} = 100 \text{ pu},$$

$$0.05 \text{ g/cc} = 0.05/1.65 \times 100$$

$$= 3.03 \text{ pu}.$$

For all practical purposes, the log scale can be marked off in increments of 3 pu starting with 0 pu at 2.65 g/cc. The error at 30 pu is only 0.3 pu, or 1%.

Combinations of Porosity Tools

Frequently a combination of porosity devices can be run into the wellbore together. Many times, these combination tools can define porosity better than any single porosity device by itself. The most commonly available pair is the density–neutron combination.

When working with well logs generated by such a tool combination, the log scales should be checked carefully before jumping to conclusions. Probably more confusion exists about density–neutron combinations and their presentations than about any other device on the market today.

Density–Neutron Combinations

The FDC log can be recorded either as a bulk density curve (ρ_b) in g/cc or as a porosity curve (ϕ_D), if certain assumptions about ρ_{ma} and ρ_f are made. The neutron log can be recorded in one of three matrix settings. The important items to note on each porosity log are the zero porosity point on each log and the porosity scale for each log. If these parameters have been chosen correctly, comparing the two logs will be like comparing apples with apples. More often than not, however, density–neutron combinations will be comparing apples with pears, because porosity scales are different or the zero points do not coincide.

A common rule is to record both logs on *compatible limestone scales*—that is, both logs will be reading the true porosity if the lithology is limestone. To accomplish this, the neutron log is run using a limestone matrix setting, and the density log is recorded as a bulk density curve and shifted so that the zero of the neutron scale coincides with a ρ_b value of 2.70 g/cc. Alternatively, and a little more accurately, ρ_b is converted into a porosity curve by using $\rho_{ma} = 2.71$ and an appropriate value of ρ_f, depending on the salinity of the mud filtrate.

For predominantly sandstone reservoirs, two methods can be used:

1. Run the neutron log on a sandstone matrix using a porosity scale of 60 to 0%. Run the density log as a ρ_b curve on a scale of 1.65 to 2.65 g/cc. The zero for both will now be at the extreme right of Track 3.
2. Alternatively, ρ_b can be run as a ϕ_D curve and scaled as for the neutron log on a 60 to 0% porosity scale.

For predominantly carbonate reservoirs, or where lithology is not well known, the best method is to use compatible limestone scales. That is, run the neutron log on a limestone matrix using a 45 to –15% porosity scale and run the density log using a ρ_b scale of 1.95 to 2.95 g/cc.

In analyzing combination density–neutron logs, some unusual combinations of scales may be encountered. If this happens, reason back to the zero point for each log and begin the analysis there.

Uses of Combination Porosity Logs

Three parameters can be determined by combining two porosity logs: porosity, lithology, and fluid content of the pore space. Regardless of the actual lithology, an extremely good approximation to true porosity can be made by combining the neutron limestone porosity (ϕ_{NLM}) with the density limestone porosity (ϕ_{DLM}) in the following manner:

$$\phi_X = (\phi_{NLM} + \phi_{DLM})/2,$$

where ϕ_X is referred to as the crossplot (Xplot) porosity.

If both logs are recorded on limestone porosity scales, and the lithology *is* limestone, both logs will read the same porosity value. If the lithology is sandstone, then ϕ_{DLM} will be too high and ϕ_{NLM} will be too low, but the combined average will be close to the actual porosity value. If gas or light hydrocarbons are present, use of the average-porosity formula will result in porosity values that are only good to a first-order approximation. Better values of porosity, when light hydrocarbons are present, are determined by the Gaymard method, which is discussed later.

DENSITY POROSITY (ϕ_D)

The simple relation

$$\phi \text{ (fractional)} = \frac{\rho_{ma} - \rho_b}{\rho_{ma} - \rho_f}$$

can be represented graphically, as in figure 22.1. The matrix density in normal reservoir rocks varies between 2.87 g/cc (dolomite) and

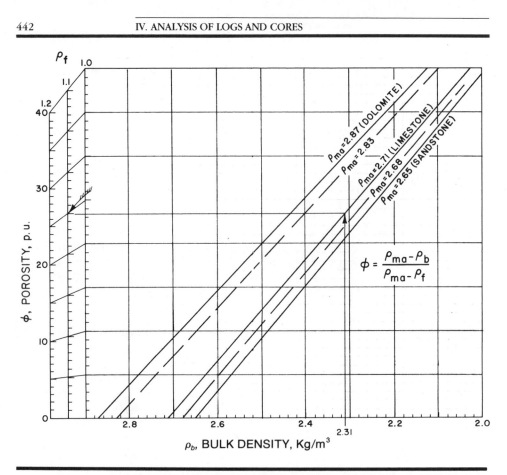

FIGURE 22.1 *Density Porosity Chart. Courtesy Schlumberger Well Services.*

2.65 g/cc (sandstone). The fluid density in normal brines ranges from 1 to 1.1 g/cc. Porosity derived from a density log is referred to as ϕ_D.

QUESTION #22.1
If ρ_b = 2.4 g/cc,
and ρ_f = 1.0 g/cc,
and the rock is sandstone, use figure 22.1 to determine

ϕ_D = ?

NEUTRON POROSITY (ϕ_N)

Compensated neutron tools are run on a matrix setting chosen by the logging engineer or the company witness. If the actual lithology coincides with the chosen matrix setting, porosities may be read directly from the log. This is seldom the case, however, and if the

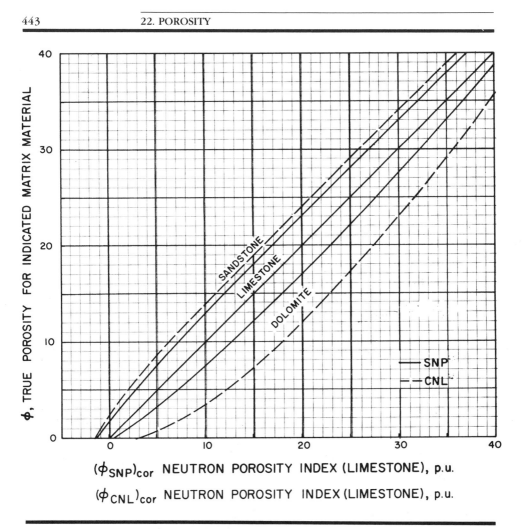

FIGURE 22.2 *Neutron Porosity Chart. Courtesy Schlumberger Well Services.*

matrix is something other than that used in running the log, the porosity reading from the log will not be correct. Also, in most instances, the lithology is not known prior to logging the well. Therefore, a standard matrix setting is normally used.

A convenient standard for the neutron log is the *neutron porosity index (limestone)*. This is the same value (ϕ_{NLM}) that the tool would have read had it been recording on a limestone scale. Figure 22.2 plots the porosity measured by the neutron tool set for a limestone matrix against the true porosity for the indicated lines of constant lithology. This chart can be used to determine the true porosity if the actual lithology is known.

QUESTION #22.2
Use figure 22.2 to find the true porosity in dolomite if the neutron porosity index (ϕ_{NLM}) is 20%.

SONIC POROSITY

Sonic porosity may be estimated from the Wyllie time-average equation:

$$\phi_S = \frac{\Delta t - \Delta t_{ma}}{\Delta t_f - \Delta t_{ma}}.$$

This equation is represented graphically in figure 22.3. The interval transit time, Δt in $\mu s/ft$, is plotted against porosity. Lines on the chart correspond to various matrix velocities in ft/s. Common reservoir rocks have matrix velocities in the range of 18,000 ft/s (sandstone) to 26,000 ft/s (dolomite). Fluid velocities depend on fluid type (oil, gas, or water) and the temperature, pressure, and salinity of the fluid; a working range is 5000 to 5300 ft/s.

Unconsolidated formations do not conform to the simple Wyllie time-average equation. A correction factor is used in these cases to extend the usefulness of figure 22.3. This correction factor is the term B_{cp}, which is also known as the compaction factor. The value of B_{cp} is usually determined for a given formation from core analysis and existing sonic logs. Alternatively, the Hunt transform may be used as outlined in chapter 16.

QUESTION #22.3
Use figure 22.3 to find the sonic porosity in limestone ($V_f = 21,000$ ft/s) if $\Delta t = 70$ $\mu s/ft$.

NEUTRON–DENSITY CROSSPLOT

Where lithology is unknown, no single porosity tool can uniquely define the porosity, which is then determined by combining two tools such as the neutron and the density. Figure 22.4 shows the neutron porosity index (ϕ_N, from the CNL) plotted against bulk density (ρ_b, from the density log). The lines of constant porosity are substantially straight lines that are parallel to the NW–SE axis. Thus, regardless of the lithology, a reading of ρ_b and ϕ_N is sufficient to define a porosity. One qualification for use of the chart is that the porosity be water filled—hydrocarbon effects will complicate the use of this chart.

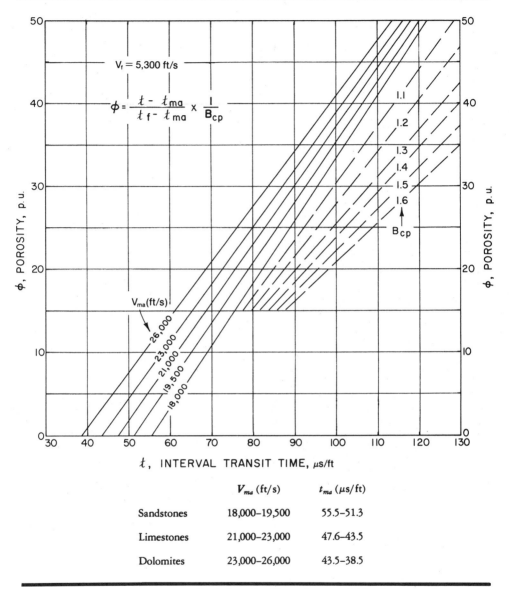

FIGURE 22.3 *Sonic Porosity Chart. Courtesy Schlumberger Well Services.*

	V_{ma} (ft/s)	t_{ma} (μs/ft)
Sandstones	18,000–19,500	55.5–51.3
Limestones	21,000–23,000	47.6–43.5
Dolomites	23,000–26,000	43.5–38.5

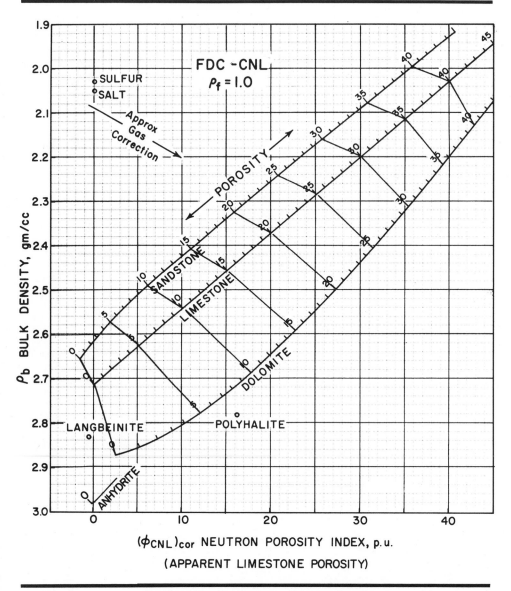

FIGURE 22.4 *Neutron–Density Crossplot Showing ϕ_X. Courtesy Schlumberger Well Services.*

A simple rule of thumb for finding a neutron–density crossplot porosity is to express ρ_b as an apparent limestone porosity, add this value to neutron limestone porosity, and divide by 2:

$$\phi_{X(N-D)} = (\phi_{N(lime)} + \phi_{D(lime)})/2.$$

QUESTION #22.4

If ρ_f =   1.0 g/cc
and ρ_b =   2.54 g/cc
and $\phi_{N(lime)}$ =   18%,

Find the corresponding crossplot porosity, ϕ_X, using figure 22.4.

QUESTION #22.5

Now convert ρ_b to $\phi_{D(lime)}$ and compute the quick-look value of ϕ_X using the formula given above.

NEUTRON–SONIC CROSSPLOT

On many occasions, a density log may be unavailable or rendered useless by washed-out hole or other operational problems. A sonic log may then be combined with a neutron log, as plotted in figure 22.5. Again, constant porosity lines proceed generally in the NW–SE direction, although the trend is not as well behaved as the trend in figure 22.4.

QUESTION #22.6

If Δt =   79 μs/ft
and $\phi_{N(lime)}$ =   18%,

use figure 22.5 to compute a crossplot porosity.

SHALY FORMATIONS

The addition of clay material to the formation implies that the properties of the matrix or the pore space have been altered. If an independent estimate is available for the bulk volume of clay material present, allowances can be made for its presence. The following development is somewhat generalized. A more complete treatment of shaly sands is to be found in chapter 27.

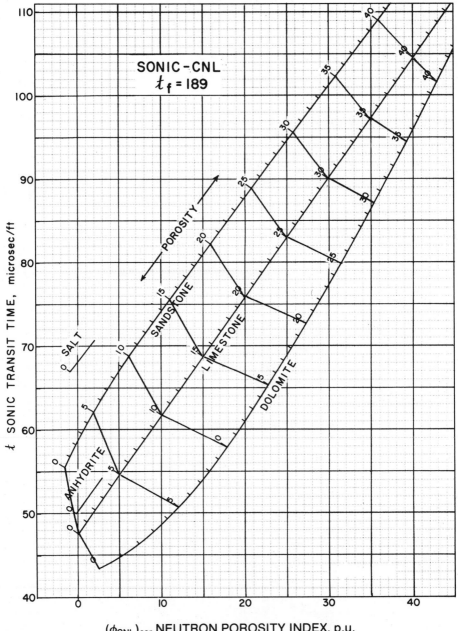

FIGURE 22.5 *Neutron–Sonic Crossplot for ϕ_X. Courtesy Schlumberger Well Services.*

If a porosity device P reads ϕ_P on the log and $\phi_{P_{sb}}$ in the foreign material (clay fillers or shale laminae), then the response of the device can be written as

$$\phi_P = \phi + (1 - \phi - V_{sb})\phi_{P_{ma}} + V_{sb}\phi_{P_{sb}},$$

where V_{sb} is the bulk volume of foreign material, $(1 - \phi - V_{sb})$ is the bulk volume matrix, $\phi_{P_{ma}}$ is the tool's response in pure matrix material, and ϕ is the actual porosity. By definition, $\phi_{P_{ma}}$ is zero. The relationship therefore reduces to

$$\phi_P = \phi + V_{sb}\phi_{P_{sb}}$$

and, by rearrangement,

$$\phi = \phi_P - V_{sb}\phi_{P_{sb}}.$$

This generalized equation may be applied to each of the porosity devices. For the density tool, for example,

$$\phi = \phi_{D_{log}} - V_{sb}\phi_{D_{sb}}.$$

Typical values of $\phi_{D_{sb}}$ will range from 20% to −10% depending on the bulk density of the extraneous materials. For example, if a density log reads 20%, for the assumed matrix, and an independent indicator (such as a gamma ray) reveals that 30% of the bulk volume is occupied by shale ($V_{sb} = 30\%$), which has an apparent density porosity of 15%, the equation can be written:

$$\phi = 20 - 0.3(15) = 15.5\%.$$

In most cases, shaly formations will appear more porous than they actually are.

For the neutron log, the equation may be written as

$$\phi = \phi_{N_{log}} - V_{sb}\phi_{N_{sb}}$$

and, for the sonic log, it may be written as

$$\phi = \phi_{S_{log}} - V_{sb}\phi_{S_{sb}}.$$

Typical values for $\phi_{N_{sb}}$ lie between 15 and 50% and typical values for $\phi_{S_{sb}}$ are in the same range.

Again, the combination of two devices can eliminate unknown qualities. For example, a simultaneous solution of the neutron and density equations for porosity in shaly formations leads to a useful result:

$$\phi_D = \phi + V_{sb}\phi_{D_{sb}} \quad \text{and} \quad \phi_N = \phi + V_{sb}\phi_{N_{sb}}.$$

Eliminating V_{sb}, these equations can be combined to give

$$\phi = \frac{\phi_{N_{sb}}\phi_D - \phi_{D_{sb}}\phi_N}{\phi_{N_{sb}} - \phi_{D_{sb}}}.$$

QUESTION #22.7
If $\phi_{N_{sb}}$ = 36%
ϕ_N = 29%,
$\phi_{D_{sb}}$ = 12%, and
ϕ_D = 23%,

use the above formula to find ϕ.

Eliminating ϕ, the following solution is found for V_{sb}:

$$V_{sb} = \frac{(\phi_N - \phi_D)}{(\phi_{N_{sb}} - \phi_{D_{sb}})}.$$

QUESTION #22.8
Using the data from question 22.7, compute V_{sb}.

SECONDARY POROSITY IN COMPLEX LITHOLOGY

Neutron and density logs respond to total porosity regardless of its form or distribution. Sonic logs, however, tend to ignore irregular porosity (such as vugs) and fracture porosity. Compressional waves passing through the formation find a path through the rock matrix around the water-filled vugs and fractures that is faster than the path through the water in them. Since the sonic tool registers the first arrival time of the compressional wave, it follows that ϕ_S will represent the primary porosity. This effect may be turned to good advantage. A comparison of ϕ_D and ϕ_S should indicate when secondary porosity is present. If ϕ_D is greater than ϕ_S, vugs and/or fractures are likely to be present in the formation.

This poses the problem of how to compute ϕ_S and ϕ_D in complex lithology when neither ρ_{ma} nor Δt_{ma} are accurately known. The solution to the problem requires use of all three porosity devices—the neutron, density, and sonic tools—and is described in detail in chapter 28.

HYDROCARBON EFFECTS

Neutron tools respond to the hydrogen index. Gas and light oils have a low hydrogen index and thus make neutron readings lower than they would be had water occupied the pore space instead of

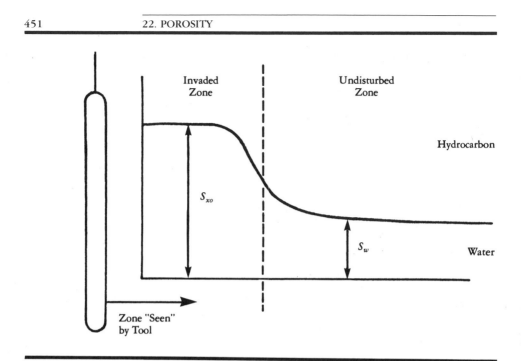

FIGURE 22.6 *Saturation Profile "Seen" by Porosity Logging Tools.*

hydrocarbons. Density tools respond to electron density and, due to hydrogen's unique Z/A ratio, gas and light oils (which contain hydrogen) appear lighter than water. Sonic tools measure the transit time of compressional waves, which travel slowly through gas. Thus, in the presence of gas and light hydrocarbons:

The neutron log indicates less than true porosity.
The density log indicates more than true porosity.
The sonic log indicates more than true porosity.

To quantify the effects of light hydrocarbons, two items must be known: (1) the volume of the pore space containing hydrocarbons in the annular volume of the formation in which the tool makes its measurement and (2) the response of the tool to hydrocarbons. In general, the porosity tools are shallow-investigation devices, that is, the majority of their measurements are made in the filtrate-invaded zone (see fig. 22.6), where the saturation is S_{xo}. Thus, the bulk volume of the hydrocarbon "seen" by the tool is $\phi(1 - S_{xo})$.

The equivalent porosity of the hydrocarbon depends on its density, as shown in figure 22.7 for neutron and density tools. A fair approximation of the functions shown in the figure are

$$\phi_{N_{hy}} = 2.2\,\rho_{hy} - 1.2\,(\rho_{hy})^2 \quad \text{and} \quad \phi_{D_{hy}} = 1.7 - 0.7\,\rho_{hy}.$$

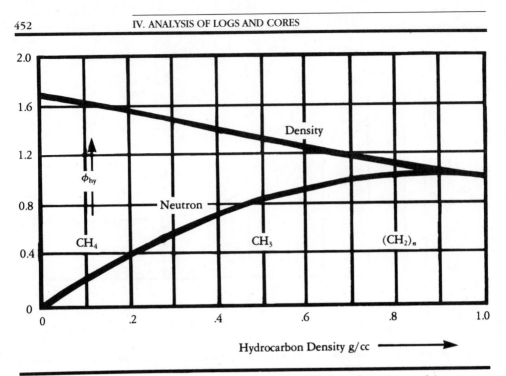

FIGURE 22.7 *Equivalent Porosities of Hydrocarbons* $\phi_{D_{hy}}$ *and* $\phi_{N_{hy}}$ *vs.* ρ_{hy}.

The response of the density tool can be described by

$$\phi_D = \phi_T [S_{xo} + \phi_{D_{hy}} (1 - S_{xo})].$$

So

$$\phi_T = \frac{\phi_D}{S_{xo} + \phi_{D_{hy}} (1 - S_{xo})}.$$

In gas-bearing formations $\phi_{D_{hy}}$ is much larger than ϕ_T and, hence, ϕ_D is larger than ϕ_T. For example, if $\phi_{D_{hy}} = 1.7$, $\phi_T = 30\%$, and $S_{xo} = 75\%$, the density log reading can be calculated as

$$\phi_D = 0.3[0.75 + 1.7(1 - 0.75)] = 0.3525$$
or   35.25% porosity.

A neutron under the same conditions would have a $\phi_{N_{hy}}$ of zero and

$$\phi_N = 0.3[0.75 + 0(1 - 0.75)] = 0.225$$
or   22.5% porosity.

The log analyst has the problem of deducing porosity from a measurement distorted by the fluid contained in the pore space. If a single porosity measurement is available, a rule of thumb can be used:

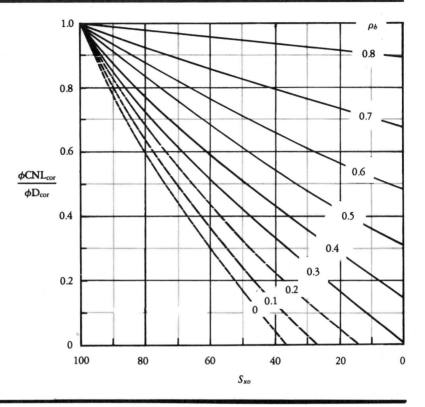

FIGURE 22.8 *Quick Determination of ρ_{by}. Courtesy Schlumberger Well Services.*

$$\phi_T = 0.8\phi_D \quad \text{in gas} \quad \text{or} \quad \phi_T = 1.25\phi_N \quad \text{in gas}.$$

A better solution could be obtained by combining the two measurements. A quick-look method for this is provided by the Gaymard (G) relationship

$$\phi_G = [(\phi_N^2 + \phi_D^2)/2]^{1/2}.$$

In the previous example with $\phi_N = 22.5\%$ and $\phi_D = 35.25\%$, the Gaymard formula gives

$$\phi_G = [(0.225^2 + 0.3525^2)/2]^{1/2} = 0.2957 \quad \text{or} \quad 29.57\%,$$

which is a close approximation to the true porosity of 30%.

More rigorous solutions can be found using iterative procedures that refine the value of porosity and S_{xo}, but these methods require complex processing. Figure 22.8 offers a quick solution to the determination of ρ_{by}. The ratio of ϕ_N/ϕ_D is plotted against S_{xo}. The curved lines are for values of constant ρ_{by}.

QUESTION #22.9
Use figure 22.8 to determine ρ_{by} when
$$\phi_N = 12.5\%,$$
$$\phi_D = 33\%, \quad \text{and}$$
$$S_{xo} = 55\%.$$

SUMMARY

In clean formations of known lithology, porosity may be determined from all three of the common porosity tools. When lithology is unknown, the safest quick-look porosity is the neutron–density crossplot porosity. When light hydrocarbons are present, the Gaymard quick-look formula should be used.

In shaly formations, a simultaneous solution of any two of the three porosity-tool response equations gives an adequate effective porosity for quick-look purposes. Complex cases, where mixed lithology, clays, and light hydrocarbons all coexist, call for a more sophisticated approach that combines the various methods, already discussed, with some new ones that are covered in chapters 27 and 28.

BIBLIOGRAPHY

Burke, J. A., Campbell, R. L., Jr., and Schmidt, A. W.: "The Litho–Porosity Crossplot," *The Log Analyst* (November–December 1969) **10**.

Gaymard, R., and Poupon, A.: "Response of Neutron and Formation Density Logs in Hydrocarbon–Bearing Formations," *The Log Analyst* (September–October 1968) **9**.

Schlumberger: "Log Interpretation Charts" (1984).

Answers to Text Questions

QUESTION #22.1
$$\phi_D = 15.2\%$$

QUESTION #22.2
$$\phi_{NLM} = 12\%$$

QUESTION #22.3
$$\phi_S = 16\%$$

QUESTION #22.4
$$\phi_X = 14\%$$

QUESTION #22.5
$$\phi_X = 14\%$$

QUESTION #22.6
$$\phi_X \quad = \quad 20\%$$

QUESTION #22.7
$$\phi \quad = \quad 20\%$$

QUESTION #22.8
$$V_{sh} \quad = \quad 25\%$$

QUESTION #22.9
$$\rho_{hy} \quad = \quad 0.2 \text{ g/cc}$$

CHAPTER TWENTY-THREE

LITHOLOGY

Logs can be used as indicators of lithology. The most useful logs for this purpose are:

Density log: ρ_b and P_e
Neutron log: ϕ_N
Sonic log: Δt
Gamma ray log: natural gamma and gamma ray spectra

With the exception of the photoelectric factor measurement, P_e, and the natural gamma ray spectral log, no single tool measurement will, by itself, give an indication of lithology. However, much useful information can be gathered by combining the measurements of more than one porosity tool. The most useful combinations are:

a. Crossplots such as ϕ_N vs. ρ_b, ϕ_N vs. Δt, and Δt vs. ρ_b
b. The M-N plot
c. The MID plot
d. Combination of the above with P_e

It is often possible to scale porosity logs in such a way that two curves, when overlaid, immediately give a visual indication of the rock type. These methods are to be encouraged. A very good picture of a geologic column may be gained by simply color coding an appropriately scaled combination of porosity logs. In mixed lithology it is essential to identify the rock type in order to pick correctly the parameters needed to perform other log analysis calculations for porosity and water saturation. Correct identification of lithology will also assist in tasks of well-to-well correlation.

THE NEUTRON–DENSITY CROSSPLOT
The neutron–density crossplot shown in figure 23.1 appears in two versions. Figure 23.1 is to be used in the case of fresh-mud filtrates with $\rho_{mf} = 1.0$, and figure 23.2 is to be used in the case of salt-mud filtrates with $\rho_{mf} = 1.1$. The differences between the two are slight. Lithology is indicated by the location of the plotted point. The positions of various nonporous minerals are shown on the charts as well as lines for the porous reservoir rocks. Shales typically fall in the area delineated by the boundaries

$$30 < \phi_N < 40 \quad \text{and} \quad 2.35 < \rho_b < 2.5.$$

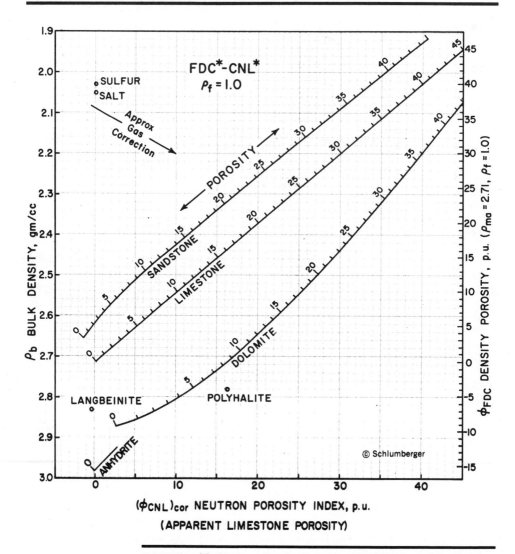

FIGURE 23.1 *The Neutron–Density Crossplot—Fresh Muds. Courtesy Schlumberger Well Services.*

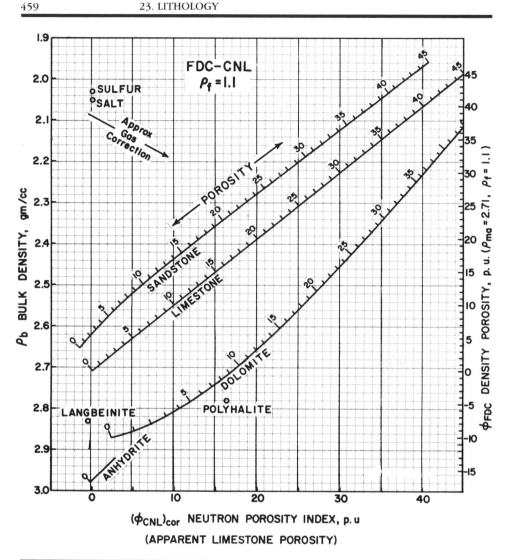

FIGURE 23.2 *The Neutron-Density Crossplot—Salt Muds. Courtesy Schlumberger Well Services.*

As an exercise, it is suggested that this rectangular shale area be boxed off on figure 23.1 and colored in.

QUESTION #23.1

ρ_b = 2.5 g/cc,

ϕ_N = 27%, and

ρ_f = 1.0 g/cc.

What lithology is indicated?

QUESTION #23.2
$$\rho_b = 2.5 \text{ g/cc},$$
$$\phi_N = 6\%, \quad \text{and}$$
$$\rho_f = 1.1 \text{ g/cc}.$$

What lithology is indicated?

The same kind of information can be gained visually from a log display on compatible limestone scales (see fig. 23.3). Note that the scale for the density log is from 2 to 3 g/cc across Tracks 2 and 3. The neutron log, in limestone porosity units, is scaled from 42 to –18% across the same tracks. The zero percent limestone point thus coincides for both devices at 4 divisions from the left of Track 3. Sandstone is indicated where the neutron curve lies to the right of the density curve. This corresponds to the area to the northwest of the limestone line on figure 23.1. Likewise, when the neutron curve lies to the left of the density curve, dolomite or shale (check the gamma ray) is indicated. This corresponds to the area to the southeast of the limestone line on figure 23.1. When the two curves coincide, limestone is indicated. Large separations, such as in anhydrite, are easily recognized.

THE NEUTRON–SONIC CROSSPLOT

Another frequently used pair are the neutron and sonic devices. Where a density log is missing or suffers from bad hole effects, this plot should be used. A neutron–sonic crossplot is shown in figure 23.4. It is used in much the same way as the neutron–density crossplot. Shales typically fall in the area delineated by the boundaries

$$30 < \phi_N < 40 \qquad \text{and} \qquad 70 < \Delta t < 100.$$

As an exercise, it is suggested that this rectangular shale area be boxed off on figure 23.4 and colored in.

QUESTION #23.3
$$\Delta t = 89 \ \mu\text{s/ft}, \quad \text{and}$$
$$\phi_N = 20.5\%.$$

Determine the lithology.

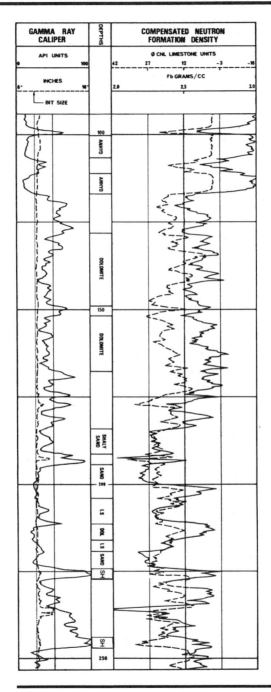

FIGURE 23.3 *Neutron and Density Curves on Compatible Limestone Scales for Lithology Identification. Courtesy Schlumberger Well Services.*

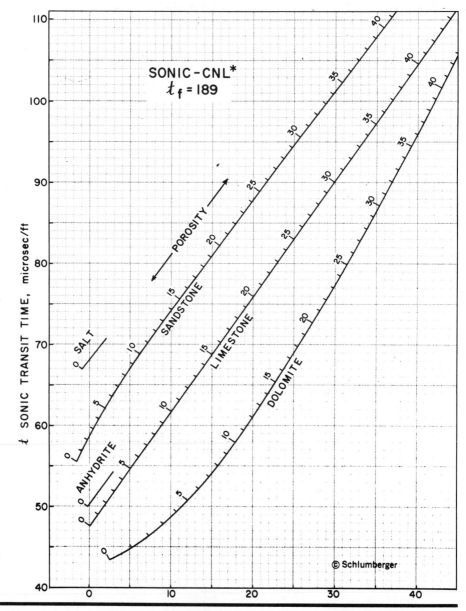

FIGURE 23.4 *The Neutron–Sonic Crossplot. Courtesy Schlumberger Well Services.*

QUESTION #23.4
Δt = 62 μs/ft, and
ϕ_N = 10%.

Determine the lithology.

THE SONIC–DENSITY CROSSPLOT
The sonic–density crossplot is particularly useful for identifying minerals in addition to reservoir rocks. This plot is illustrated in figure 23.5.

QUESTION #23.5
ρ_b = 2.1 g/cc, and
Δt = 65 μs/ft.

What mineral is this?

QUESTION #23.6
ρ_b = 2.98 g/cc, and
Δt = 50 μs/ft.

What mineral is this?

THE *M-N* PLOT
Sometimes three, not two, porosity devices are available. If a three-dimensional graph could be built with *x, y,* and *z* axes corresponding to the neutron, density, and sonic responses, then identifiable minerals would occupy unique points in space. Cases where a lithologic mixture exists could be more easily interpreted. For example, a mixture of sand and dolomite could appear as limestone on the neutron–density and neutron–sonic plots but would be identified by the sonic–density plot. Various attempts have been made to resolve the problem of reducing three log readings to a two-dimensional crossplot. One of the first was the *M-N* plot. This requires that the two parameters *M* and *N* be defined as

$$M = \frac{\Delta t_f - \Delta t}{(\rho_b - \rho_f) \times 100} \quad \text{and} \quad N = \frac{\phi_{Nf} - \phi_N}{\rho_b - \rho_f}.$$

Note that ϕ_N needs to be in fractional units and that ϕ_{Nf} is assumed to be 1.0. Effectively, these two definitions are algebraic methods of

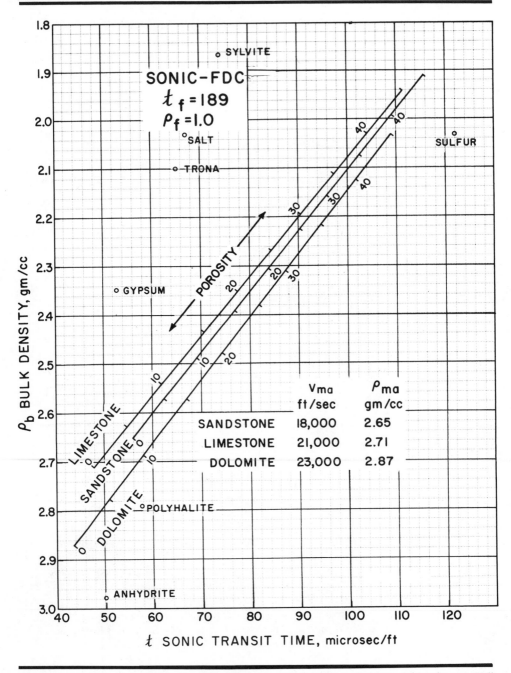

FIGURE 23.5 *The Sonic–Density Crossplot. Courtesy Schlumberger Well Services.*

finding the slope of a line that passes through a plotted point and the 100% porosity point. Since any pure reservoir rock will plot on a line on the crossplots, and the slope of this line is substantially constant, then that slope is a characteristic of the rock type. Having defined M and N and determined them for a given point on the log, their values may be plotted on an M-N plot as shown in figure 23.6.

QUESTION #23.7
Assume fresh-mud filtrate with

$$\rho_f = 1.0 \text{ g/cc} \quad \text{and}$$
$$\Delta t_f = 189 \ \mu s/ft.$$

a. If $\Delta t = 69 \ \mu s/ft$ and
   $\quad \rho_b = 2.5 \text{ g/cc}$, find M.
b. Where does this plot on the Δt-ρ_b plot?
c. If $\phi_N = 10\%$, find N.
d. Where does this plot on the ϕ_N-ρ_b plot?
e. Plot M and N on the M-N plot.
f. What does this point reveal about the lithology mixture?

The M-N plot suffers from a number of shortcomings. For example, it is not truly porosity independent in the case of dolomite since the neutron response is not linear. Another annoying feature is that the plotted point depends on ρ_{mf}, the filtrate density. Finally, the actual values of M and N for common minerals and reservoir rocks are not easily remembered and have no particular significance in themselves. Although the M-N plot is still used by some analysts, it has largely been superseded by another plot that accomplishes the same end result more elegantly—the MID plot.

THE MATRIX IDENTIFICATION PLOT (MID)
The MID plot requires three porosity tools for input. Its main advantages over the M-N plot include its independence from porosity and mud type and the fact that it uses meaningful parameters directly related to rock properties that are easily remembered. The neutron and density logs are combined to define an apparent matrix density, ρ_{maa}. The neutron and sonic logs are combined to define an apparent matrix travel time Δt_{maa}. These two parameters are then crossplotted to define lithology mixtures.
The value of ρ_{maa} can be computed from the equation

$$\phi_X = [\phi_{D(lime)} + \phi_{N(lime)}] / 2$$

Whence,

$$\rho_{maa} = \frac{\rho_b - \phi_X \rho_f}{1 - \phi_X} \quad \text{and} \quad \Delta t_{maa} = \frac{\Delta t - \phi_X \Delta t_f}{1 - \phi_X}$$

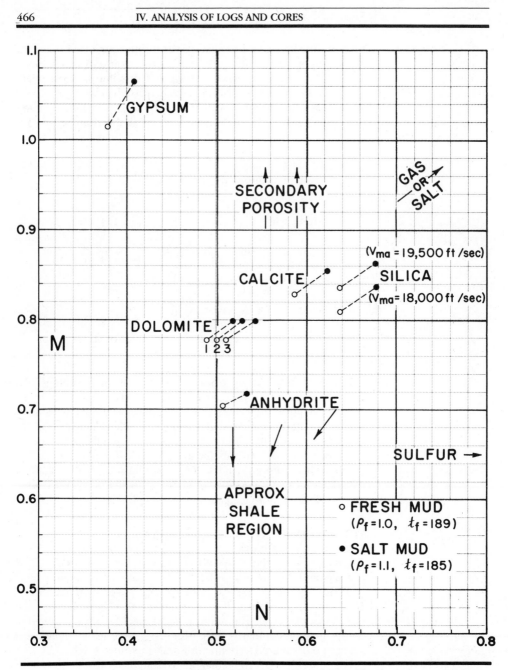

FIGURE 23.6 *The M-N Plot for Mineral Identification. Courtesy Schlumberger Well Services.*

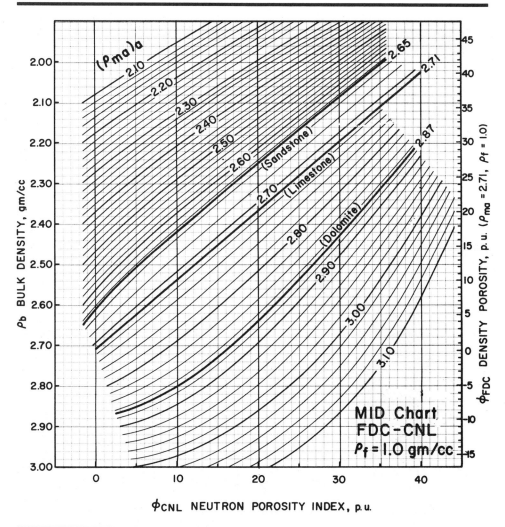

FIGURE 23.7 *Determination of ρ_{maa} from Neutron and Density Logs in Fresh Muds. Courtesy Schlumberger Well Services.*

However, in practical log evaluation, these values can be found more simply from charts. For example, figure 23.7 defines ρ_{maa} for any ρ_b, ϕ_N pair and figure 23.8 defines Δt_{maa} for any Δt, ϕ_N pair. Figure 23.9 should be used in place of Figure 23.7 when $\rho_f = 1.1$ (salt-mud filtrate). Once ρ_{maa} and Δt_{maa} have been determined, they are crossplotted on the MID-plot chart shown in figure 23.10.

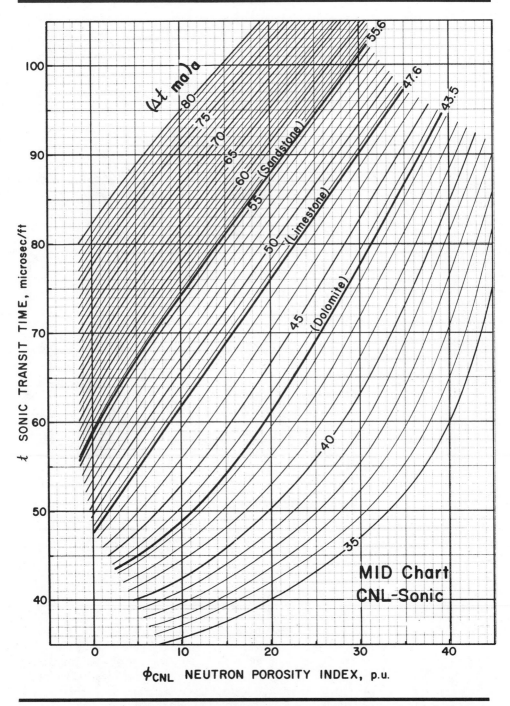

FIGURE 23.8 *Determination of* Δt_{maa} *from Neutron and Sonic Logs. Courtesy Schlumberger Well Services.*

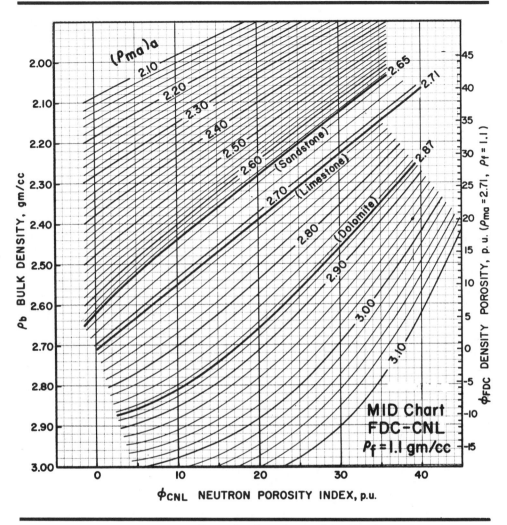

FIGURE 23.9 *Determination of ρ_{maa} from Neutron and Density Logs in Salt Muds. Courtesy Schlumberger Well Services.*

QUESTION #23.8

$\rho_b = 2.5$ g/cc ($\rho_f = 1.0$),
$\phi_N = 21\%$, and
$\Delta t = 72$ μs/ft.

a. Find ρ_{maa}.
b. Find Δt_{maa}.
c. Plot these two values on figure 23.10.
d. What lithology mix does this point indicate?

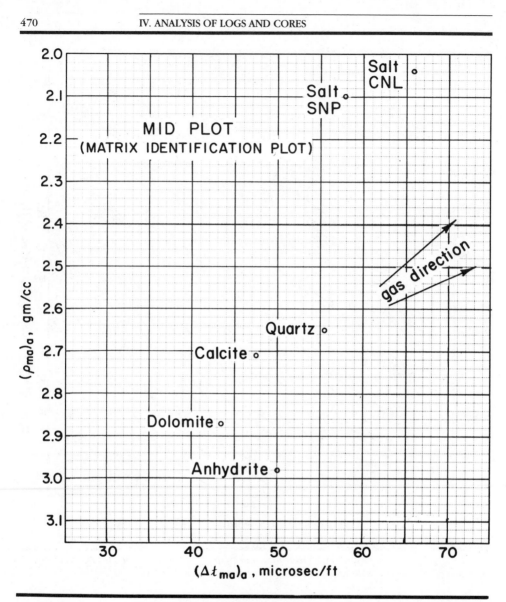

FIGURE 23.10 *The Matrix Identification (MID) Plot. Courtesy Schlumberger Well Services.*

When using either the MID plot or the *M-N* plot, it is useful to have available a reference table similar to table 23.1. This table summarizes the values of ρ_{ma}, Δt_{ma}, M, and N for common minerals and reservoir rocks.

THE LITHO–DENSITY LOG
AND THE PHOTOELECTRIC FACTOR P_e

Details and the measurement principle of the LDT are given in chapter 17 together with a quick-look guide to interpretation of the

TABLE 23.1 *Lithology Indicators for Common Minerals*

Mineral	ρ_b	ρ_{ma}	ϕ_{CNL}	Δt_{ma}	M	N
Sand	2.65	2.648	-1.5	55.5	0.84	0.64
Limestone	2.71	2.710	0.0	47.5	0.85	0.60
Dolomite	2.87	2.876	+2.5	43.5	0.80	0.53
Anhydrite	2.96	2.977	0.0	50.0	0.72	0.52
Halite	2.17	2.032	0.0	67.0	1.24	1.02
Sylvite	1.98	1.863	0.0	74.0	1.41	1.23
Gypsum	2.32	2.351	60.0	52.0	1.05	0.31

TABLE 23.2 *Density and Photoelectric Index for Common Minerals*

	P_e	ρ_{maa}	U_{maa}	ρ_b	U
Dolomite	3.14	2.88	9.11	2.88	9.05
Limestone	5.08	2.71	13.78	2.71	13.78
Sandstone	1.81	2.65	4.79	2.65	4.79
Magnesite	0.83	3.00	2.50	2.98	2.47
Anhydrite	5.05	2.98	15.06	2.98	15.06
Gypsum	3.99	3.69	18.76	2.35	9.38
Halite	4.65	2.36	12.44	2.04	9.49
Sylvite	8.51	2.25	23.08	1.86	15.83
Chlorite*	6.30	3.39	23.63	2.76	17.39
Illite*	3.45	2.92	10.97	2.52	8.69
Kaolinite*	1.83	2.96	6.14	2.41	4.41
Montmorillonite*	2.04	2.89	7.288	2.12	4.32
Muscovite	2.40	2.97	7.35	2.82	6.77
Biotite*	6.27	3.10	19.80	2.99	18.75
Glauconite*	6.37	3.05	21.52	2.54	16.18
Coal, bituminous	0.17	1.99	0.87	1.24	0.21
Barite	267	4.09	1091	4.09	1091
Hematite	21.5	5.27	113.5	5.18	111.3
Fresh water	0.36			1.00	0.36
Salt water (330 ppk)	1.64			1.19	1.95
Oil	0.12			0.88	0.11

*Average Values.

P_e curve. The more detailed discussion that follows is designed to shed more light on the usefulness of the P_e measurement when combined with other log readings.

Table 23.2 lists typical values of densities and photoelectric indices for the most common minerals. A more complete list is given in the appendix to this chapter. In most lithologies, combinations of minerals exist and the overall photoelectric index is not a linear function of the P_e values of the individual components. A new term, the volumetric photoelectric absorption index (U), must be calculated. This index is the product of electron density and the photoelectric factor.

$$U = \rho_e P_e .$$

An approximation is usually made by using bulk density ρ_b rather than electron density ρ_e. In a formation of complex lithology, the measured volumetric photoelectric absorption index is the sum of the individual volumetric photoelectric absorption indices weighted by their relative proportions in the formation:

$$U = U_1 V_1 + U_2 V_2 + \cdots,$$

where

U_1 = volumetric photoelectric absorption index of mineral 1,
V_1 = volumetric fraction of mineral 1 in the formation, etc.

For a porous single mineral shaly formation containing hydrocarbons, a general equation can be written as:

$$U = \underset{\substack{\text{matrix} \\ \text{term}}}{(1 - \phi V_{sh})U_{ma}} + \underset{\substack{\text{water} \\ \text{term}}}{\phi S_{xo} U_f} + \underset{\substack{\text{hydrocarbon} \\ \text{term}}}{\phi (1 - S_{xo}) U_{hy}} + \underset{\substack{\text{shale} \\ \text{term}}}{V_{sh} U_{sh}}.$$

It can be seen from the data in table 23.2 that the absorption coefficient U_f of fresh water is substantially lower than any of the matrix coefficients U_{ma} and can therefore be neglected without introducing a major error. Only if very salty muds are used would the term need to be included. The hydrocarbon contribution can also be neglected, because U_{hy} is less than 0.12. If the shale content of the formation is included in the matrix, the foregoing equations can be combined to produce the relationship

$$U_{ma} = \frac{P_e \rho_e}{1 - \phi}. \quad *$$

This equation can be solved using the nomogram shown in figure 23.11. For a quick-look interpretation, the porosity ϕ can be taken from a neutron–density crossplot.

In complex lithology (a mixture of two to three minerals), a second physical parameter is needed to define the individual minerals and their volume percentage in the formation. The apparent matrix density ρ_{maa} derived from the neutron–density crossplot may be used or ρ_{maa} can be calculated using the standard formula:

$$\rho_{maa} = \frac{\rho_b - \phi_X \rho_f}{1 - \phi_X},$$

where

*More strictly $U_{ma} = \dfrac{P_e \rho_e - \phi U_f}{1 - \phi}.$

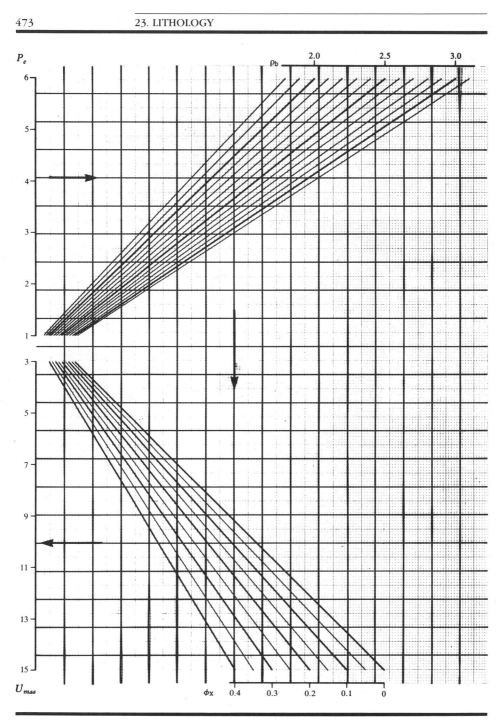

FIGURE 23.11 *Nomogram to Determine U_{maa} from P_e, ρ_b, and ϕ_X. Courtesy Schlumberger Well Services.*

ρ_f = fluid density of the invaded zone, and

ϕ_X = apparent porosity from density and neutron data.

The apparent volumetric photoelectric absorption index of the matrix, U_{maa}, can be calculated for any mineral composition if the apparent porosity is used:

$$U_{maa} = \frac{P_e \rho_b}{1 - \phi_X}.$$

Once the values of ρ_{maa} and U_{maa} have been calculated, they may be crossplotted against one another to help in identifying lithology. Figure 23.12 shows such a crossplot, made over a depth interval that includes anhydrite, dolomite, and shaly sand.

Using the pure data to obtain the matrix parameters, an overlay can be constructed (see fig. 23.13) which, when placed over the crossplot, indicates the most probable mineral composition. Of course, geological knowledge and cuttings analysis are a substantial help in selecting the main contributing components.

When the main mineral components of the formation have been identified, a plot can be made to determine their relative proportions. Figure 23.14 is an example of a plot valid for a quartz-calcite-dolomite composition, with a grid already established. Continuous computation of the relative proportions of up to three minerals is possible using this approach. Wellsite computation and display can be illustrated by an example: A section logged with neutron, density, and P_e curves is first analyzed by a U_{maa}-ρ_{maa} crossplot (fig. 23.15), which indicates the three main components of the formation to be magnesite, dolomite, and anhydrite. The relative proportions of the three minerals at each depth are computed according to where the data points fall within the defined triangle. Figure 23.16 shows the corresponding playback over the same interval. The average grain (matrix) density (RHGA) is displayed in Track 1; the lithology is displayed in the depth track; and the photoelectric absorption index (PEF), density porosity (DPL), average porosity (PHIA), and neutron porosity (NPL) are displayed in Tracks 2 and 3. The porosity values are scaled in limestone units. Because of the statistical nature of the measurements, not all of the data points will plot inside the selected solution triangle (see fig. 23.15) defined by the three minerals. When a point falls outside the triangle, a flag is raised on the left edge of Track 1, and only one or two minerals will be displayed.

LIMITATIONS

Both gas in the formation pore space and barite in the mud have a marked effect on the LDT tool measurements. These can be detected on a U_{maa}-ρ_{maa} crossplot (see fig. 23.17). Because of its

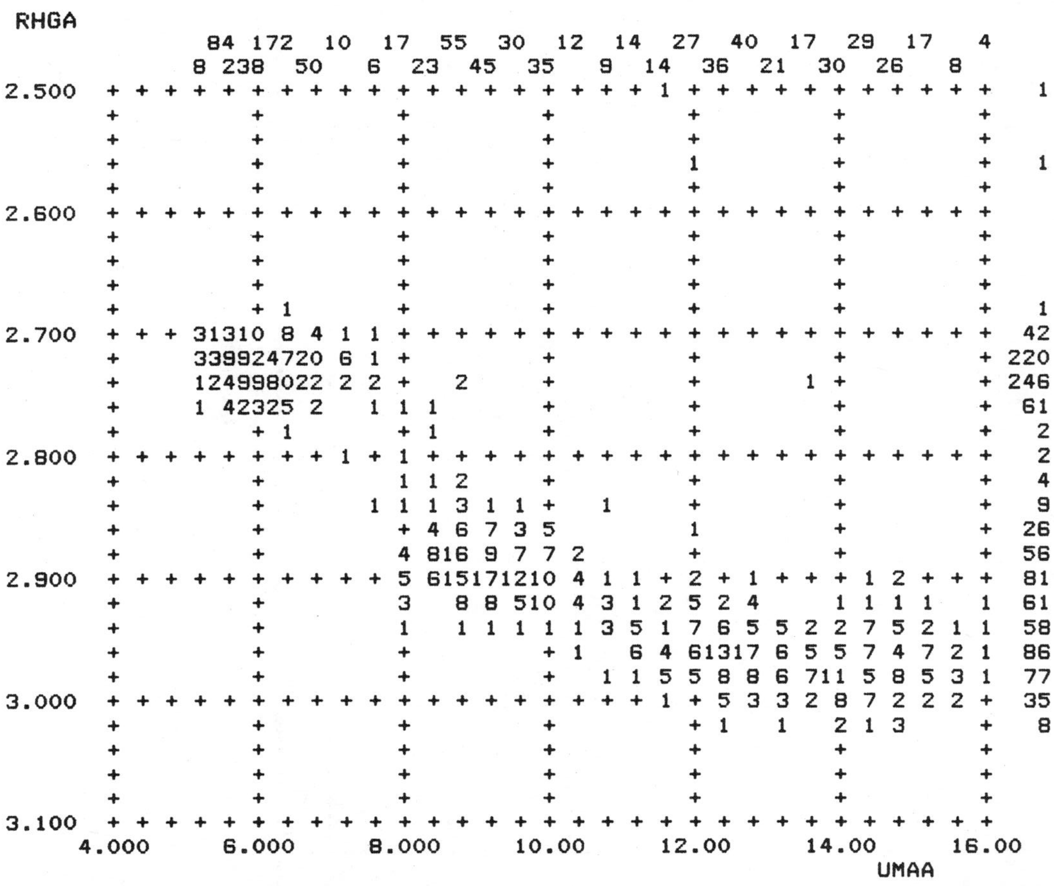
RHGA
2.500
2.600
2.700
2.800
2.900
3.000
3.100
4.000
6.000
8.000
10.00
12.00
14.00
16.00
UMAA

FIGURE 23.12 *Crossplot of Apparent Volumetric Photoelectric Absorption Index of the Matrix against Apparent Matrix Density. Courtesy Schlumberger Well Services.*

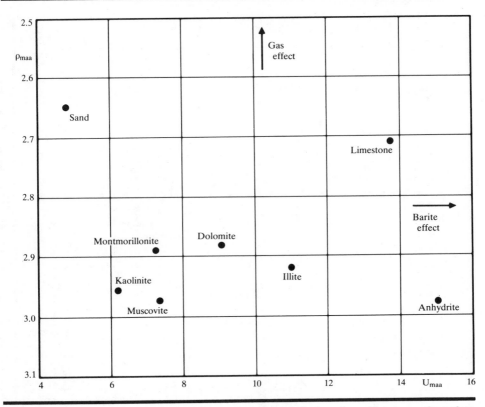

FIGURE 23.13 *Chart to Determine Lithology from U_{maa}-ρ_{maa} Crossplots. Courtesy Schlumberger Well Services.*

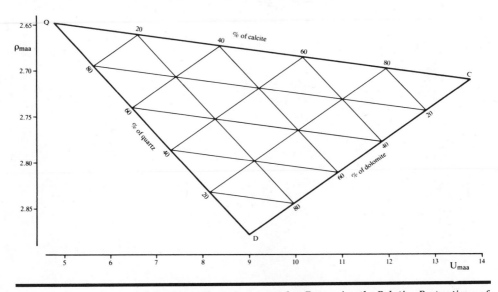

FIGURE 23.14 *Chart Constructed to Determine the Relative Proportions of Quartz, Dolomite, and Calcite. Courtesy Schlumberger Well Services.*

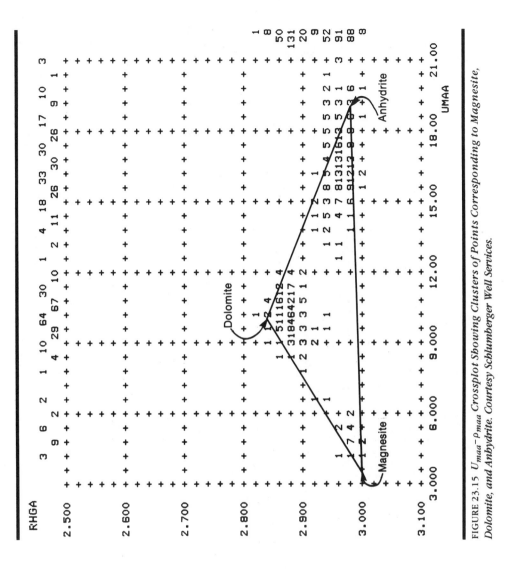

FIGURE 23.15 $U_{maa}-\rho_{maa}$ Crossplot Showing Clusters of Points Corresponding to Magnesite, Dolomite, and Anhydrite. Courtesy Schlumberger Well Services.

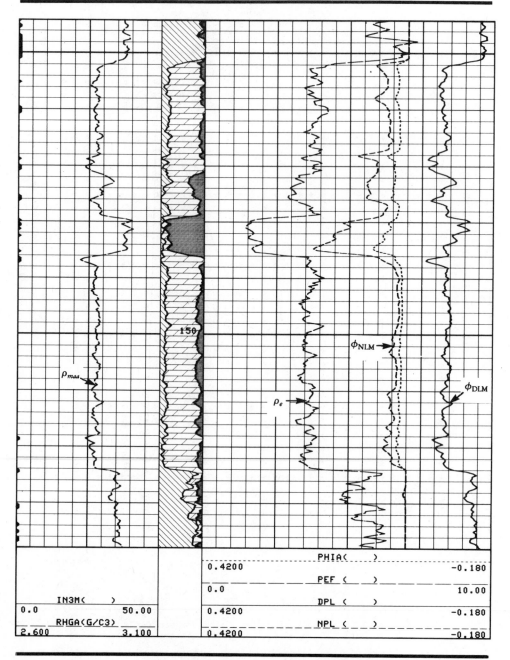

			PHIA( )	
		0.4200		-0.180
			PEF ( )	
		0.0		10.00
IN3M( )			DPL ( )	
0.0	50.00	0.4200		-0.180
RHGA(G/C3)			NPL ( )	
2.600	3.100	0.4200		-0.180

FIGURE 23.16 *Example of Wellsite Computation in Mixed Lithology. Courtesy Schlumberger Well Services.*

RHGA

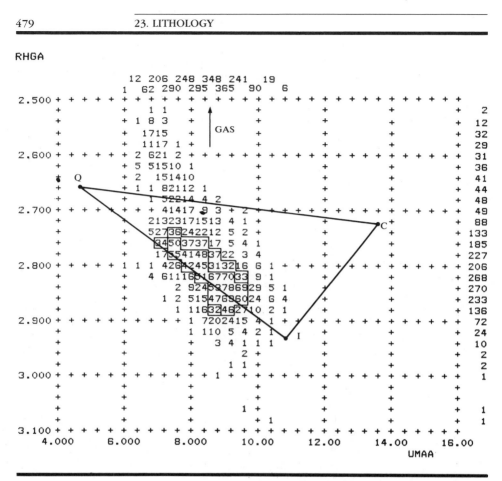

FIGURE 23.17 *Influence of Gas on the $U_{maa}-\rho_{maa}$ Crossplot: Main Effect is on the Apparent Matrix Density. Courtesy Schlumberger Well Services.*

high atomic number, barite moves the U_{maa} points toward the right (higher U_{maa}). Gas influences the ρ_{maa} values but has little effect on U_{maa} and so moves the points upward. By picking the U_{maa} and ρ_{maa} values of the minerals according to the crossplots rather than their theoretical values, the relative proportions of the three minerals can still be determined in some cases.

APPENDIX: *Photoelectric Factors of Elements and Minerals*

Name	Formula	Molecular Weight	P_e	Z_{eq}	ρ	ρ_e	ρ_b	c
A. Elements	H	1.000	0.00025	1				1.984
	C	12.011	0.15898	6				0.9991
	O	16.000	0.44784	8				1.0000
	Na	22.991	1.4093	11				0.9566
	Mg	24.32	1.9277	12				0.9868
	Al	26.98	2.5715	13	2.700	2.602	2.596	0.9637
	Si	28.09	3.3579	14				0.9968
	S	32.066	5.4304	16	2.070	2.066	2.022	0.9979
	Cl	35.457	6.7549	17				0.9589
	K	39.100	10.081	19				0.9719
	Ca	40.08	12.126	20				0.9980
	Ti	47.90	17.089	22				0.9186
	Fe	55.85	31.181	26				0.9311
	Cu	63.5	43.5	29				
	Sr	87.63	122.24	38				0.8673
	Zr	91.22	147.03	40				0.8770
	Ba	137.36	493.72	56				0.8154
B. Minerals								
Anhydrite	$CaSO_4$	136.146	5.055	15.69	2.960	2.957	2.977	0.9989
Barite	$BaSO_4$	233.366	266.8	47.2	4.500	4.011		0.8913
Calcite	$CaCO_3$	100.09	5.084	15.71	2.710	2.708	2.710	0.9991
Carnallite	$KCl\text{-}MgCl_2\text{-}6H_2O$	277.88	4.089	14.79	1.61	1.645	1.573	1.0220
Celestite	$SrSO_4$	183.696	55.13	30.4	3.960	3.708		0.9363
Corundum	Al_2O_3	101.96	1.552	11.30	3.970	3.894		0.9808
Dolomite	$CaCO_3\text{-}MgCO_3$	184.42	3.142	13.74	2.870	2.864	2.877	0.9977
Gypsum	$CaSO_4\text{-}2H_2O$	172.18	3.99	14.07	2.320	2.372	2.350	1.0222
Halite	$NaCl$	58.45	4.65	15.30	2.165	2.074	2.040	0.9580
Hematite	Fe_2O_3	159.70	21.48	23.45	5.240	4.987		0.9518
Ilmenite	$FeOTiO_2$	151.75	16.63	21.87	4.70	4.46		0.9489
Magnesite	$MgCO_3$	84.33	0.829	9.49	3.037	3.025	3.049	0.9961
Magnetite	Fe_3O_4	231.55	22.08	23.65	5.180	4.922		0.9501
Marcasite	FeS_2	119.98	16.97	21.96	4.870	4.708		0.9668
Pyrite	FeS_2	119.98	16.97	21.96	5.000	4.834		0.9668
Quartz	SiO_2	60.09	1.806	11.78	2.654	2.650	2.648	0.9985
Rutile	TiO_2	79.90	10.08	19.02	4.260	4.052		0.9512
Siderite	$FeCO_3$	115.86	14.69	21.09	3.94		3.89	
Sylvite	KCl	74.557	8.510	18.13	1.984	1.916	1.862	0.9657
Zircon	$ZrSiO_4$	183.31	69.10	32.45	4.560	4.279		0.9383
C. Liquids								
Water	H_2O	18.016	0.358	7.52	1.000	1.110	1.000	1.1101
Salt water	(120,000 ppm)		0.807	9.42	1.086	1.185	1.080	1.0918
Oil	$CH_{1}\text{-}6$		0.119	5.53	0.850	0.948	0.826	1.1157
	CH_2		0.125	5.61	0.850	0.970	0.880	1.1407

APPENDIX: *Photoelectric Factors of Elements and Minerals (continued)*

Name	Formula	Molecular Weight	P_e	Z_{eq}	ρ	ρ_e	ρ_b	c
D. Miscellaneous								
Berea sandstone			1.745	11.67	2.308	2.330	2.305	0.9993
Pecos sandstone			2.70	13.18	2.394	2.414	2.395	1.0000
Average shale			3.42	14.07	2.650	2.645	2.642	0.998
Anthracite coal	C:H:O=93:3:4		0.161	6.02	1.700	1.749	1.683	1.0287
Bituminous coal	C:H:O=82:5:13		0.180	6.21	1.400	1.468	1.383	1.0485

BIBLIOGRAPHY

Burke, J.A., Campbell, R.L., and Schmidt, A.W.: "The Litho-Porosity Cross-plot," *Trans.*, SPWLA 10th Annual Logging Symposium, May 1969.

Burke, J.A., Curtis, M.R., and Cox, J.T.: "Computer Processing of Log Data Enables Better Production in Chaveroo Field," SPE paper 1576, presented at the 41st Annual Meeting, Dallas, Oct. 2–5, 1966.

Nugent, W.H., Coates, G.R., and Peebler, R.P.: "A New Approach to Carbonate Analysis," *Trans.*, SPWLA 19th Annual Logging Symposium, El Paso, June 13–16, 1978.

Poupon, A., Hoyle, and Schmidt: "Log Analysis in Formations with Complex Lithologies," *J. Pet. Tech.* (August 1971) **24**.

Raymer, L.L., and Biggs, W.P.: "Matrix Characteristics Defined by Porosity Computations." *Trans.*, SPWLA 4th Annual Logging Symposium (1963).

Savre, Wayland, C.: "Determination of a More Accurate Porosity and Mineral Composition in Complex Lithologies with the Use of Sonic, Neutron, and Density Surveys." *J. Pet. Tech.* (September 1963) **16**, 945–959.

Schlumberger: "Well Evaluation Conference—Iran" (1976).

Schlumberger: "Log Interpretation Charts" (1977).

Schlumberger: "Litho Density Tool Interpretation," M 086108, (1982).

Schlumberger: "Well Evaluation Developments, Continental Europe," (1982).

Schmidt, A.W., Land, A.G., Yunker, J.D., and Kilgore, E.C.: "Applications of the Coriband Technique to Complex Lithologies," paper Z, *Trans.*, SPWLA, 12th Annual Logging Symposium, Dallas, May 2–5, 1971.

Tilly, H.P., Gallagher, B.J., and Taylor, T.D.: "Methods for Correcting Porosity Data in a Gypsum-Bearing Carbonate Reservoir," *J. Pet. Tech.* (October 1982) **35**, 2449–2454.

Tixier, M.P., and Alger, R.P.: "Log Evaluation of Non-Metallic Mineral Deposits," SPWLA 9th Annual Logging Symposium, New Orleans, 1968.

Answers to Text Questions

QUESTION #23.1
Dolomite

QUESTION #23.2
Sandstone

QUESTION #23.3
Sandstone

QUESTION #23.4
Limestone (or is it sand and dolomite?)

QUESTION #23.5
Trona

QUESTION #23.6
Anhydrite

QUESTION #23.7
a. $M = 0.8$
b. $N = 0.6$
f. sand-dolomite mixture

QUESTION #23.8
a. $\rho_{maa} = 2.80$
b. $\Delta t_{maa} = 46.0$
d. Lime-dolomite mix

CHAPTER TWENTY-FOUR

R_w DETERMINATION

Until this point in the discussions, the value of R_w has been assumed. However, in many cases its value cannot be assumed and must be found. Various methods are available for determining the resistivity of the formation water, R_w:

Direct measurement of a water sample
Computation from chemical analysis of a water sample
Use of the SP
The R_{wa} technique
The ratio technique
Various types of crossplots
Use of F overlays

The following is a detailed discussion of each of these means for finding the value of R_w.

DIRECT MEASUREMENT
Direct measurement of R_w from a water sample requires two things: an uncontaminated representative sample of formation fluid and a resistivity cell. The sample could come from:

a. a drillstem test,
b. a wireline formation tester, or
c. a produced sample collected at the separator.

In all cases, collection of the water sample should be made in a clean, dry glass container that is stoppered to avoid evaporation. In cases (a) and (b), care should be taken to assure that the recovered sample is actual formation water and not mud filtrate. Nitrate tracer purposely placed in the drilling mud is an easy way to distinguish drilling mud or filtrate-contaminated samples from the formation water. Nitrates are never found in nature, so the presence of nitrate in a recovered water sample is indicative of contamination by mud filtrate. Normally, a concentration of 100 ppm of NO_3 is maintained in the mud. Most mud companies are equipped to measure nitrate concentrations at, and below, this concentration.

Resistivity cells are found in laboratories, logging trucks, and mud loggers' briefcases. Mud loggers like the battery-operated type, which invariably have dead batteries, are usually out of calibration, and have very condensed scales that inhibit accurate measurement.

Also, they operate on DC current and are prone to polarization effects. The resistivity cells in logging trucks are more sophisticated, accept larger samples of water, work on AC current, and can be calibrated before each reading. For these reasons, they are preferred when measuring R_w values.

The measurement will be in $\Omega \cdot m$ at a given temperature. At 75°F (a normal ambient temperature), fresh waters will have a resistivity from half an ohm to several ohms and brines will have a resistivity of a tenth of an ohm or less. Seawater normally has a resistivity of 0.2 $\Omega \cdot m$. These figures are approximate. Conversion of a resistivity at one temperature to equivalent resistivity at another temperature can be made by using figure 24.1. Note that this chart assumes a sodium chloride solution and may not be strictly valid for waters containing ions other than sodium and chloride.

R_w COMPUTED FROM CHEMICAL ANALYSIS

If Na^+ and Cl^- ions are the only ions present in a solution, computation of an R_w value is a simple matter. Figure 24.1 can be used by entering the chart with a total NaCl concentration and a temperature and exiting with an R_w value in $\Omega \cdot m$.

QUESTION #24.1

An NaCl solution is shown by chemical analysis to contain 20,000 ppm NaCl. What is its resistivity at 175°F?

Titration test results are frequently quoted as ppm of chlorides. This is merely an indication of the chemistry of the method used. To convert ppm of chlorides to the equivalent ppm of NaCl, the value of chlorides in ppm is multiplied by the factor of 1.65.

QUESTION #24.2

A recovered DST water sample titrated out at 87,000 ppm chlorides.

a. What would the equivalent ppm of NaCl be?
b. At 75°F, what resistivity would this solution exhibit?

When working with formation waters that have ions other than Na^+ and Cl^-, the effect these ions have on the electrical behavior of an electrolyte must be considered. The two factors that must be taken into consideration are (1) the charge carried by the ion and (2) its mobility through an electrolyte in the presence of an electric field.

Obviously, a bivalent or trivalent ion carrying multiple charges (for example, calcium and magnesium ions carry two positive

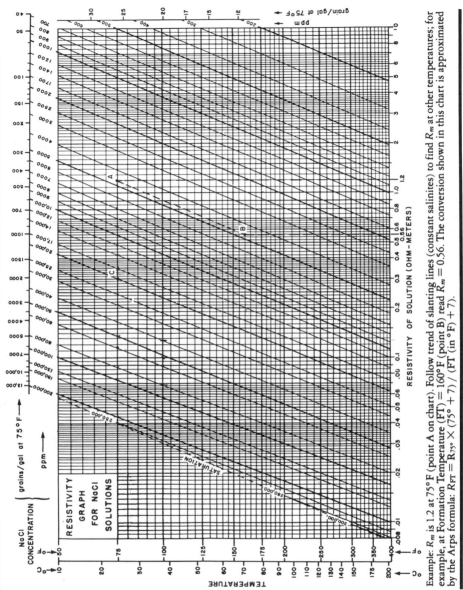

Example: R_m is 1.2 at 75°F (point A on chart). Follow trend of slanting lines (constant salinities) to find R_m at other temperatures; for example, at Formation Temperature (FT) = 160°F (point B) read R_m = 0.56. The conversion shown in this chart is approximated by the Arps formula: $R_{FT} = R_{75°} \times (75° + 7) / (FT (in °F) + 7)$.

FIGURE 24.1 *Resistivity Nomograph for NaCl Solutions. Courtesy Schlumberger Well Services.*

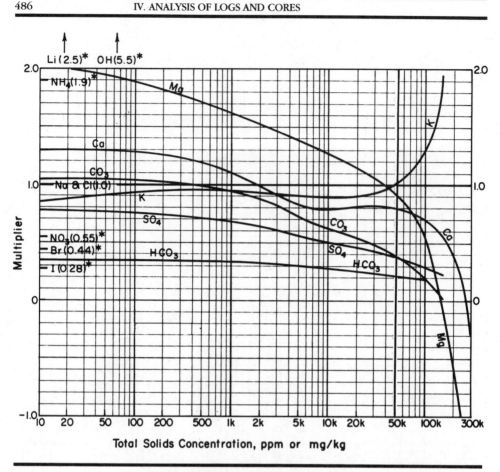

FIGURE 24.2 *Multipliers and Total-Solids Concentration for Various Ions.*
Courtesy Schlumberger Well Services.

charges each) will contribute more conductivity than a monovalent
ion such as Na^+ or Cl^-. However, these multivalent ions normally are
surrounded by oppositely charged ions—for example, Ca^{++} may be
surrounded by a couple of OH^- (hydroxide) ions. These attached
ions tend to slow down the Ca^{++} ion in its progress through the
electrolyte. So even though a Ca^{++} ion delivers more charges per
trip, it may end up making the trip through the electrolyte so slowly
that a monovalent ion would do the charge-carrying job better
because it moves more quickly.

These two opposing effects, charge-carrying capability and
mobility (speed), are very sensitive to the total ionic concentration
in the electrolyte. At low concentrations, the bivalent ions can move
around freely and are better charge carriers than the monovalent
ions. At high concentrations, the reverse is true. A practical method
of dealing with this problem is to use multipliers for each ion to
compute its effective charge-carrying power relative to Na^+ and Cl^-.
See figure 24.2, where the y axis gives the value of the multiplier
and the x axis the total-solids concentration in ppm. Lines on the

chart show the behavior of each type of common ion found in naturally occurring waters.

Consider the following example: Chemical analysis reveals that a water sample contains the following ions and respective concentrations:

Ion	Symbol	Concentration (ppm)
Sodium and chloride	Na^+ and Cl^-	25,000
Carbonate	CO_3^{--}	2000
Sulfate	SO_4^{++}	5000
Bicarbonate	HCO_3^-	10,000
Magnesium	Mg^{++}	4000
Total		46,000

The first step is to find the total concentration in ppm—which is 46,000 for this water. On figure 24.2, a line can be drawn vertically through the point on the x axis at 46,000 (46K) ppm. Where this line intersects the corresponding line for each ion present in the mixture, the multiplier is read. The next step is to multiply each ion's concentration by its own multiplier:

Ion	Concentration		Multiplier		Effective Concentration
Na^+ and Cl^-	25,000	×	1.00	=	25,000
CO_3^{--}	2000	×	0.40	=	800
SO_4^{++}	5000	×	0.37	=	1850
HCO_3^-	10,000	×	0.20	=	2000
Mg^{++}	4000	×	0.95	=	3800
Total	46,000				33,450

The effective concentrations are then totaled. For this example water, the total ion concentration is 33,450 ppm. This value is the ppm of NaCl that would have the same electrical resistivity as the actual mixture of ions present in the formation water.

The final step is to determine the resistivity of the electrolyte solution containing these 33,450 ppm of equivalent NaCl ions. For example, the R_w at 200°F is about 0.075 Ω·m (fig. 24.1). If the actual concentrations of the ions had been used (without taking into account the individual ability of each ion to conduct electricity), an R_w of 0.056 Ω·m would have been determined, resulting in an error of 25%.

Significant errors in determining R_w values for formation waters frequently occur (a) in formations at or near evaporite and/or carbonate formations, and (b) in shallow freshwater formations, especially near mountains. When drilling in an area where evaporite and carbonate formations occur, for example, a recovered water sample may be classified as a "saturated brine"; R_w will be calculated at 0.015 Ω·m on the assumption that the only ions present are Na^+ and Cl^-; and, using this low R_w value, log evaluation will indicate the presence of oil in numerous formations. Therefore, numerous drillstem tests are planned. Then, one after the other, all of the formations produce only water during the DST's. After much headscratching, it dawns on the eager engineer that something went wrong.

To understand the effect of carbonates on R_w, the behavior of Mg^{++} and Ca^{++} ions must be considered. At a total concentration of 200,000 ppm, the Mg^{++} multiplier is a *negative* value (-0.65) and the Ca^{++} multiplier is only 0.35. Assuming this solution had the following concentrations of ions, the effective NaCl ppm is computed using the appropriate multipliers:

Ion	Concentration	Multiplier		Effective Concentration
Na^+ and Cl^-	100,000	$\times$ 1.00	=	100,000
Ca^{++}	50,000	$\times$ 0.35	=	17,500
Mg^{++}	50,000	$\times$ -0.65	=	-32,500
Total	200,000			85,000

The effective NaCl equivalent concentration is only 85,000 ppm. For a formation temperature of 240°F, the correct R_w is 0.029 $\Omega \cdot m$ instead of the 0.015 originally computed. The effect of carbonates on the computation of water saturation is dramatic. Formations where water saturation was calculated to be 40% using an R_w value of 0.015 will have a calculated value of 56% using an R_w value of 0.029.

The problem with freshwater-bearing formations is different. A frequently found component in fresh formation water is the bicarbonate ion, HCO_3^-. Its multiplier remains less than 1.00 throughout the range of normal formation waters. For example, consider a freshwater sample reported as having a total ion content of 5000 ppm. Based on an assumption of NaCl content only, such a solution would be expected to have a resistivity of 0.56 $\Omega \cdot m$ at 150°F. If half the ions present are bicarbonate ions, however, the situation changes radically:

Ion	ppm	Multiplier		Equivalent ppm NaCl
HCO_3^-	2500	$\times$ 0.3	=	750
Na^+ and Cl^-	2500	$\times$ 1.0	=	2500
Total	5000			3250

The resistivity of a 3250 ppm NaCl solution at 150°F is 0.85 $\Omega \cdot m$ or 52% higher than would be expected from the assumption that sodium and chloride were the only ions present.

The bivalent cations Mg^{++} and Ca^{++} can also occur in freshwaters. In dilute solutions, these ions can cause difficulties because they are very good charge carriers and build massive SP deflections. Thus, they cause R_w values to appear too low.

In summary, sodium chloride is not the only substance found in formation waters. Other ions, even in quite small concentrations, can alter the resistivity characteristics of the water. To account adequately for these other ions, a chemical analysis is required. The effects of bivalent cations at high concentrations (total ppm) is to make a solution less conductive (raise R_w). At low total concentrations, bivalent cations have the reverse effect and make a solution more conductive (lower R_w). Bicarbonates always reduce conductivity (raise R_w).

R_w FROM THE SP

This subject has been dealt with extensively in chapter 9. However, one aspect of R_w determination from the SP may be enlarged on here with advantage. It is common practice to drill wells using potassium-based muds mainly for control of problem shales. In these cases, the conventional SP charts and nomograms are no longer applicable since the electrochemical properties of KCl are not the same as those of NaCl. When faced with this situation, the analyst should use figure 24.3 to estimate R_{mfe} and figure 24.4 to estimate the R_{mfe}/R_{we} ratio from the SP deflection.

QUESTION #24.3

A well is drilled with KCl mud.

R_{mf} at formation temperature of 200°F is 0.28.

The SP development is –40 mV.

Calculate R_w at formation temperature assuming connate water is a NaCl solution.

THE R_{wa} METHOD

The R_{wa} technique relies on Archie's relationship:

$$S_w^n \ = \ F \ \times \ R_w/R_t.$$

With a little rearranging this equation can be written as

$$R_w \ = \ R_t \ \times \ S_w^n/F.$$

If S_w is assumed to be 1 (100% water-saturated formation), then the apparent value of R_w is called R_{wa}:

$$R_{wa} \ = \ R_t/F.$$

In a 100% water-saturated (wet) zone, R_{wa} will have a low value and in an oil-bearing zone R_{wa} will have a high value. The lowest value found for R_{wa} will be where the original assumption of $S_w = 1$ is valid, in which case $R_{wa} = R_w$.

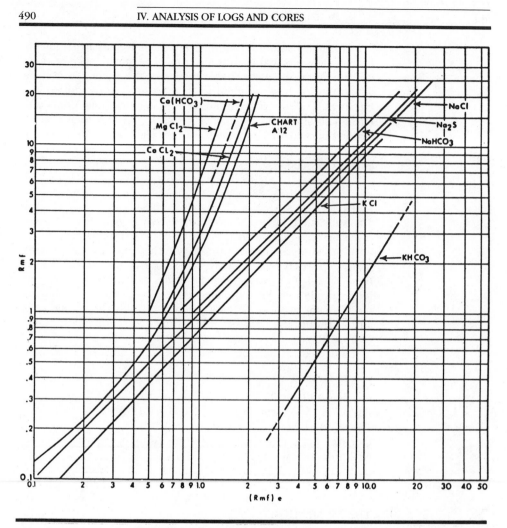

FIGURE 24.3 *Conversion of R_{mf} to R_{mfe} for Potassium Mud Filtrates at 25°C. Reprinted by permission of the SPWLA from Cox and Raymer 1976. SPWLA 17th Symposium.*

Remembering the relationship between F and ϕ,

$$F = a/\phi^m,$$

substitution into Archie's equation results in an expression for R_{wa} in terms of porosity:

$$R_{wa} = \phi^m \times R_t/a.$$

The simplest form of this equation, for use on a pocket calculator, is

$$R_{wa} = \phi^2 \times R_t,$$

where m is assumed to equal 2 and a is assumed to equal 1.

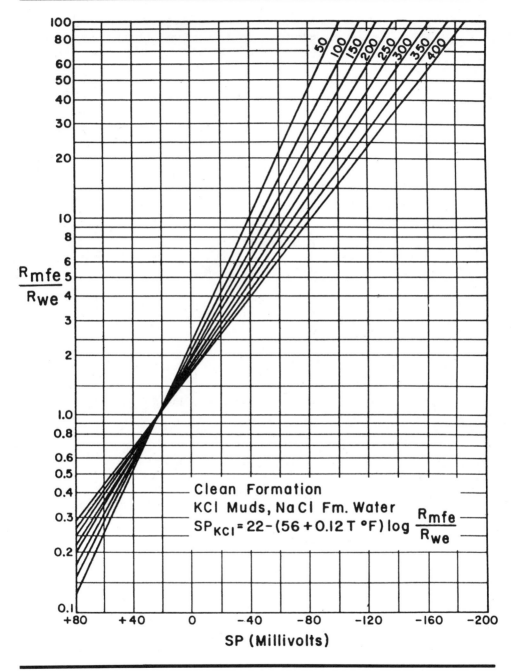

FIGURE 24.4 *SP vs. R_{mfe}/R_{we} for KCl Muds. Reprinted by permission of the SPWLA from Cox and Raymer 1976. SPWLA 17th Symposium.*

The question arises, which porosity device should one use? What about shale effects and the influence of filtrate invasion on R_t? Usually the choice will be between a neutron porosity log and a density porosity log. Shale will, in general, cause neutron porosity to read higher than true porosity; and the presence of shale will, in general, reduce the resistivity of a formation. Thus, a combination of the two is to some extent self-compensating. A neutron-derived R_{wa} value is referred to as an R_{wa-N} value. In pure shale beds, R_{wa} may compute *lower* than R_w. Distinguishing these spuriously low values of R_{wa} from the legitimate measurements requires some independent indicator of shale, such as a gamma ray curve.

QUESTION #24.4

The following porosities and resistivities are noted as depth increases through a zone.

Level	Porosity	R_t	R_{wa}
1	23	19.5	
2	17	36.0	
3	19	18.0	
4	22	6.5	
5	18	5.5	
6	20	4.0	
7	16	6.5	

Compute R_{wa} at each level assuming $R_{wa} = (\phi^2 R_t)$ and pick a minimum value for R_w.

The R_{wa} technique is most easily applied by computing R_{wa} over long intervals of logs and reviewing the analog print of the resulting curve. This approach calls for a computer. Modern computerized logging trucks can generate this curve during the logging run. The R_{wa} plot is a very useful quick-look tool. Not only can R_w be simply found but quick estimates of S_w can be made as well. Minor rearrangement of Archie's equation will suffice to show that

$$S_w^n = R_w / R_{wa}.$$

Thus, if the logged section includes both a water-bearing interval and a hydrocarbon-bearing interval, it is a simple matter to compute water saturation.

QUESTION #24.5

a. Inspect the logs shown in figure 24.5.
   Identify a wet interval and read the minimum value of R_{wa}.
b. Find a hydrocarbon-bearing interval and compute S_w.

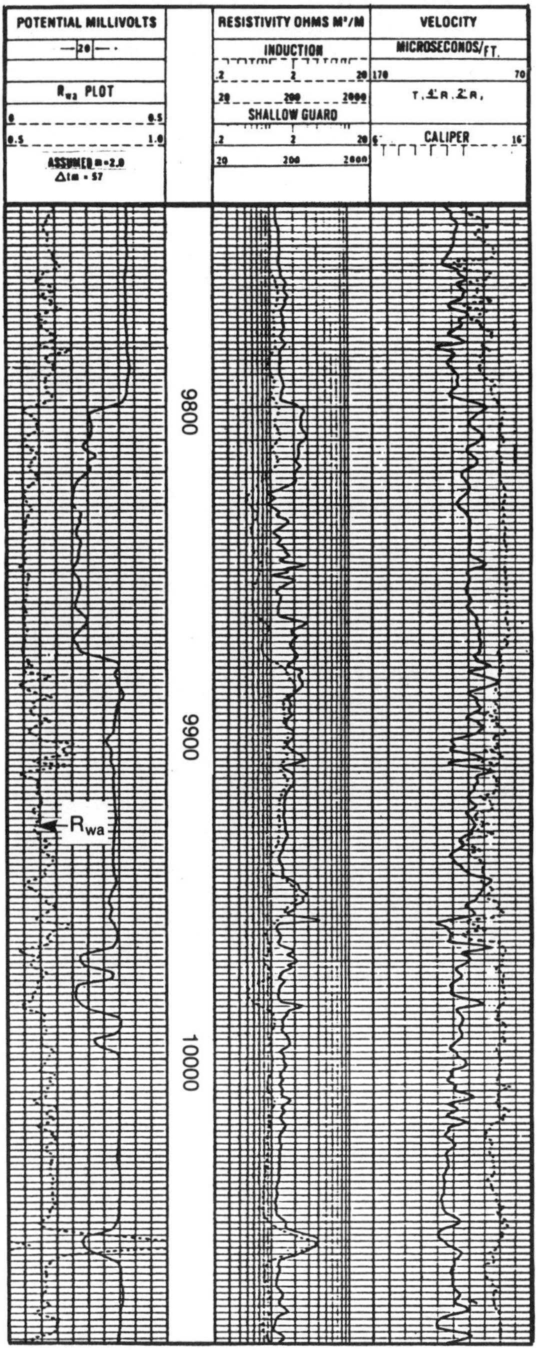

FIGURE 24.5 *Example of Continuous R_{wa} Plot.*

THE RATIO TECHNIQUE

The ratio technique is rather easy to use because it eliminates the need to know every variable. The equation for water saturation in the undisturbed portion of the formation is

$$S_w^n = F(R_w/R_t) ;$$

and in the zone invaded by the drilling-mud filtrate, the equation is

$$S_{xo}^n = F(R_{mf}/R_{xo}) .$$

If only 100% water-saturated zones are considered, then $S_w = S_{xo} = 1$. By taking a ratio of the two above equations, the need to know F is eliminated and

$$R_{mf}/R_w = R_{xo}/R_t ,$$

thus,

$$R_w = R_{mf}(R_t/R_{xo}) .$$

If reliable values of R_t, R_{xo}, and R_{mf} are available, a value of R_w can be determined.

Provided invasion effects do not materially affect the R_t and R_{xo} measurements determined by the deep and shallow investigation resistivity devices, the technique works well and has the advantage of being independent of porosity.

QUESTION #24.6

If R_{mf} (at 80°F) = 0.15 Ω·m,

T_{form} = 175°F,

R_t = 2.5 Ω·m, and

R_{xo} = 1.5 Ω·m,

find R_w at formation temperature.

CROSSPLOTS

The use of crossplots can also give good indications of R_w. Again, Archie's relation can be modified to produce a workable system. Some analysts favor plots on linear axes and others plots on logarithmic axes. Both types will be covered here.

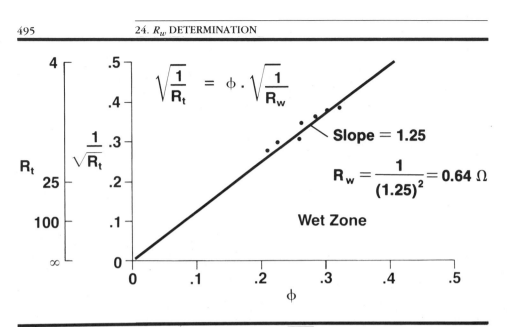

FIGURE 24.6 *Plot of ϕ against $\sqrt{1/R_t}$ in Wet Zone.*

Linear Crossplots

Archie's equation may be rearranged to read

$$1/R_t = S_w^n \phi^m / a R_w.$$

If $n = m = 2$, then,

$$\sqrt{1/R_t} = S_w \phi / \sqrt{a R_w}.$$

Thus, in a 100% water-bearing interval, a plot of $\sqrt{1/R_t}$ versus ϕ will produce a straight-line of slope $\sqrt{1/a \; R_w}$, as shown in figure 24.6.

Since the parameter $\sqrt{1/R_t}$ is rather an awkward animal to handle, special sheets of graph paper can be prepared with a reciprocal root scale already marked (see fig. 24.7). The practical procedure is straightforward; points are taken at regular intervals over the gross thickness of a formation and plotted for both oil- and water-bearing sections. A water-bearing line is drawn from the origin $\sqrt{1/R_t} = 0$, $\phi = 0$, through the most northwesterly points plotted. The slope of this line determines R_w. Note that only water-bearing points will fall on a line. Points for which S_w is less than 1 will fall below the line.

R_w is determined by calculating the square of the porosity value (expressed as a fraction) corresponding to an R_t value of 1. If this does not fall on the graph conveniently, R_w is equally well determined by calculating ten times the square of the porosity corresponding to an R_t value of 10.

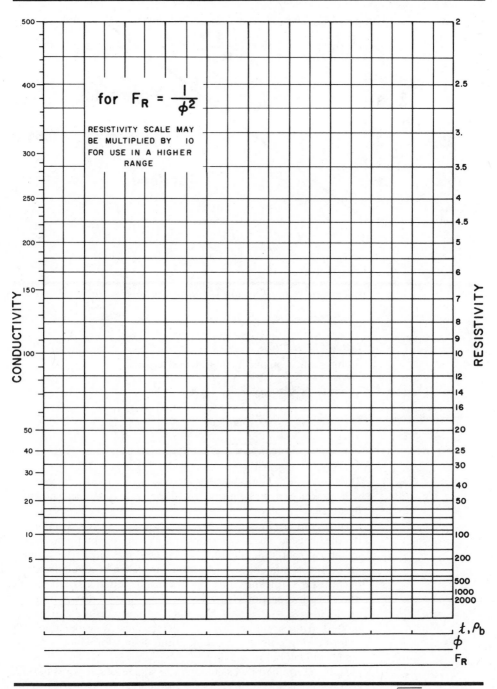

FIGURE 24.7 *Specially Scaled Graph Paper for Plotting $\sqrt{1/R_t}$ vs. ϕ.*
Courtesy Schlumberger Well Services.

QUESTION #24.7
Plot the following points on figure 24.7 and deduce R_w.

ϕ (%)	R_t ($\Omega \cdot m^2/m$)
19.0	4.2
15.0	6.7
12.0	10.4
17.0	5.2
13.0	8.9
18.0	4.6
14.0	7.7
16.0	5.9

This kind of plot has two other uses, which will be discussed later—namely, saturation determination and matrix response for the porosity tool involved.

Logarithmic Plots

Archie's equation may equally well be written in logarithmic form such that

$$n \log S_w = \log a + \log R_w - m \log \phi - \log R_t$$

which, by rearrangement, yields

$$\log R_t = -m \log \phi + \log a R_w - n \log S_w .$$

A plot on log-log paper of R_t versus ϕ will, therefore, give a straight line in water-bearing points (where $S_w = 1$ and hence $n \log S_w = 0$) with a slope of $-m$ and an intercept of $a R_w$. Figure 24.8 shows an appropriately scaled sheet of log-log graph paper on which the points used in question 24.6 have been plotted. Note that, on this type of plot, oil- or gas-bearing points will lie to the northeast of the 100% water line.

F-OVERLAY TECHNIQUE

Archie's equation can be expressed in a logarithmic form:

$$S_w^n = F R_w / R_t$$

$$n \log S_w = \log F + \log R_w - \log R_t .$$

If the value of $S_w = 1$, the value of $\log S_w = 0$. Thus,

$$\log R_w = \log R_t - \log F .$$

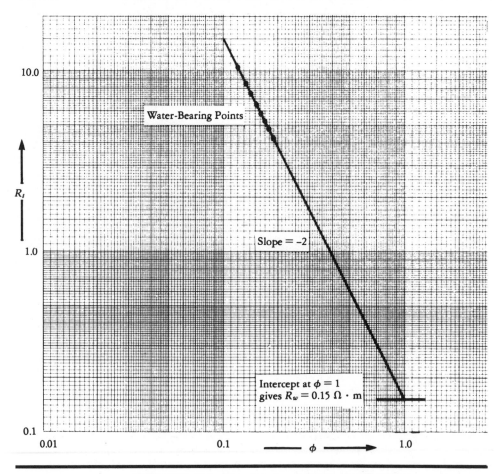

FIGURE 24.8 *Plot of Log ϕ vs. Log R_t for R_w Determination.*

This is most conveniently solved graphically using a logarithmic F curve generated from a porosity tool (usually a neutron or density log) and a logarithmic R_t curve (usually R_{ILD} or R_{LLd}). In a wet zone, these curves will overlay one another provided they are shifted relative to one another by an amount equal to log R_w. Thus, the act of normalizing the F curve to the R_t curve effectively calculates R_w, as explained in chapter 21.

R_w CATALOGS
R_w catalogs are another possible source of values for R_w. They have been compiled for several geographical areas around the world, by various groups (such as the API and the SPWLA), from data collected by various oil and service companies and from commercial laboratory data. Normally, they are useful as guides to what one may expect for values of R_w in various formations. For the explorationist in a wildcat situation, an R_w catalog will not be of any great use.

BIBLIOGRAPHY

Cox, J.W., and Raymer, L.L.: "The Effect of Potassium Salt Muds on Gamma Ray, and Spontaneous Potential Measurement," *Trans.,* SPWLA, 17th Annual Logging Symposium, Denver, June 9–12, 1976.

Desai, K.P., and Moore, E.J.: "Equivalent NaCl Concentrations from Ionic Concentrations," *The Log Analyst* (May–June 1969) **10**.

Schlumberger: "Log Interpretation Charts" (1972).

Schlumberger: "Log Interpretation Charts" (1979).

Welex: "Openhole Services," G-6003.

Answers to Text Questions

QUESTION #24.1
0.13 $\Omega \cdot m^2/m$

QUESTION #24.2
a. 143,550 ppm NaCl
b. 0.59 $\Omega \cdot m^2/m$

QUESTION #24.3
0.065 $\Omega \cdot m^2/m$ at 200°F

QUESTION #24.4
0.160 $\Omega \cdot m^2/m$

QUESTION #24.5
a. Wet zone 9830 to 9875 ft
R_w = 0.05 $\Omega \cdot m^2/m$
b. Pay zone 10,052 to 10,060 ft
R_{wa} = 0.55 $\Omega \cdot m^2/m$
S_w = 30%

QUESTION #24.6
0.12 $\Omega \cdot m^2/m$

QUESTION #24.7
0.15 $\Omega \cdot m^2/m$

CHAPTER TWENTY-FIVE

WATER SATURATION

This chapter will review in detail various methods of determining the water saturation of a rock system. Water saturation can be used to determine whether hydrocarbon production is probable and to calculate the volume of oil in a unit volume of rock (e.g., how many barrels per acre).

Water saturation (S_w) is defined as the fraction of the pore volume occupied by water. This is shown schematically in figure 25.1. The actual bulk volume of water in a unit volume of porous rock is given by the product ϕS_w and the bulk volume of hydrocarbons is given by $\phi(1 - S_w)$. This latter value is also referred to as the hydrocarbon pore volume or HCPV.

QUESTION #25.1
A rock sample has a volume of 100 cc. It contains 30 cc of pore space, of which 10 cc are occupied by water.

a. What is the porosity (ϕ)? $\quad\quad\quad\quad\quad\quad\quad\quad\quad\quad \phi$ =
b. What is the water saturation (S_w)? $\quad\quad\quad\quad\quad\quad S_w$ =
c. What is the hydrocarbon pore volume? $\quad\quad\quad$ HCPV =

METHODS AVAILABLE TO DETERMINE S_w
Many methods are available for finding the water saturation of a rock sample. They include:

Core analysis
Dielectric measurements
Nuclear measurements
Archie's equation
The ratio technique (SP vs. R_{xo}/R_t)
Crossplots
F overlays
Shaly sand methods

Core analysis is covered in detail in chapter 26; however, it is worthwhile to mention here, in the context of water saturation determination, that in general only native-state cores can be relied on for representative S_w figures. It is, perhaps, equally important to mention the role of permeability measurements from cores. As will

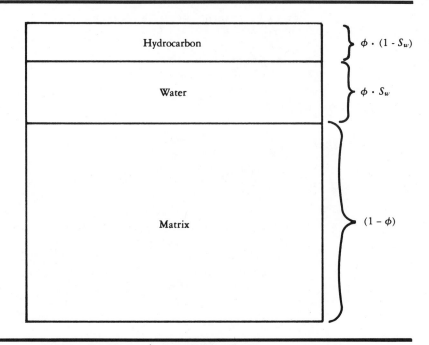

FIGURE 25.1 *Definition of Water Saturation.*

be discussed in chapter 27, many low-resistivity pay sections with very high apparent water saturations do in fact produce water-free hydrocarbons. In general, these are formations of low permeability. These troublesome formations can be evaluated properly through adequate core analysis.

Dielectric measurements include log measurements, such as the EPT covered in chapter 15, and measurements made on cores using lab apparatus. These methods will not be covered here in detail.

Nuclear methods include log measurements from pulsed neutron logs (TDT, NLL, TMD, etc.) of the thermal neutron-capture cross section of the formation and inelastic gamma ray measurements from carbon/oxygen logs. These measurements are commonly made in cased holes and are covered in Bateman (1985).

The other five methods will be discussed in detail along with suggestions on when each should be used.

THE BASIC METHOD—ARCHIE'S EQUATION

Archie combined three measurable observations into one equation. By saturating a rock sample with salt solutions of different salinities he found that the resistivity of the wet (water saturated) rock (R_0) was related to the resistivity of the saturating water (R_w) by the relation

$$R_0 = FR_w,$$

where F is called the formation factor. This formation factor was found to vary predictably as the rock porosity changed, according to

$$F = a/\phi^m,$$

where a is a constant and m is known as the cementation exponent. Lastly, Archie found that rocks at less than 100% water saturation with resistivity R_t obey the rule:

$$S_w^n = R_0/R_t,$$

where n is the saturation exponent. A combination of these relationships gives Archie's equation:

$$S_w^n = a R_w/\phi^m R_t.$$

The values of a, m, and n, when they are unknown, can be set to the following generally accepted values (although, for serious petrophysical analysis, they should be determined by core analysis):

Sandstones: $a = 0.81$ $m = 2$ $n = 2$ }
    or $a = 0.62$ $m = 2.15$ $n = 2$ } or less
Carbonates: $a = 1$ $m = 2$ $n = 2$ or more

Thus, three commonly used versions of Archie's equation are:

Sandstones: $$S_w = \left(\frac{0.81 R_w}{\phi^2 R_t}\right)^{1/2}$$

    or $$S_w = \left(\frac{0.62 R_w}{\phi^{2.15} R_t}\right)^{1/2}$$

Carbonates: $$S_w = \left(\frac{R_w}{\phi^2 R_t}\right)^{1/2}$$

The analyst should avoid a mind-set that locks the conventional values for a, m, and n into any and all log analysis evaluations. Some crossplot methods allow one or another of these parameters to be deduced, but there is no substitute for rigorous core analysis to pin down the precise values required for each and every reservoir unit.

THE SATURATION EXPONENT n

While n is usually assumed to be 2, its exact value may vary depending on the wettability of the rock. Oil-wet systems usually exhibit an n greater than 2. Some water-wet systems show an n less than 2. To find n, a core sample is prepared at a number of different water saturations, and for each saturation the ratio of R_0 to R_t is

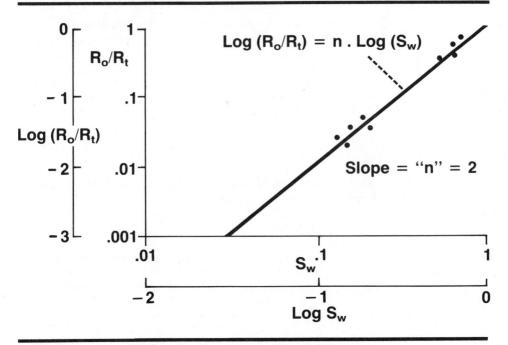

FIGURE 25.2 *Determination of Saturation Exponent* n.

measured. By taking the logarithm of Archie's equation, it is transposed into

$$n \log S_w = (R_0/R_t)$$

Therefore, a plot of log S_w versus log (R_0/R_t) should produce a straight line of slope n. Figure 25.2 illustrates the method.

QUESTION #25.2
A core from the Fubar formation was measured in the laboratory to have the following resistivities and saturations:

S_w	R_t
100%	1.00
75%	1.73
50%	3.73
25%	14.0

Plot these results on the log-log graph paper of figure 25.3 and find the value of n.

It is worthwhile to explore the sensitivity of S_w to the value of n. Suppose a calculation of S_w is made using $n = 2$ and the resulting value of S_w is 0.5, or 50%. We can then deduce that

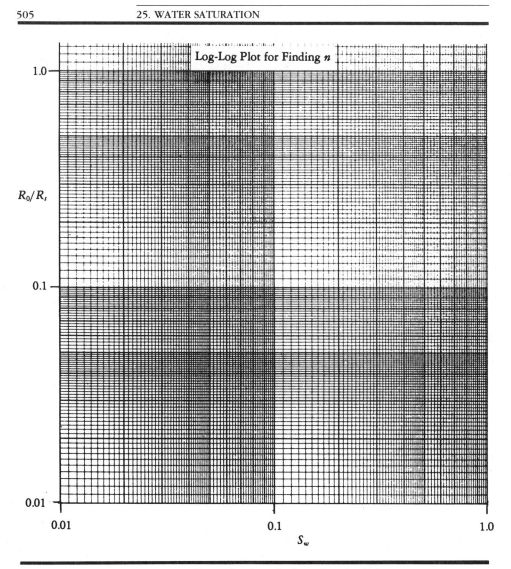

FIGURE 25.3 *Determination of* n *from Core Analysis.*

$$S_w = (FR_w/R_t)^{1/2} = 0.5$$

and hence

$$(FR_w/R_t) = 0.5^2 = 0.25.$$

Having found that FR_w/R_t is 0.25, we can now play a "what-if" game and see the results of raising n to, say, 2.2:

$$S_w = (FR_w/R_t)^{1/2.2} = (0.25)^{1/2.2} = 0.533 \text{ or } 53.3\%.$$

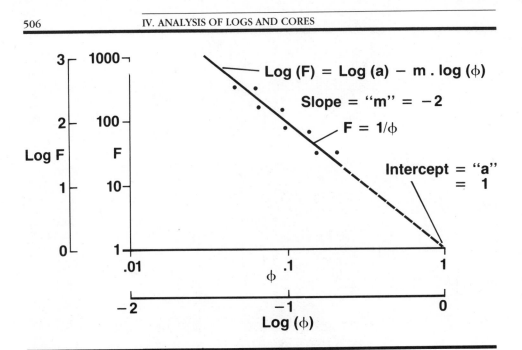

FIGURE 25.4 *Determination of* a *and* m.

Lowering *n* to, say, 1.7 gives

$$S_w = (0.25)^{1/1.17} = 0.442 \text{ or } 44.2\%.$$

In summary, other factors remaining the same, raising *n* raises S_w; and, conversely, lowering *n* lowers S_w.

FORMATION FACTOR/POROSITY—FINDING *a* and *m*

Each rock type has its characteristic formation factor versus porosity relationship. Core measurements to determine this relationship require a range of core samples of different porosities. At each porosity, a measurement is made of R_0, the resistivity of the rock at 100% water saturation. If the value of R_w is known, the formation factor *F* can be deduced by using the definition $F = R_0/R_w$. Note that the solution used to saturate the core should be of the same NaCl concentration as is the connate water found in the formation. By taking the logarithm of both sides of the equation relating formation factor to porosity, the result is

$$\log F = \log a - m \log \phi.$$

Thus, a plot of log *F* versus log ϕ should give a line of slope $-m$ and an intercept at $\phi = 1$ of *a* (see fig. 25.4).

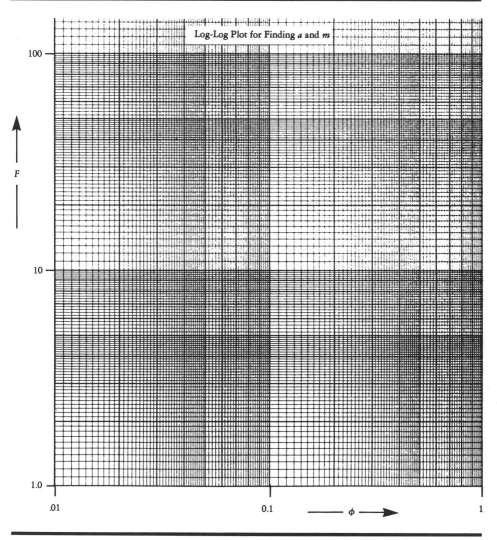

FIGURE 25.5 *Determination of a and m from Core Analysis.*

QUESTION #25.3
Use the following measurements of F and ϕ to deduce a and m
by plotting the points on Figure 25.5.

ϕ	F
32	9.8
27	14.1
22	21.6
17.5	35.0
12	77.3

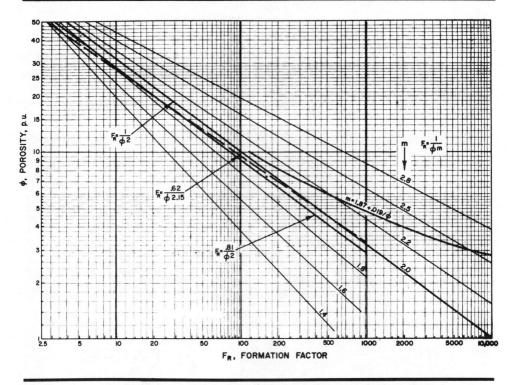

FIGURE 25.6 *Formation Factor/Porosity Relationships. Courtesy Schlumberger Well Services.*

Various formation factor/porosity relationships have been proposed, such as the Shell formula for low-porosity carbonates, which boosts the value of m at low porosities according to the relationship $m = 1.87 + 0.019/\phi$. This relationship is included, along with various other F-to-ϕ relationships, in figure 25.6 for comparison purposes.

DETERMINATION OF S_w IN CLEAN FORMATIONS

Archie Technique

Archie's equation may be solved by use of either a calculator or a nomogram of the sort illustrated in figure 25.7. A line from R_w drawn through the porosity value indicates an R_0 value for the particular F-to-ϕ relation used to build the nomogram (fig. 25.7 used $F = 0.62/\phi^{2.15}$). From the R_0 value, another line is drawn through the R_t value to find S_w.

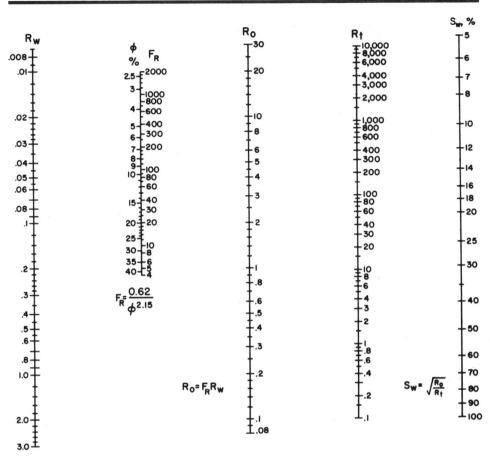

FIGURE 25.7 *Nomogram for Water Saturation in Clean Formations. Courtesy Schlumberger Well Services.*

QUESTION #25.4
Use figure 25.7 to find S_w when

$$\phi = 20\,\%,$$
$$R_w = 0.1\ \Omega \cdot m^2/m, \quad \text{and}$$
$$R_t = 60\ \Omega \cdot m^2/m.$$

Check the value from this nomogram with a calculator.

Archie's equation should be used when all the parameters necessary to solve it are known. It is the conventional method of finding S_w. However, there are some circumstances where, owing to lack of a particular parameter, an alternative method may help to compute S_w.

The Ratio Technique

Also known as the Rocky Mountain method, the ratio technique was invented by Maurice Tixier. If Archie's equation is written for both the flushed zone and the undisturbed zone, the result is

$$S_w^2 = FR_w/R_t \quad \text{and} \quad S_{xo}^2 = FR_{mf}/R_{xo}.$$

Dividing the second equation by the first, the formation factor is eliminated. This is useful if no porosity log is available.

$$\left(\frac{S_{xo}}{S_w}\right)^2 = \frac{R_{mf}}{R_w} \times \frac{R_t}{R_{xo}}$$

If some relationship between S_{xo} and S_w is assumed to exist, the need to know S_{xo} can be eliminated. A commonly used approximation is

$$S_{xo} = S_w^{1/5}$$

which, by substitution, leads to

$$\left[\frac{(S_w)^{0.2}}{S_w}\right]^2 = S_w^{8/5} = \frac{R_{mf}}{R_w} \times \frac{R_t}{R_{xo}}.$$

The relationship R_{mf}/R_w can be related to the SP value by the equation

$$SP = -K \log (R_{mf}/R_w).$$

Therefore, instead of R_{mf}/R_w, the entity $10^{-SP/K}$ may be used. Then the equation to determine water saturation can be written

$$S_w = \left(\frac{R_{xo}}{R_t} \times 10^{-SP/K}\right)^{5/8}$$

This is a useful equation since it offers a method of computing S_w from a resistivity log with an SP even if no porosity log is available and the values of R_w or R_{mf} are not known. It can be solved, provided R_{xo}, R_t, and SP are known, either by calculator or graphically using figure 25.8.

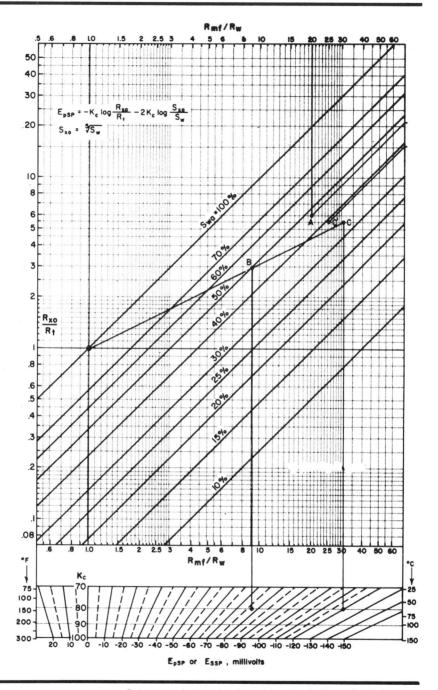

FIGURE 25.8 *Saturation Determination Using the Ratio Method. Courtesy Schlumberger Well Services.*

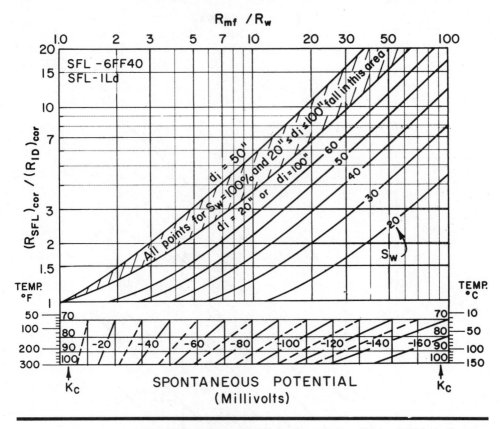

FIGURE 25.9 *Ratio Method Using R_{ID} and R_{SFL} (20 in. d_i 80 in.). Courtesy Schlumberger Well Services.*

QUESTION #25.5
Given that the hole caved in and was lost after the resistivity log was run and no porosity log is available, find S_w given that

$$SP = -105 \text{ mV},$$
$$T = 200°F \ (K = 90), \text{ and}$$
$$R_{xo}/R_t = 5.$$

This method assumes that the R_{xo}/R_t ratio is known. This value would normally be found from a deep resistivity device (deep induction or deep laterolog tool) and a shallow device (SFL or MSFL). However, invasion effects normally reduce the apparent value of R_{xo}/R_t taken from the logs in this fashion. An alternative method for determining the value of R_{xo}/R_t is shown in figure 25.9, which was prepared for the induction–SFL combination. The figure takes into account filtrate invasion effects.

QUESTION #25.6

R_{SFL} = 40 Ω·m²/m,
R_{ID} = 10 Ω·m²/m,
T = 200°F, and
SP = –115 mV.

Find S_w =

This R_{xo}/R_t ratio method can be extended to produce a continuous overlay on a log. If the logging unit is so equipped, the surface panels compute the R_{xo}/R_t ratio, which is scaled according to K (temperature) and the 5/8 power. This pseudo SP is then superimposed on the real SP curve. Where the two curves overlay, S_w is 100%. Where the two curves separate, S_w is less than 100% and the zone is immediately identified for further study. An example of this kind of presentation is given in chapter 21.

Crossplots

Crossplot methods are useful when many points need to be analyzed together—for example, where computer processing of digital log files is available. However, these same methods can be applied manually using convenient charts or plain graph paper.

Archie's equation can be rearranged to the following equation:

$$\frac{1}{R_t} = \frac{S_w^n \phi^m}{a\, R_w}.$$

If n and m are set equal to 2 and a is equal to 1, this equation reduces to

$$(1/R_t)^{1/2} = \phi\, S_w/(R_w)^{1/2}.$$

Inspection of this equation reveals that a plot of $(1/R_t)^{1/2}$ against ϕ will produce a straight line of slope $S_w/(R_w)^{1/2}$ (see fig. 25.10). On the y axis, the reciprocal root of R_t is plotted on a *linear* scale. For the sake of clarity, the actual values of R_t are also given on a *non-linear* scale. On the x axis, porosity is plotted on a *linear* scale. As a result of these scales, all points for a given water saturation lie on a straight line passing through the origin.

Use of this plot for finding R_w was explained in chapter 24. By plotting only water-bearing points, the S_w 100% line is established and its slope defines R_w. By extending the plot to include hydrocarbon-bearing points, estimates of S_w can be made as well.

This crossplot has another interesting application. If ρ_b or Δt rather than porosity is plotted on the x axis, it is possible to determine the matrix point provided there is a fair range of wet

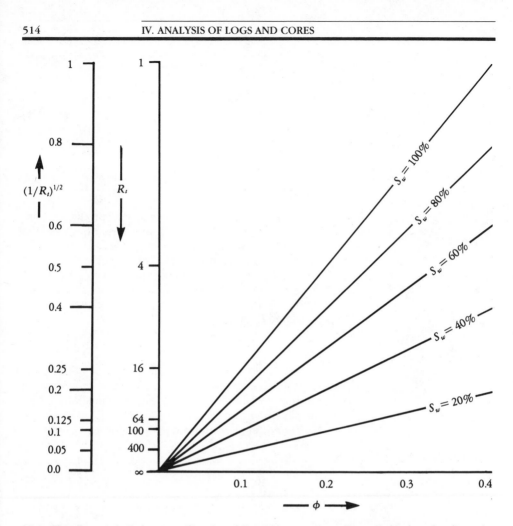

Note: This figure is built for a specific value of R_w and is not universally applicable for other formation water resistivities.

FIGURE 25.10 *Resistivity–Porosity Crossplot for Saturation Determination.*

points at different porosities. The following question illustrates this
method.

QUESTION #25.7
Readings from sonic and induction logs at various points through an
interval of interest give:

Depth	R_t	Δt
1	200	70
2	150	74
3	400	62
4	125	76
5	70	66
6	30	70
7	14	68
8	13	64
9	4.5	78
10	20	60
11	16	62
12	9	68
13	5	76

Use figure 25.11 and scale the x axis in μs/ft so that *porosity
increases* to the *right*. From the plot find:

a. R_w
b. Δt_{ma}
c. S_w in the pay

(Assume that $\Delta t_f = 188\ \mu$s/ft.)

F Overlay

The *F* overlay is a powerful interpretation technique for quick
analysis of resistivity and porosity logs. An *F* log can be overlaid
with the deep resistivity log (both logs on a logarithmic scale) so
that the two logs coincide in a clean, wet zone. The departure
between the two curves may then be used to determine S_w using an
appropriately marked scaler, as described in chapter 21. When the *F*
curve has been normalized in a wet zone (coincidence of the *F*
curve with the deep resistivity curve in a clean wet zone), a useful
side benefit is a quick method of finding R_w. Where the $F = 100$
line overlays the resistivity log scale, that value of R_t will equal
$100 \times R_w$.

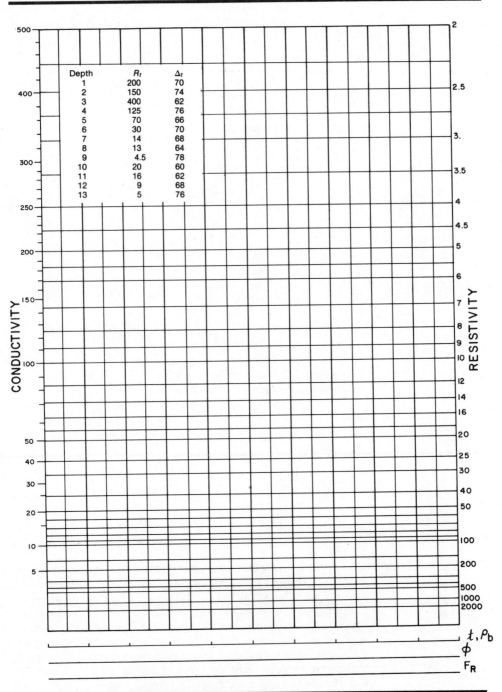

Depth	R_t	Δ_t
1	200	70
2	150	74
3	400	62
4	125	76
5	70	66
6	30	70
7	14	68
8	13	64
9	4.5	78
10	20	60
11	16	62
12	9	68
13	5	76

FIGURE 25.11 *Resistivity–Porosity Crossplot* ($F = 1/\phi^2$). *Courtesy Schlumberger Well Services.*

SUMMARY

1. Use the *Archie technique* when the formation is clean and when a deep resistivity log and a porosity log are available and the value of R_w is known.
2. Use the *ratio technique* where a porosity log is missing, but where deep and microresistivity logs are available together with an SP log.
3. Use *crossplots* to reconnoiter a porosity and resistivity log if the value of R_w and/or the matrix type are questionable.
4. Use the *F overlay* technique for quick analysis when a logarithmic F and deep resistivity logs are available and a wet zone is present where the two curves may be normalized.

BIBLIOGRAPHY

Bateman, R.M.: *Cased-Hole Log Analysis and Reservoir Performance Monitoring,* IHRDC, Boston, 1985.

Poupon, A., Loy, M.E., and Tixier, M.P.: "A Contribution to Electric Log Interpretation in Shaly Sands," *J. Pet. Tech.* (June 1954) 6.

Schlumberger: "Log Interpretation Charts" (1977).

Answers to Text Questions

QUESTION #25.1
a. ϕ =   30%
b. S_w =   33.3%
c. HCPV   =   0.2

QUESTION #25.2
n = 1.9

QUESTION #25.3
a = 0.9
m = 2.1

QUESTION #25.4
S_w = 23%

QUESTION #25.5
S_w = 50%

QUESTION #25.6
S_w = 50%

QUESTION #25.7

a. R_w $=$ $0.25\,\Omega\cdot\text{m}^2/\text{m}$

b. Δt_{ma} $=$ $44\ \mu\text{s/ft}$

c. S_w $=$ 20%

CHAPTER TWENTY-SIX

CORE ANALYSIS

Rock samples suitable for laboratory analysis may come from a variety of sources, for example:

Cuttings
Sidewall cores and/or plugs
Conventional cores

The sophistication of the analysis used depends on the source of the sample. At worst, a good idea of the rock type and porosity can be obtained; at best, a vast range of rock and fluid properties can be measured, including:

Porosity
Fluid saturation
Permeability
Relative permeability
Wettability
Capillary pressure
Pore-throat distribution
Grain-size distribution
Grain density
Mineral composition
Electrical properties
Effects of overburden stress
Sensitivity to fluids
Hydrocarbon analysis

Inasmuch as most of these properties of a rock/fluid system are of vital interest to the formation evaluator, it behooves us to learn more about core analysis methods and the application of their results to log analysis in particular and formation evaluation in general.

SAMPLE SELECTION AND PREPARATION

The analyst may be able to select only samples that originate from the depth interval of interest. To some extent selection is made at the time the sample is taken; for example, sidewall cores are usually shot at preselected depths determined from wireline logs. Cuttings are usually collected at 2-, 5-, 10-, or 20-ft intervals (see chapter 4) and should be properly lagged and identified as to source (depth). Whole cores are usually cut to driller's depth, which may be at odds

519

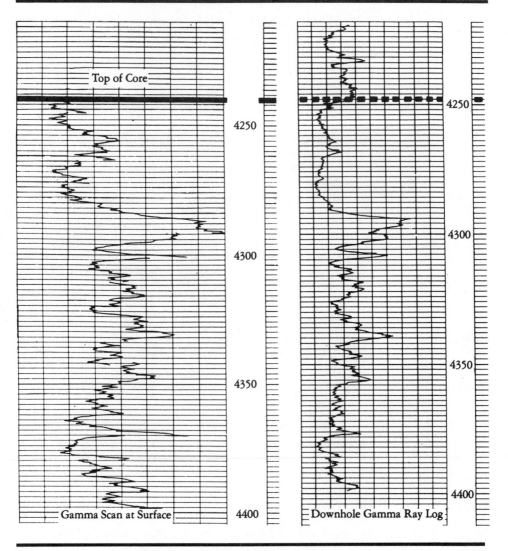

FIGURE 26.1 *Comparison of Core Gamma Scan to Wireline Openhole Gamma Ray Log.*

with wireline logging depths. Thus a good starting point for whole-core analysis is a gamma ray scan of the core. The core is laid out in its shipping container and moved relative to a gamma ray counter that records a graph of radioactivity versus distance traveled along the core. This record may then be compared directly with a wireline gamma ray log (see fig. 26.1).

QUESTION #26.1
Inspect the logs of figure 26.1 and estimate the depth difference between core depths and log depths.

When cutting conventional cores, it is wrong to assume that the only formations of interest are the clean reservoir rocks. Useful data may be extracted from shales as well, and the temptation to high-grade the core at the wellsite by discarding shale sections should be resisted. The core gamma scan, for example, would be useless if the radioactive section of the core were missing. The type of analysis to be used determines whether a plug cut from the whole core or the whole core itself is used. Plugs are 1 to 1½ in. in diameter and 1 to 3 in. long. Whole cores are normally 5 in. in diameter and up to 10 in. long.

Fresh Cores are cores cut with water- or oil-based muds and are preserved without any cleaning. *Native-state cores* are cores cut with lease crude as the coring fluid to minimize changes in rock wettability. *Restored cores* are cores that are cleaned and dried prior to any testing—wettability and fluid distributions are changed in these samples. Cores may be epoxy coated, jacketed in heat-shrinkable tubing or metal (for unconsolidated cores), or coated with lucite.

MEASUREMENT OF BASIC ROCK PROPERTIES

Porosity

There are two components in a porous rock system—the grain volume (V_G) and the pore volume (V_p). The sum of the two gives the bulk volume (V_B).

$$V_B = V_G + V_P.$$

Porosity is defined as the ratio of the pore volume to the bulk volume:

$$\phi = V_p/V_B.$$

It can be measured in a number of ways, for example:

$$\phi = (V_B - V_G)/V_B \quad \text{or} \quad \phi = V_P/(V_G + V_P).$$

That is, provided any two of the three entities are measured, porosity can be deduced. Among the commonly employed methods of measuring porosity are:

The *summation of fluids* method, in which the volumes of water, oil, and gas are independently measured and then summed to give V_P. V_B is deduced from the dimensions of the core.

The *Boyle's law* method, in which the core is cleaned and dried. Air or other gas is allowed to fill the pore space. When the sample is connected to another chamber filled with gas at a different pressure, the gas in the core pore space expands. The final pressure in the system allows deduction of V_P by use of Boyle's (Charles') law. Again, V_B is deduced from the dimensions of the core.

The *Washburn-Bunting* method, which uses vacuum extraction and collection of gas from the pore space is somewhat similar to the Boyle's law method; it measures V_P.

Liquid restoration simply involves filling the pore space with a liquid of known density and measuring the weight increase. V_P is then deduced by dividing the weight increase by the known liquid density.

Grain-density methods require that both bulk volume and dry weight of the sample be determined first. Then the sample is crushed to grain-size particles and V_G is measured. V_P is then deduced as the difference between V_B and V_G. A side benefit of this method is the estimation of grain density from the knowledge of dry weight and grain volume. The disadvantage of the method is the physical destruction of the core sample.

The accuracy of core porosity measurements is claimed to be within one half pu. Methods that require the core to be cleaned and dried are subject to error when the core contains hydrated clay material. Retorting such a sample may drive off water of hydration and thus the porosity measured may be larger than effective porosity. Use of a humidity-controlled oven to dry samples will alleviate this problem.

As can be expected, the use to which the porosity value is put should be modified by the analyst's knowledge of the heterogeneity of the reservoir. A core plug is very localized, and core-plug porosities may be higher than whole-core porosities if the whole core includes portions of rock of a lower porosity. The selection of places from which to take a plug from a whole core is somewhat subjective and usually the "best looking spot" is picked.

Sidewall core plugs can produce porosities that are either higher or lower than true porosities, depending on the porosity range. Figure 26.2 illustrates this effect by plotting sidewall core porosity against conventional core porosity. In general low-porosity formations tend to fracture when sidewall core bullets strike them, and this creates additional pore volume.

Porosity measurements on cuttings may be made on samples as small as a cubic centimeter or less. Such measurements compare well to core-plug porosities (within one pu).

Permeability

Permeability (k) is measured by causing a fluid to flow through the core sample. Measurements required to deduce permeability include the cross-sectional area A, the length L, the pressure drop Δp, the flow rate q, and the fluid viscosity μ. Darcy's equation can be applied to deduce permeability:

$$k = q\mu L / A\Delta p$$

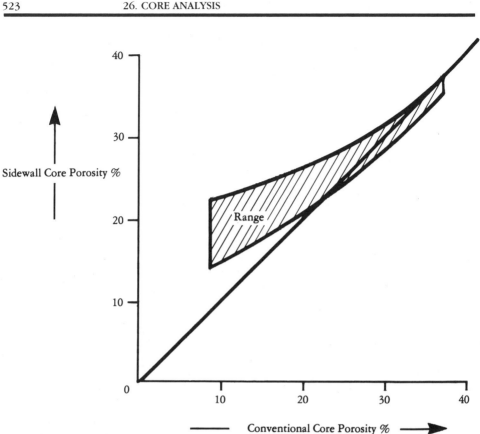

FIGURE 26.2 *Comparison of Conventional and Sidewall Core Porosity.*

TABLE 26.1 *Klinkenberg Correction Factors*

Air Permeability (md)	Correlation Factor	Equivalent Liquid Permeability (md)
1000	0.95	950
100	0.88	88
10	0.78	7.8
1	0.68	0.68
0.18	0.66	0.12

Frequently, the fluid used is air. Gases and liquids do not behave in the same fashion in these types of flow conditions, however. Klinkenberg (1941) studied the gas slippage involved and determined that gas permeability is a function of both the gas composition and the mean pressure in the core. It is common, therefore, to convert air permeabilities to equivalent liquid permeabilities using the Klinkenberg correction. Air permeabilities measure higher than liquid permeabilities. Table 26.1 lists Klinkenberg corrections for air. When reading a core analysis report, always check to see

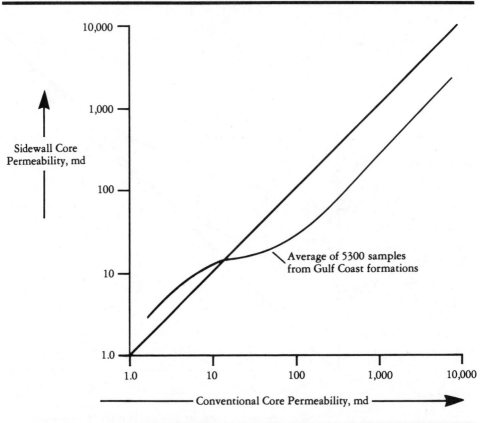

FIGURE 26.3 *Comparison of Conventional and Sidewall Core Permeability.*

whether permeability was measured with air or brine and whether the values reported are corrected for Klinkenberg effect or not.

Like the porosity measurements from sidewall cores, permeability measurements from sidewall cores may be higher or lower than the true value. Figure 26.3 shows a comparison of conventional core permeability and sidewall core permeability.

Relative Permeability, Wettability, and Sensitivity

Tests for relative permeability and wettability require samples from native-state core, if possible, to avoid chances that the rock wettability was altered as a result of the coring procedure. Surprisingly, wettability tests, although they use lease crude, employ a polished silica surface to simulate the formation rock. Wettability tests depend to some extent on the time the wetting fluid remains in contact with the surface being tested. For these reasons a certain degree of skepticism is allowed when reviewing contact-angle tests from laboratories.

It is a good idea to request that sensitivity tests be carried out at the same time permeability is measured, to see if fresh water in any way damages the permeability of the core. It is quite common to find that fresh water dislodges fines from the surfaces of the grains and that these fines subsequently bridge together and create obstructions in pore throats. Permeability can usually be restored by flushing with saline water. Clay sensitivity and treatment is worth checking into as well. Depending on the type of clay present, an appropriate treatment can be recommended. What works for one situation may create problems in another. For example, hydrofluoric acid will cause calcium fluoride precipitates in rocks with calcite present, but hydrochloric acid works fine. Chlorite is liable to produce iron hydroxide precipitates when faced with oxygen-rich systems with a pH greater than 4. Other clay types are susceptible to treatment with HCl and HF acids if freshwater systems have already caused damage.

Fluid Saturation

Gas saturation can be measured in two ways. One, a sidewall or conventional core sample is subjected to injection of mercury at about 750 psi, which causes all the gas present to go into solution with the liquids in the pore system. The volume of mercury injected gives the fraction of V_P occupied by gas. Two, a whole core sample is subjected to rapid evacuation of all free gas. Water is then forced into the sample, and the increase in weight is used to measure the fraction of V_P that held gas.

Oil saturation is measured by a retorting process in which the core sample is heated to between 450°F and 1200°F (according to the technique used) and the oil distilled off is collected in a calibrated condenser. This volume of oil represents the fraction of V_P that contained oil; from this, oil saturation may be deduced.

Water saturation is measured at the same time as oil saturation when the retorting methods are employed. Oil and water recovered by the distillation are separated and their volumes are measured individually. Other methods, such as the Dean-Stark distillation method, use a solvent, toluene for example, to take the oil into solution. The recovered water volume is divided by V_P to give water saturation.

EFFECTS OF OVERBURDEN PRESSURE

Although most basic rock properties are measured at atmospheric conditions, in some cases it is important to make measurements (especially those of porosity and permeability) at simulated reservoir conditions. Both rock matrix and pore fluids are compressible, and measurements at standard conditions therefore give overly optimistic values for ϕ and k. The tests illustrated in figure 26.4 show, for example, that increasing overburden pressure can reduce permeability by as much as 80%—depending, of course, on the type of material tested.

Note that permeability is an intrinsic rock property independent of the fluid in the pore space. When it comes to discussing reduced

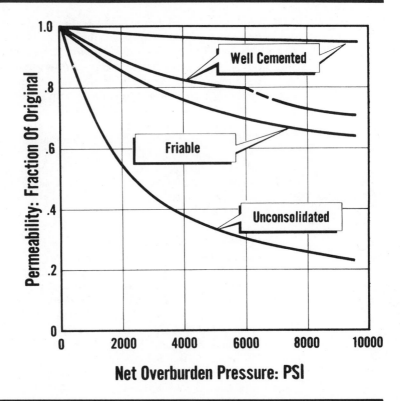

FIGURE 26.4 *Effects of Net Overburden Pressure on Permeability Measurements.*

porosity caused by changes in pressure, however, the total rock/fluid system must be considered since both the rock matrix and the fluids in the pore space are compressible. Thus, both overburden and pore pressure come into play. Compressibilities are expressed in vol./vol./psi. It would not be uncommon to find a 10% reduction in porosity resulting from compression—that is, if a sample in the lab measured 20%, the porosity at depth might be 18%. Lab measurements of oil, water, and rock compressibility can be made and the exact pore reduction factor deduced if reservoir pore and overburden stress are known.

Other pressure-sensitive parameters include acoustic velocity and formation factor. It is not uncommon to raise lab-measured values of these parameters by 20 to 30% in order to compensate for the effects of pressure.

MEASUREMENTS OF CAPILLARY PRESSURE
A number of techniques are used to measure capillary pressure, and they all depend on forcing a fluid into the core and measuring how much fluid goes in at a fixed pressure. Techniques include injection

TABLE 26.2 *Contact Angle and Interfacial Tension for Various Fluid Pairs*

Fluids	Contact Angle θ (degrees)	Interfacial Tension (dynes/cm)	$T \cos \theta$
Laboratory:			
Air-water	0	72	72
Air-mercury	140	480	367
Air-oil	0	24	24
Oil-water	30	48	42
Average Reservoir:			
Water-oil	30	30	26
Water-gas	0	50	50

of mercury (which renders the core sample useless for further analysis), use of a centrifuge, and pressure saturation in a closed cell. Both imbibition and drainage situations can be simulated. The end results will consist of a set of paired readings of saturation and pressure. Since the contact angles and interfacial tensions of the fluids used in the lab may differ from those that exist in the reservoir, it is useful to be able to convert the lab measurements to equivalent reservoir capillary pressures and hence to a plot of water, oil, and gas saturations in the reservoir as a function of height above the free water level. Table 26.2 may be consulted for the appropriate conversion factors. Remembering from chapter 6 that height above the free-water level may be expressed as

$$h = p_c / (\rho_w - \rho_{by}),$$

and that

$$p_c = 2T \cos \theta / r,$$

it is only necessary to take care of units of measurement and the factor

$$\frac{T \cos \theta_{\text{lab}}}{T \cos \theta_{\text{reservoir}}}$$

in order to convert from surface measurements to downhole conditions. For example, air-mercury measurements of the pressure required to attain a given saturation in a core plug will be about 5 times higher than that measured using air-brine ($367/72 = 5.097$, see table 26.2), and air-mercury pressures will be about 14 times that of a reservoir water-oil case ($367/26$). If p_c is measured in psi and reservoir fluid density is in g/cc, h (in feet) can always be deduced using

$$h = 2.3 p_c / \Delta \rho.$$

Since capillary pressure is directly related to the capillary radius, the capillary pressure data can also be displayed in terms of the distribution of pore-throat radii:

$$r = c \times 2T \cos \theta / p_c,$$

where

r is in microns,
T is in dynes/cm,
p_c is in psi, and
c is a constant (to take care of units) that is equal to 1.45×10^{-5}.

The relationship between capillary pressure and pore-throat distribution is illustrated schematically in figure 26.5.

Sieve analysis is a useful technique for gauging the distribution of grain size in the formation. This distribution will usually compare with the pore throat distribution. It can be of importance in designing gravel packs for unconsolidated formations.

PETROGRAPHIC AND OTHER MEASUREMENTS

Petrographic studies of cores provide descriptions of texture, rock type, minerology, etc. Techniques employed include:

Thin sections
X-ray diffraction
Scanning electron microscope (SEM)
Cathodoluminescence

Thin sections (usually a 1-by-1-in. square) are prepared from any core sample source. Prior to cutting and polishing the section, the core may be injected with colored plastic in order to fill pore spaces. The section is examined under the microscope to determine grain sizes, distribution (sorting), depositional history, and so on.

X-ray diffraction measurements yield quick and accurate analysis of both matrix and clay types. The normal procedure is to separate a crushed rock sample into clay fines (particles with a diameter less than $4 \, \mu$) and large fragments using a centrifuge, so that matrix and clay portions may be analyzed separately. Each fraction is then sampled by the x-ray diffraction technique in order to "fingerprint" the crystal structure present. Only 5 or 10 g of material are required to make the measurement. Results are normally quoted by weight percent of the whole rock. A typical report might read:

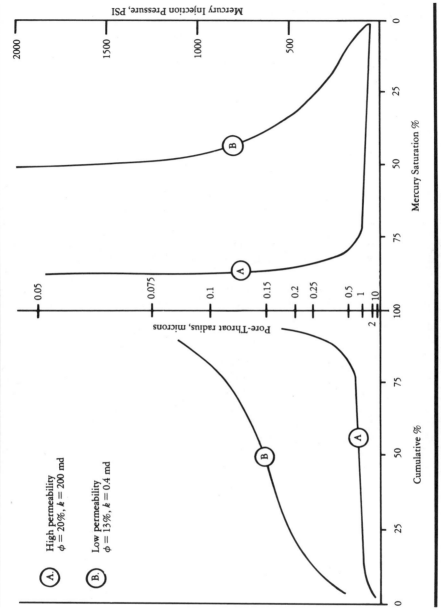

FIGURE 26.5 Interrelationships between Capillary Pressure and Pore Throat Distribution.

Bulk Rock Analysis by Weight %

Quartz	42
Feldspar	28
Calcite	21
Fubarite	6
Kaolinite	1
Chlorite	TR
Illite	TR
Montmorillonite	TR

Kaolinite 1

Chlorite, Illite, Montmorillonite (TR) } clays constitute 3% by weight of total rock

Clay Fractions

Kaolinite	31
Chlorite	10
Illite	55
Montmorillonite	4

} % of clay present

X-ray diffraction should be considered accurate to within 5 or 10%.

Scanning electron microscope (SEM) studies allow a much closer look at individual grains all the way down to clay-sized particles. Samples are first prepared by a special process that plates the grain surfaces with a conductive film such as carbon, gold, or palladium by use of a sputter or evaporative coater. Then, under high vacuum, the sample is scanned using a beam of electrons. Very high magnification is possible (up to 40,000×) with this technique. It is usually combined with a KEVEX-type system that also measures elemental composition using the EDX (energy dispersive x-ray) technique (see figure 26.6). For a very complete "dictionary" of the appearance and distribution of clays and other minerals the reader is referred to the AAPG SEM Petrology Atlas (Welton 1984). Samples of SEM photos are given in figure 26.7 (and also fig. 27.6).

Cathodoluminescence is a phenomenon that is observed when a thin section of rock (30 to 100 μ thick) is bombarded with electrons in a vacuum. Different minerals glow in different colors depending on mineral composition and the depositional and geochemical history of the rock.

Hydrocarbon Analysis

Pyrochromatic analysis of residual hydrocarbons may be carried out by heating the core sample and passing the vaporized hydrocarbons directly to a chromatographic column. The chromatograph analysis gives the percentage of each component of the hydrocarbon. If the cumulative percentage is plotted versus the hydrocarbon number, the resulting curve may be compared with "type" curves for gas, oil, and nonproductive formations (fig. 26.8).

ELECTRICAL AND RELATED MEASUREMENTS

Electrical measurements on cores are conducted using specialized cells that allow such parameters as R_0 and R_t to be found at various saturations. Cells of the four-electrode type should be used. Such

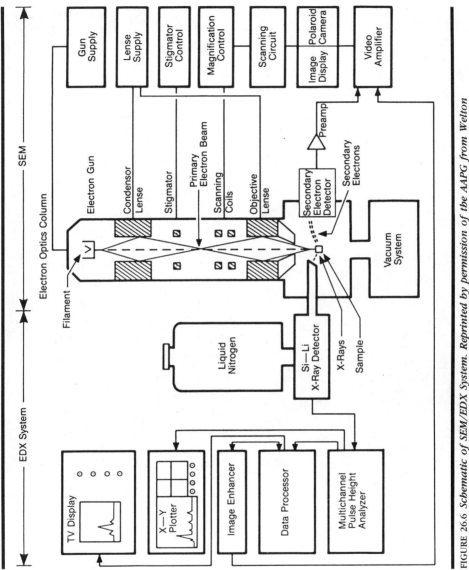

FIGURE 26.6 *Schematic of SEM/EDX System. Reprinted by permission of the AAPG from Welton 1984; modified from H. M. Beck, Schematic drawing of SEM/EDX system, unpublished.*

531

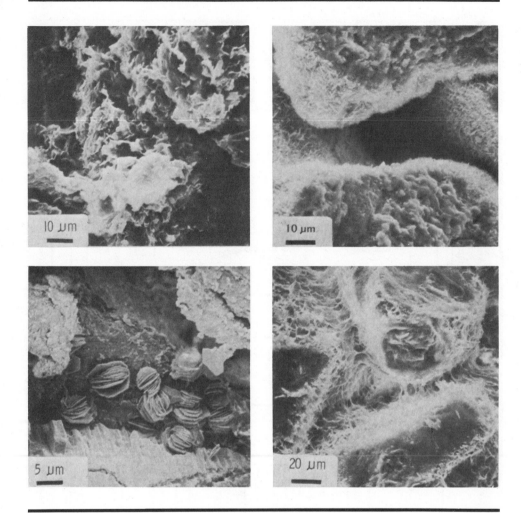

FIGURE 26.7 *Examples of SEM Photos of Clays in Sandstone Formations.*
(a) *Smectite;* (b) *Authigenic Kaolinite Partially Filling Pore Space*
(ϕ = 20.3%, k = 7.7 md); (c) *Authigenic Chlorite Lining Pores*
(ϕ = 14%); (d) *Authigenic Illite Lining Pores and Filling Pore Throats*
between Sand Grains (ϕ = 11.3%, k = 7.7 md). Reprinted by permission
of the SPE-AIME from Pittman and Thomas 1979, figs. 6-9, p. 1378.
© 1979 SPE-AIME.

measurements allow deduction of formation factor and saturation
exponents. Specifically, the constants *a*, *m*, and *n* in the Archie
relationships may be indirectly measured. Details are covered in
chapter 25.

In shaly cores, cation exchange capacity (CEC) can be measured
by dehumidifying the core sample and comparing the humidified
weight to the dehumidified weight. Considerable controversy exists
regarding the validity of CEC measurements on cores, since different
techniques give different results.

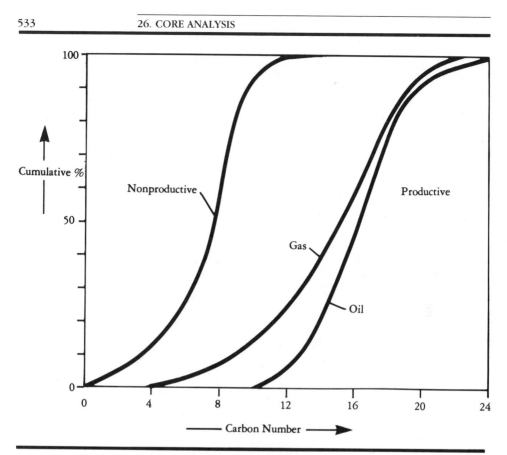

FIGURE 26.8 *Chromatographic "Type" Curves for Residual Hydrocarbon in Core Samples. After Maness and Price 1979.*

SUMMARY

It behooves the formation evaluator to acquire an intimate knowledge of core analysis techniques. Only then can coring programs be properly designed and the results of core analysis correctly used. It is hoped that this brief discussion of a vast topic has at least sensitized the reader to some of the pitfalls of coring and core analysis while, at the same time, opening up for consideration some of the possibilities that exist to enhance other forms of formation evaluation.

BIBLIOGRAPHY

Amott, Earl: "Observation Relating to the Wettability of Porous Rock," *Trans.,* AIME (1959) **216.**

Bobex, Mattax, and Denekas: "Reservoir Rock Wettability—Its Significance and Evaluation," *Trans.,* AIME (1958) **213.**

Dupuy, M., Morineau, Y., and Simandoux, P.: "About the Importance of Small Scale Heterogeneities on Fluid Flow in Porous Media," paper presented at the 58th Annual AIChE Meeting, Dallas, Feb. 8, 1966.

Fatt, I., and Davis, D. F.: "Reduction in Permeability with Overburden Pressure," *Trans.*, AIME (1952) **195**.

Geffen, T. M., Owens, W. W., Parrish, D. R., and Morse, R. A.: "Experimental Investigation of Factors Affecting Laboratory Relative Permeability Measurements," *Trans.*, AIME (1951) **192**.

Hall, H. N.: "Compressibility of Reservoir Rocks," *Trans.*, AIME (1953) **198**.

Hassler, G. L., Brunner, E., and Deahl, T. J.: "The Role of Capillarity in Oil Production," *Trans.*, AIME (1951) **192**.

Jones, F. O., Jr.: "Influence of Chemical Composition of Water on Clay Blocking of Permeability," *J. Pet. Tech.* (April 1964) **16**.

Keelan, D. K.: "Core Analysis for Aid in Reservoir Description," *J. Pet. Tech.* (November 1982), 2483–2491.

Kieke, E. M., and Hartmann, D. J.: "Scanning Electron Microscope Application to Formation Evaluation," *Trans. Gulf Coast Assoc. of Geol. Socs.* (1973) **23**.

Klinkenberg, L. J.: "API Drilling and Production Practice." (1941) 200.

Maness, M., and Price, J. G. W.: "Well Formation Characterization by Residual Hydrocarbon Analysis," *J. Pet. Tech.* (January 1979) 118–120.

Monaghan, P. H., et al.: "Laboratory Studies of Formation Damage in Sands Containing Clays," *Trans.*, AIME (1959) **216**.

Osoba, J. S., et al.: "Laboratory Measurements of Relative Permeability," *Trans.*, AIME (1951) **192**.

Pittman, E. D., and Thomas, J. B.: "Some Applications of Scanning Electron Microscopy to the Study of Reservoir Rocks," *J. Pet. Tech.* (November 1979) 1375–1380.

Stewart, C. R., and Spurlock, J. W.: "Analysis of Large Core Samples," *Oil and Gas J.* (September 17, 1952) **51**.

Warren, J. E., et al.: "An Evaluation of the Significance of Permeability Measurements," *J. Pet. Tech.* (August 1961) **31**.

Welton, Joann, E.: *SEM Petrology Atlas,* The American Association of Petroleum Geologists, Tulsa, OK (1984).

Wyllie, M. R., Gregory, A. R., and Gardner, G. H. F.: "An Experimental Investigation of Factors Affecting Elastic Wave Velocity in Porous Media," *Geophysics* (July 1958) **23**.

Answer to Text Question

QUESTION #26.1
Approximately 8 ft

CHAPTER TWENTY-SEVEN

FORMATION EVALUATION IN SHALY SANDS

OBJECTIVES

The problem the log analyst faces when confronted by shaly sands is formidable. The objective is to define porosity, hydrocarbon content, permeability, and productivity from logs and any other data that may be available, in a situation where no adequate model exists and where the fundamental electrical behavior of the rock system is poorly understood. The problem arises either where clay material is present with quartz matrix and formation waters are fresh—in which case conventional log analysis overestimates water saturation—or where formation waters are relatively salty and abnormally low formation resistivities are observed in pay zones. Both cases lead to bypassed production unless adequate steps are taken to identify these difficult zones.

The fundamental reason behind the problem lies in the failure of the classical Archie relationships to hold under the conditions encountered in shaly sands. The postulate that C_0, the wet rock conductivity, is linearly related to C_w, the water conductivity, is found to be false in shaly sands. Figure 27.1 shows that, as formation water conductivity increases, C_0 increases through a nonlinear zone until it reaches a plateau at some conductivity offset X above the clean-sand line. The form of the C_0 versus C_w plot is a function of the shaliness of the sand. Figure 27.2 shows that the apparent formation factor F_a ($= C_w/C_0$) is decreased with increasing shale content. Note that at high values of C_w (salty formation waters), F_a approaches the classical Archie formation factor. The problem, therefore, is to explain the source of the "excess" conductivity, X.

Over the years, many workers have tackled this problem. In general the approach has been first to model the behavior of the 100%-wet shaly sand system and then to transpose the model to include nonconductive hydrocarbons as well. A very large body of published material is available on the subject, and the reader is referred to the SPWLA reprint volume on shaly sand (1982).

CLAY TYPES AND DISTRIBUTIONS

In discussions of shaly sands, the distinction is usually made between the terms *clay* and *shale*. In this discussion, "clay" or "dry clay" will be used to refer to dehydrated shale and "shale" will be used when referring to hydrated dry-clay materials. The potter works with wet clay but you eat off china (dry clay). In some of the literature, the terms are used interchangeably, which can confuse the issue.

535

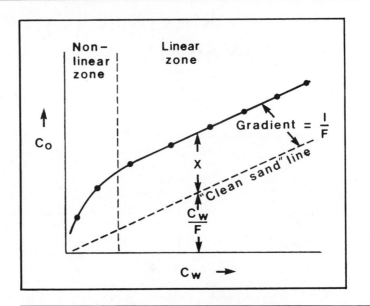

FIGURE 27.1 C_0 vs. C_w for Shaly Sand. Reprinted by permission of the SPWLA from Worthington 1985, Vol. 26, No. 1, The Log Analyst.

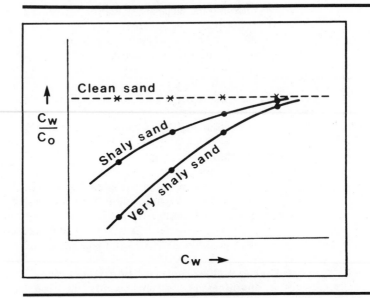

FIGURE 27.2 C_w/C_0 as a Function of C_w and Shale Content. Reprinted by permission of the SPWLA from Worthington 1985, Vol. 26, No. 1, The Log Analyst.

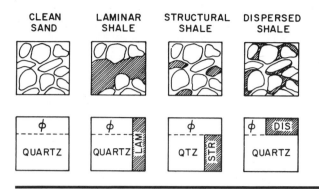

FIGURE 27.3 *Shale Distribution Models. Courtesy Schlumberger Well Services.*

Shale effects depend on:

1. the distribution of the clay material,
2. the type of clay material,
3. the amount of clay material,
4. the salinity of the formation water, and
5. the water saturation.

Distribution

Figure 27.3 illustrates three different ways clay materials may be distributed. *Dispersed shale* is an inexact term used to describe clay overgrowths on sand grains. These clay particles reduce porosity and permeability within the pore structure of the sandstone. *Laminated shale* can occur as layers of compacted clay, mudstone, and/or siltstone and meets this model definition provided it has zero effective porosity. *Structural shale* is a term used to describe the random replacement of grains of primary matrix material with fragments of lithified reworked shale.

Clay Types

There are two ways of defining shales. One is by grain size and the other is by minerological description. For example, the standard definitions by grain size (diameter) are:

Sand: 0.05 to 2 mm (50 to 2000 μ)
Silt: 0.004 to 0.05 mm (4 to 50 μ)
Clay: less than 0.004 mm (less than 4 μ)

Mineralogical analysis defines the common clay minerals as montmorillonite (smectite), illite, kaolinite, and chlorite. These are various molecular arrangements of alumino-silicates with various quantities of quartz and feldspar and are further subclassified by their origin. *Detrital clays* are deposited with the sandstone at the time the sediments are laid down. *Authigenic clays* appear as precipitates from solution at a later time. A schematic of the molecular building blocks and their various arrangements to form clay crystals is shown in figure 27.4.

The most important aspect of these minerals is their ability to hold adsorbed water on their grain surfaces. Table 27.1 lists the specific areas of porous formations and figure 27.5 generalizes the relationship between grain size and grain surface area. As grain diameter decreases, so too does pore-throat radius. As noted in chapter 6 this decrease in pore-throat radius is accompanied by an increase in capillary pressure and hence an increase in the amount of water that can be imbibed into the system. When silt- and clay-sized particles are present, microporosity accounts for a large percentage of the total porosity. Under these circumstances, irreducible water saturation can be very high. To give a visual understanding of where and how the micropore system develops, figure 27.6 shows SEM photos of the major clay types.

Crystal surfaces have what are known as *exchange sites,* where ions can temporarily reside because of the charge imbalance on the external surface of the clay's molecular building blocks. These exchange sites effectively offer an electrical path through the clay by means of surface conductance. A dry clay is an insulator, of course, but a wet clay is not. The cation exchange capacity (CEC) can be measured to quantify the conductivity of wet clays. It will come as no surprise to find that the larger the specific surface area the larger the CEC—that is, the clay type per se is not important, only its specific surface area. Even quartz sand grains, if sufficiently small, exhibit surface conductance and can be ascribed a CEC value. Figure 27.7 shows the relation between CEC and surface area.

Unfortunately, log measurements do not allow a direct measurement of either CEC or surface area. Table 27.2 lists the properties of, and log responses to, the common clay minerals. It can be seen from the table that there is no general correlation between the CEC or specific area numbers and any *one* clay indicator. Perhaps the best hope for determining clay type from logs is to note that the least electrically active clays have a large neutron–density difference and the most active a small one.

MODELS

There must be some physical model for analysis of shaly sand logs in order to produce quantitative results that are of practical use. A number of these models will be considered including:

Laminated shales and sands
Dispersed clay systems
Structural shales
Electrochemical models

The following is a discussion of each model, together with some commonly used equations for determining effective porosity and water saturation for each model described.

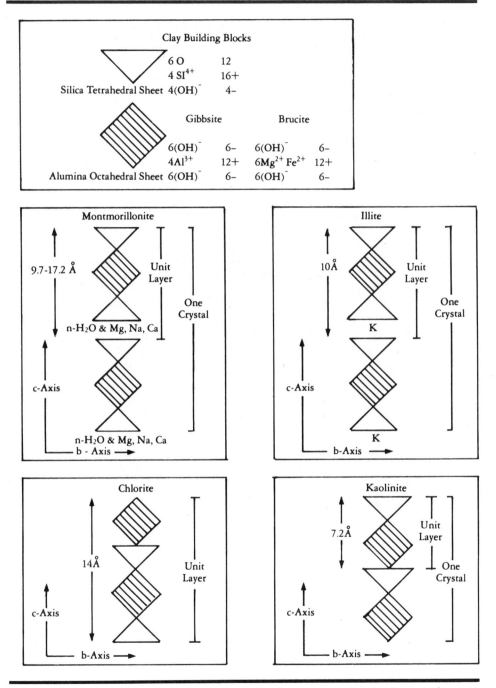

FIGURE 27.4 *Schematic of Clay Crystal Structure. After Moore 1960.*

TABLE 27.1 *Specific Areas of Porous Formations*

Formation	m²/cc	Acres/cu ft
Montmorillonite	900	6300
Illite	280	1960
Kaolinite	50	350
Sands (100 grains)	—	0.1 to 0.2

Laminated Shale Model

The model for analysis of laminated shale proposes a multilayer sandwich of alternate layers of clean sand and shales (lithified clay materials). The thickness of each layer is small in relation to the vertical resolution of the porosity and resistivity devices used to log it. Thus, both porosity and resistivity devices "see" an "average" that is not a true indicator of the properties of the clean-sand laminae. Figure 27.8 illustrates schematically that as the proportion of shale laminae (V_{lam}) increases, the amount of clean sandstone is reduced. The porosity in the clean sand remains unchanged, but there is progressively less and less of the clean sand present.

In the laminated-shale model, porosity measured by the density tool will be expressed by

$$\phi_D = V_{lam}\phi_{D_{sh}} + (1 - V_{lam})\phi_e,$$

where $\phi_{D_{sh}}$ is the apparent density porosity in the shale and ϕ_e is the true porosity in the clean sand. A similar equation can be written for the neutron log:

$$\phi_N = V_{lam}\phi_{N_{sh}} + (1 - V_{lam})\phi_e.$$

Combining these two equations results in equations for true porosity and shale content in a laminated shale:

$$\phi_e = \frac{\phi_{N_{sh}}\phi_D - \phi_{D_{sh}}\phi_N}{(\phi_{N_{sh}} - \phi_{D_{sh}}) - (\phi_N - \phi_D)},$$

$$V_{lam} = \frac{\phi_N - \phi_D}{\phi_{N_{sh}} - \phi_{D_{sh}}}.$$

This is represented graphically in figure 27.9.

QUESTION #27.1

In a laminated sandstone/shale sequence,

$\phi_N = 31\%$, and

$\phi_D = 19\%$.

If $\phi_{N_{sh}} = 40\%$, and

$\phi_{D_{sh}} = 10\%$,

find ϕ_e and V_{lam}.

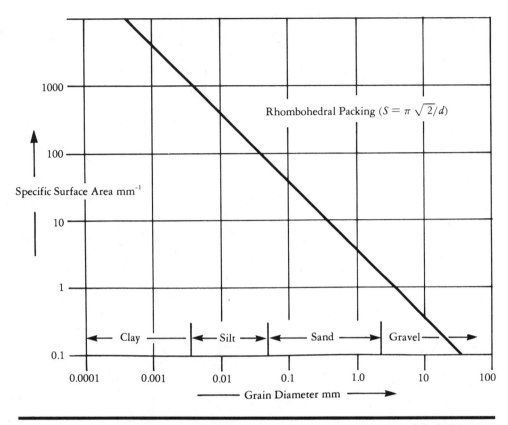

FIGURE 27.5 *Specific Grain Surface Area as a Function of Grain Diameter for Granular Formations.*

The model for water saturation in laminated shale considers that there are two resistances in parallel, that of the shale and that of the sandstone. This may be looked at schematically (fig. 27.10) as a measurement of an R_t that is the parallel resistivity of the sand and shale fractions—that is, total rock conductivity is the weighted sum of the conductivities of its components:

$$C_t = C_{\text{sand}}(1 - V_{\text{lam}}) + C_{sh} V_{\text{lam}}.$$

If C_{sand} is considered the conventional Archie conductivity, then

$$C_{\text{sand}} = \frac{\phi_e^m S_w^n}{aR_w} \quad \text{and} \quad C_{sh} = \frac{1}{R_{sh}}.$$

Hence, S_w will be related to the other parameters by

$$\frac{1}{R_t} = \frac{(1 - V_{\text{lam}}) \times \phi_e^m S_w^n}{a R_w} + \frac{V_{\text{lam}}}{R_{sh}},$$

FIGURE 27.6 *SEM Photos of the Major Clay Types and Associated Micropore Structure. Reprinted by permission of the SPE-AIME from Wilson 1982, fig. 1, p. 2872. © 1982 SPE–AIME.*

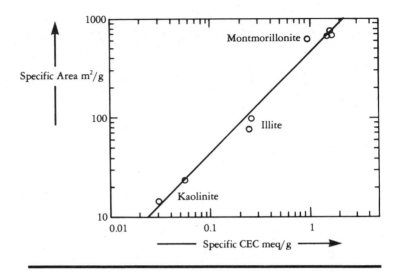

FIGURE 27.7 *Relation between Specific CEC and Specific Area for API Standard Clays. After Patchett 1975.*

TABLE 27.2 *Properties of, and Log Responses to, Common Clay Materials*

	Chlorite	Kaolinite	Illite	Montmorillonite
ρ_{ma} g/cc	2.6–2.96	2.6–2.7	2.6–2.9	2.2–2.7
$(\rho_{ma})_{av}$	2.79	2.63	2.65	2.53
CEC meq/100 g	10–40	3–15	10–40	80–150
GR (relative)	1	9	100	5
Σ_{ma} cu	25.0	14.0	17.2	11.2
P_e	6.33	1.84	3.55	2.3
$(\phi_{N_{sh}})_{av}$	52	37	30	44
$(\phi_{D_{sh}})_{av}$	-7	15.0	8	32
$\phi_{N_{sh}} - \phi_{D_{sh}}$	59	22	22	12
ϕ_T	29.5	26.0	19.0	38.0

where R_{sh} is the shale resistivity. With rearrangement we have

$$S_w^n = \left[\frac{1}{R_t} - \frac{V_{lam}}{R_{sh}} \right] \frac{a R_w}{\phi_e^m (1 - V_{lam})} \ .$$

If ϕ_e is available from a neutron–density crossplot, it should be used. If only one porosity device is available, the density porosity is preferred and ϕ_e may be approximated as $\phi_D/(1 - V_{lam})$, where V_{lam} is estimated from another indicator (GR or SP).

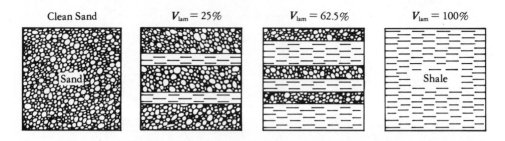

FIGURE 27.8 *The Laminated-Shale Model.*

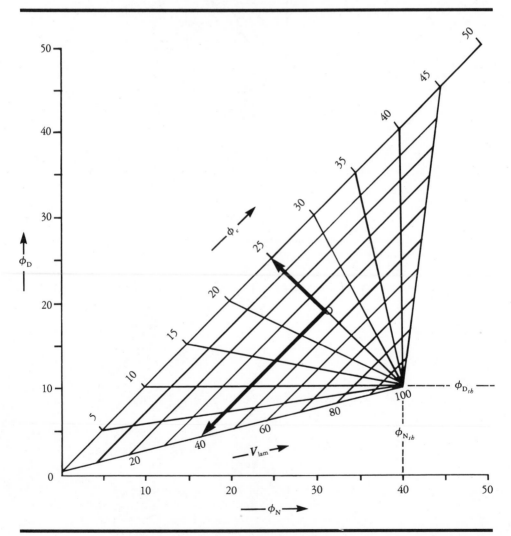

FIGURE 27.9 *Neutron–Density Crossplot in a Laminated Sand/Shale Sequence.*

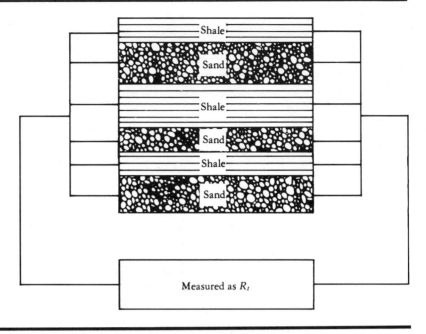

FIGURE 27.10 *Electrical Model of Laminated Shaly Sand.*

QUESTION #27.2

In the example above, where $\phi_N = 31\%$, $\phi_D = 19\%$, $\phi_{N_{sb}} = 40\%$, and $\phi_{D_{sb}} = 10\%$, ϕ_e was found to be 25% and V_{lam} to be 40%.

a. If $R_t = 2.31$, $R_{sb} = 1$, and $R_w = 0.1$, find S_w using the laminated-model saturation equation. (Assume $a = 1$, $m = n = 2$.)
b. Use the density porosity and solve for an Archie clean-sand value of S_w.
c. Use $\phi_e = \phi_D/(1 - V_{lam})$ and then apply the Archie equation to find S_w.

The method runs into difficulties when the resistivity device "sees" the sandwich not as a parallel system, but as a series or hybrid system. Indeed, depending on the geometry of the laminated sand, the electrode placement on the resistivity tool, and its mode of measurement, it may in fact "see" the sand and shale layers in the sandwich in series rather than in parallel.

It should be noted that the hydrocarbon pore volume should be determined using

$$\text{HCPV} = \phi_e(1 - S_w)(1 - V_{lam})h,$$

since, in each foot of formation, only a fraction $(1 - V_{lam})$ will

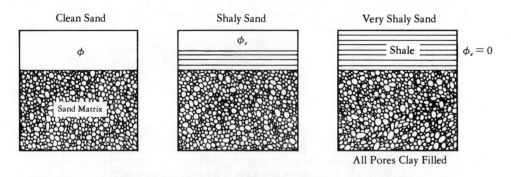

FIGURE 27.11 *The Dispersed Clay Model.*

actually be sandstone at porosity ϕ_e and saturation S_w.

The laminated model should be used where the individual laminations are small with respect to the vertical resolution of the logging device. A good rule of thumb would be to use laminated logic whenever bed thickness is less than 1 m.

Dispersed Shale Model

The dispersed shale model proposes that clay fines and clay overgrowths on the sand grains progressively replace pore space. They bring with them very large surface areas on which large quantities of water are absorbed. Thus, high water saturations are likely to be calculated if Archie's clean model is employed.

Figure 27.11 illustrates schematically that the dispersed clays replace porosity. The maximum value of V_{dis} is equal to the original porosity, but the volume of matrix material (sandstone) remains unchanged.

In the dispersed shale model, porosity measured by the density log will be expressed by

$$\phi_{\mathrm{D}} \;=\; \phi_e \;+\; V_{\mathrm{dis}}\phi_{\mathrm{D}_{sb}}$$

and for the neutron log, porosity is expressed by the equation

$$\phi_{\mathrm{N}} \;=\; \phi_e \;+\; V_{\mathrm{dis}}\phi_{\mathrm{N}_{sb}}$$

By combining these two equations, an expression for true porosity can be developed:

$$\phi_e \;=\; \frac{\phi_{\mathrm{N}_{sb}}\phi_{\mathrm{D}} \;-\; \phi_{\mathrm{D}_{sb}}\phi_{\mathrm{N}}}{\phi_{\mathrm{N}_{sb}} \;-\; \phi_{\mathrm{D}_{sb}}} \quad \text{and} \quad V_{\mathrm{dis}} \;=\; \frac{\phi_{\mathrm{N}} \;-\; \phi_{\mathrm{D}}}{\phi_{\mathrm{N}_{sb}} \;-\; \phi_{\mathrm{D}_{sb}}}$$

These equations are solved graphically using figure 27.12.

The electrical model of the dispersed clay system considers the total porosity to be filled with a mixture of clay slurry of resistivity R_{dis} and free water and hydrocarbons, if any. Thus, the total

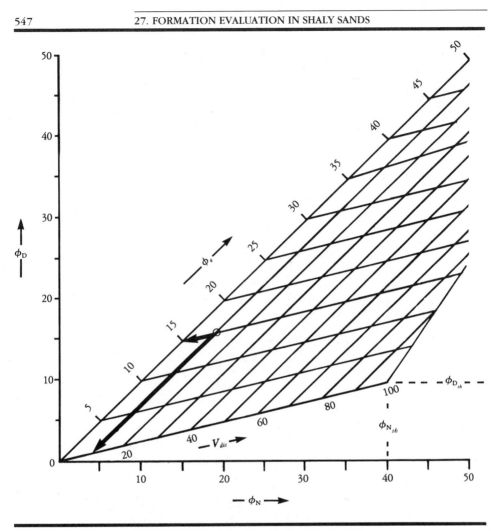

FIGURE 27.12 *Neutron–Density Crossplot in Dispersed Clays.*

formation conductivity is considered to be the sum of an Archie term referred to the total porosity (i.e., both the freely interconnected pores and the slurry-filled pores) and a clay conductivity term that depends both on water saturation and the clay fraction (see fig. 27.13). Thus, for the dispersed case,

$$C_t \;=\; \frac{\phi_T^2 \, S_{wT}^2}{a \, R_w} \;+\; \frac{\phi_T \, S_{wT} \, V_{\text{dis}}}{a}\left(\frac{1}{R_{\text{dis}}} \;-\; \frac{1}{R_w}\right)$$

and

$$S_{we} \;=\; 1 \;-\; \frac{\phi_T}{\phi_e} \times (1 \;-\; S_{wT}).$$

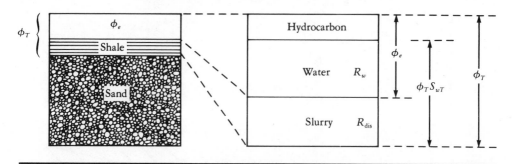

FIGURE 27.13 *Dispersed Clay Electrical Model.*

For practical purposes ϕ_T, ϕ_e, and V_{dis} can be calculated from the neutron–density crossplot (some prefer to take ϕ_T from the sonic log). R_{dis} may be calculated at the shale point as $R_{sb}\phi^2_{Tsb}$

QUESTION #27.3

Assuming a dispersed clay model, solve for ϕ_e and V_{dis} for a case where $\phi_N = 19\%$ and $\phi_D = 16\%$, if $\phi_{N_{sb}} = 40\%$ and $\phi_{D_{sb}} = 10\%$.

Then solve the dispersed saturation equation for S_{wT} and S_{we} given $R_{sb} = 2$, $R_w = 0.1$, and $R_t = 8$.

Structural-Shale Model

The structural shale model proposes that the grains of the sand matrix are progressively replaced with grains of shale. To the extent that the replacement grains may have a different grain density and hydrogen index, this process will have an effect on neutron and density response. The maximum theoretical fraction of shale in this case is $(1 - \phi_e)$ (see fig. 27.14). The response of the neutron and density tools can be written as

$$\phi_D = \phi_e + (1 - \phi_e)\, V_{\text{str}}\, \phi_{D_{sb}}$$

$$\phi_N = \phi_e + (1 - \phi_e)\, V_{\text{str}}\, \phi_{N_{sb}}$$

which gives

$$\phi_e = \frac{\phi_D\, \phi_{N_{sb}} - \phi_N\, \phi_{D_{sb}}}{\phi_{N_{sb}} - \phi_{D_{sb}}}$$

and

$$V_{\text{str}} = \frac{\phi_N - \phi_D}{\phi_{N_{sb}}(1 - \phi_D) - \phi_{D_{sb}}(1 - \phi_N)}.$$

This is illustrated graphically in figure 27.15.

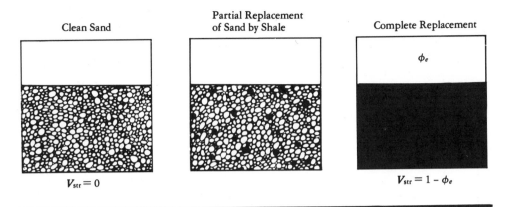

FIGURE 27.14 *Structural Shale Model.*

The electrical model for the structural case assumes that in addition to the Archie term a simple shale-conductivity term can be added such that

$$C_t = \frac{\phi_e^m S_w^n}{a R_w} + \frac{V_{str}}{R_{sb}}$$

whence

$$S_w^n = \left(\frac{1}{R_t} - \frac{V_{str}}{R_{sb}} \right) \frac{a R_w}{\phi_e^m}.$$

When in doubt as to the actual distribution of shale/clay in the formation, an alternative approach is to use one of the so-called *total-shale* relationships that is nonspecific to the model selected. A commonly used formula is

$$\frac{1}{R_t} = \frac{\phi_e^m S_w^n}{a R_w (1 - V_{sb})} + \frac{V_{sb} S_w}{R_{sb}} \quad \text{(modified total shale)}.$$

None of these relations is based on any sound experimental basis, but the most widely used is probably the modified total-shale relationship.

Waxman-Smits Technique

Physical chemists can measure the ability of a crystal surface to absorb water by finding the number of sites available for ionic exchanges. This is referred to as the cation exchange capacity (CEC). Different materials have different CEC values. Quartz in the form of sandstone has practically none, but illite and montmorillonite, because of their high specific (surface) areas, have higher CEC values.

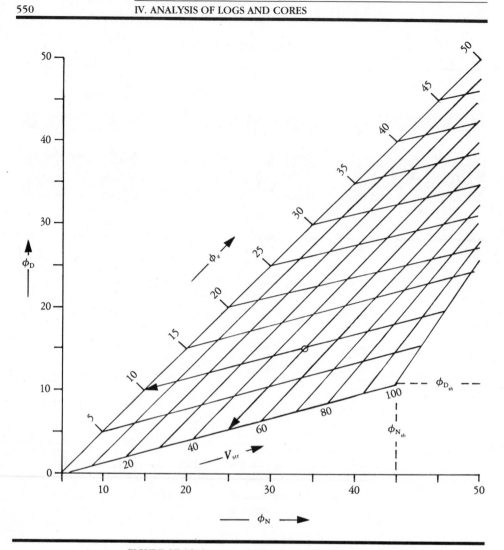

FIGURE 27.15 *Structural Model Neutron–Density Crossplot.*

Waxman and Smits derived an equation to express S_w in shaly sands as a function of the cation exchange capacity of the clay disseminated in the sand:

$$\frac{1}{R_t} = \frac{S_w^2}{F^*R_w} + \frac{B\,Q_v\,S_w}{F^*}\,,$$

where both F^* and S_w refer to total interconnected pore space; B is a constant that is dependent on the value of R_w; and Q_v is a constant determined by multiplying the volume of clay and the CEC. One commonly used approximation for Q_v is

$$Q_v = \frac{\text{CEC}\,(1 - \phi)\,\rho_{ma}}{100\,\phi}.$$

This method, while undoubtedly based on sound laboratory measurement and good petrophysical principles, lacks one vital factor for day-to-day application, namely, the availability of CEC values from conventional log measurements. Correlations for a given field can be made between core measurements of CEC and other log measurements (GR, ϕ_N, ϕ_D, etc.). However, no practical means exist to use this method if only well log data are available. Modifications of the Waxman-Smits method by Juhasz have made it more directly applicable to measurements from logs.

Dual Water Model

The dual water model proposes that there are two distinct waters to be found in the pore space. Close to the surface of the grains is found *bound* water of resistivity R_{wB}. This water is fresher (more resistive) than the remaining water further away from the grain surface. This *far* water of resistivity R_{wF} is more saline (less resistive) than the bound water and is free to move in the pores. The model suggests that the amount of bound water is directly related to the clay content of the formation. Thus, as clay volume increases, the portion of the total porosity occupied by bound water increases. This is illustrated schematically in figure 27.16.

For a water-bearing shaly sand, the model proposes a total bulk volume of water $\phi_T S_{wT}$, which is made up of a bulk volume of bound water $\phi_T S_{wB}$ and a bulk volume of free water $\phi_e S_{we}$, where ϕ_T is a *total porosity* and ϕ_e is the effective porosity. The resistivity model for this mixture of waters requires that the effective mixed-water resistivity be such that

$$\frac{1}{R_{\text{fluid mixture}}} = \frac{(1 - S_{wB})}{R_{wF}} + \frac{S_{wB}}{R_{wB}}$$

from whence

$$R_0 = \frac{R_{wF}\,R_{wB}}{[R_{wB} + S_{wB}\,(R_{wF} - R_{wB})\,\phi_T^2]}.$$

In hydrocarbon-bearing sections, the model proposes a bulk volume of bound water, $\phi_T S_{wB}$; a bulk volume of hydrocarbon, $\phi_T(1 - S_{wT})$; and a bulk volume of free water $\phi_e\,S_{we}$. The total conductivity of such a system after taking into account both clay and hydrocarbon leads to the relation

$$C_t = \frac{C_w\,S_w^2}{F_o} + \frac{(C_{wB} - C_w)\,V_Q\,Q_v\,S_w}{F_o},$$

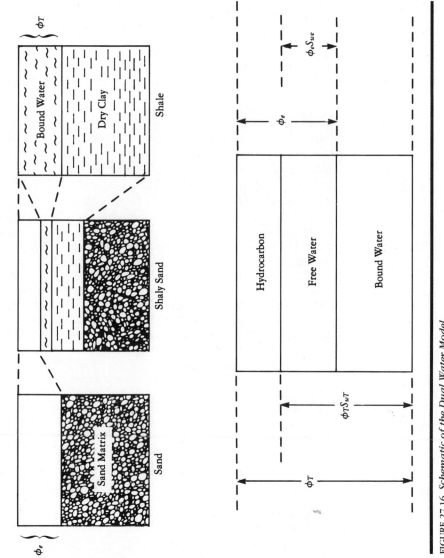

FIGURE 27.16 *Schematic of the Dual Water Model.*

where C_{wB} is the conductivity of the bound water and the term $V_Q Q_v$ is effectively the product of the amount of clay (bound) water and its ability to conduct. F_o refers to total interconnected porosity. In practice, ϕ_T is taken as the neutron–density crossplot porosity; R_{wF} is considered equal to a conventional R_w; and R_{wB} is estimated from porosity and resistivity readings in neighboring shales.

Service Company Products

Most service companies' shaly sand computer analysis methods are well documented in the literature and will not be reviewed here in detail. However, it is worthwhile to be familiar with the way in which these canned "answer" products are presented. Typical of these products are:

Saraband (Schlumberger)
Cyberlook (Schlumberger)
VOLAN (Schlumberger)
CLASS & CLAYS Epilog (Dresser)

An example of Schlumberger's VOLAN is shown in figure 27.17

QUESTION #27.4
From figure 27.17 determine in the upper zones:

a. S_w
b. Hydrocarbon density
c. Permeability to oil
d. Matrix density

QUICK-LOOK COMPUTER ANALYSIS
In the field, this type of analysis may be useful provided the analyst is in control of the parameters selected to perform the computation. As an example, figures 27.18 and 27.19 show a Schlumberger *Cyberlook*. Figure 27.18 shows a first pass from which R_{wF} and R_{wB} may be picked. Figure 27.19 shows a final pass with computed results. Although based on the dual-water theory, a number of approximations and assumptions are built in to this wellsite routine.

QUESTION #27.5
From the log of figure 27.18, pick values for:

a. R_{wF} =
b. R_{wB} =
c. GR_{cl} =
d. GR_{sh} =

Do you see any evidence of hydrocarbon?

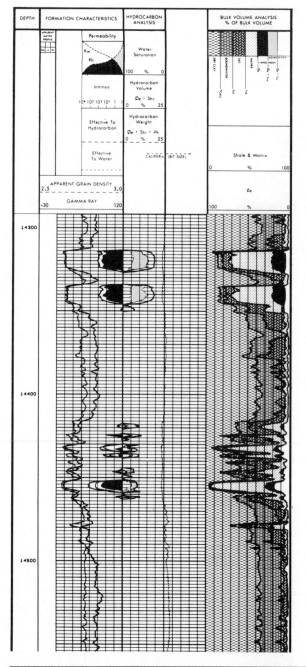

Track 1—*Formation Characteristics*

If a litho-density log is available, there will be data in the first two large divisions. The small track will be fractional volumes of up to three constituent matrix minerals.

GR—Solid line
ρ_{ga}—Dotted line
K—Intrinsic permeability
K_h—Relative permeability to hydrocarbon

Track 2—*Hydrocarbon Analysis*

Water Saturation (S_w)— Fraction of pore volume filled with formation water.

Hydrocarbon Volume ($\phi \cdot S_{hr}$)— Residual hydrocarbon per bulk volume where S_{hr} is residual hydrocarbon saturation.

Hydrocarbon Weight ($\phi \cdot S_{hr} \cdot \rho_h$)— Weight of residual hydrocarbon per bulk volume where ρ_h is the density of the hydrocarbon. The two curves, $\phi \cdot S_{hr}$ and $\phi \cdot S_{hr} \cdot \rho_h$, converge in oil zones because the density of oil is close to unity. In light hydrocarbon zones the two curves diverge.

The ratio of $\phi \cdot S_{hr} \cdot \rho_h$ to $\phi \cdot S_{hr}$ is the hydrocarbon density.

The values of hydrocarbon density derived from the computation appear on the tabular listing.

Track 3—*Differential Caliper*

Track 4—*Bulk Volume Analysis*
V_{dc}—Volume of dry clay
ϕ_{wb}—Volume of bound water (ϕ_t-ϕ_{wf})
V_{si}—Volume of silt
V_{ma}—Volume of matrix
ϕ_e—Effective porosity corrected for effects of hydrocarbons
$\phi_e \cdot S_{wt}$—Water-filled effective porosity

FIGURE 27.17 *Schlumberger's VOLAN Presentation. Courtesy Schlumberger Well Services.*

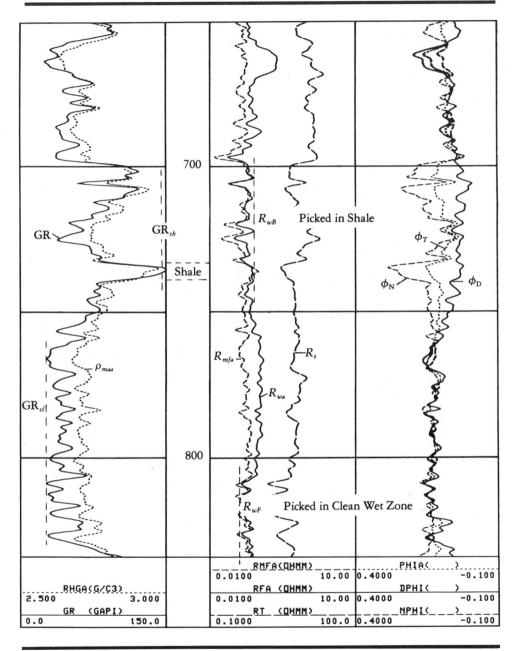

FIGURE 27.18 *Cyberlook Pass-One Presentation for Parameter Selection. Courtesy Schlumberger Well Services.*

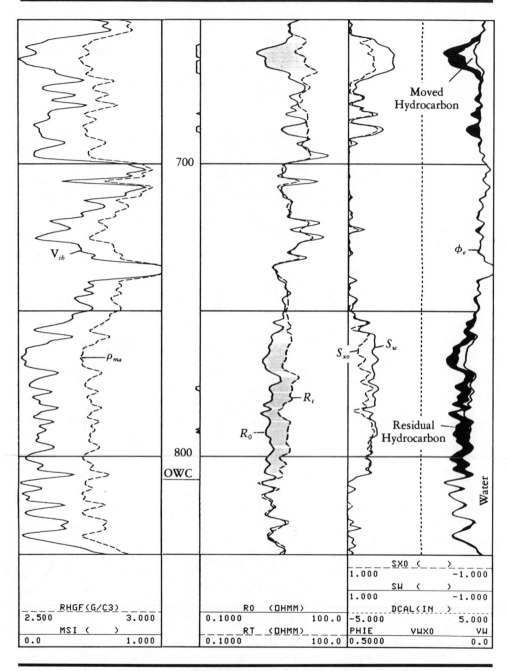

FIGURE 27.19 *Cyberlook Final Presentation in a Shaly Sand Using the Dual Water Model. Courtesy Schlumberger Well Services.*

PRACTICAL LOG ANALYSIS IN SHALY SANDS

For the analyst with access only to a calculator, many of the more sophisticated algorithms for dealing with shaly sands would appear to be out of reach. What then can the analyst do to assess quickly the potential of a shaly sand? The following method is suggested as a first approximation.

First, it should not be assumed that all clay minerals are radioactive, nor that all sandstones are nonradioactive. Kaolinite, for example, is potassium free and will therefore look like sand if only the gamma ray is used as a shale indicator. Many sandstones, on the other hand, have a high potassium feldspar content and are thus radioactive. The method suggested here should cover both eventualities by combining both gamma ray and neutron–density logs for shale measurement.

Step 0: Compute R_w in clean wet zone.
Compute R_{wB} in a shale using $R_t \phi_T^2$.

Step 1: Compute V_{sh} from the gamma ray in the conventional fashion:

$$V_{sh} = \frac{GR - GR_{cl}}{GR_{sh} - GR_{cl}}.$$

Step 2: Correct ϕ_N and ϕ_D using

$$\phi_{N_{cor}} = \phi_N - V_{sh}\phi_{N_{sh}} \quad \text{and}$$

$$\phi_{D_{cor}} = \phi_D - V_{sh}\phi_{D_{sh}}.$$

If the corrected values lie such that $\phi_{N_{cor}} < \phi_{D_{cor}}$, then one of two things could be the cause. Either light hydrocarbons are influencing ϕ_N and ϕ_D or there is radioactive matrix, such as K-feldspar. In either case, compute ϕ_e as $[(\phi_{N_{cor}}^2 + \phi_{D_{cor}}^2)/2]^{1/2}$. If $\phi_{N_{cor}} > \phi_{d_{cor}}$, then go to Step 3, else jump to Step 4.

Step 3: Compute ϕ_e from the neutron and density using

$$\phi_e = \frac{\phi_D\phi_{N_{sh}} - \phi_N\phi_{D_{sh}}}{\phi_{N_{sh}} - \phi_{D_{sh}}}$$

and reset V_{sh} to $\dfrac{\phi_N - \phi_D}{\phi_{N_{sh}} - \phi_{D_{sh}}}$.

Step 4: Compute $\phi_T = (\phi_N + \phi_D)/2$.

Step 5: Compute $S_{wT} = \left[\dfrac{R_w}{R_t \phi_t^2} - \dfrac{R_w}{R_{wB}} \left(1 - \dfrac{\phi_e}{\phi_T} \right)^2 \right]^{1/2}$.

Step 6: Compute $S_{we} = 1 - \dfrac{\phi_T}{\phi_e} (1 - S_{wT})$.

LOW-RESISTIVITY PAY

Many shaly sand formations give all appearances on logs of being water bearing. Resistivities may be less than $1\Omega\cdot m$ and neutron and density logs give little indication of any porosity development. Figure 27.20 gives an example of a suite of logs from a well in the U.S. Gulf Coast. After studying these logs for a while, try to guess which is the productive interval. When you can bear the suspense no longer, turn the page to discover the productive interval and the test results (fig. 27.21). A second example is from two North Sea wells. The logs are presented in Figure 27.22. Again, try to guess the test results from the various marked (gray crosshatching) intervals. The unbelievable, but true, results are shown in figure 27.23.

SUMMARY

Log analysis in shaly sands is difficult. Anyone who tells you otherwise is out of date. The easy-to-find pay has mostly been found already. What was not economical to explore and produce a few years ago is now the target of ambitious drilling programs. There is no doubt at all that more and more shaly reservoirs will have to be evaluated and produced. It is worth pondering that, in all probability, many prolific producers have already been plugged and abandoned owing to lack of adequate formation evaluation.

The key to successful log analysis in these troublesome formations lies in the experimental sciences. The observations are to hand. What is lacking is an adequate theory to explain the observations and adequate logging tools to make the necessary measurements. Figure 27.24 shows the key observation—that the C_w/C_o versus C_w plot is dependent on both C_w and shaliness, CEC, specific surface area, or whatever you care to call it.

The present author proposes a model in which total porosity, ϕ_T, is defined as the difference between 1 and the sum of the dry clay and matrix fractions. The model also defines an effective porosity, ϕ_e, as the difference between ϕ_T and the volume of water held in the very smallest pores, or microporosity, associated with the dry clay fraction. Conductivity is viewed as dependent on the bulk volume of water (of whichever type) raised to the power m such that

$$C_o = C_w (\phi_T)^m + C_{wB} (\phi_T - \phi_e)^m.$$

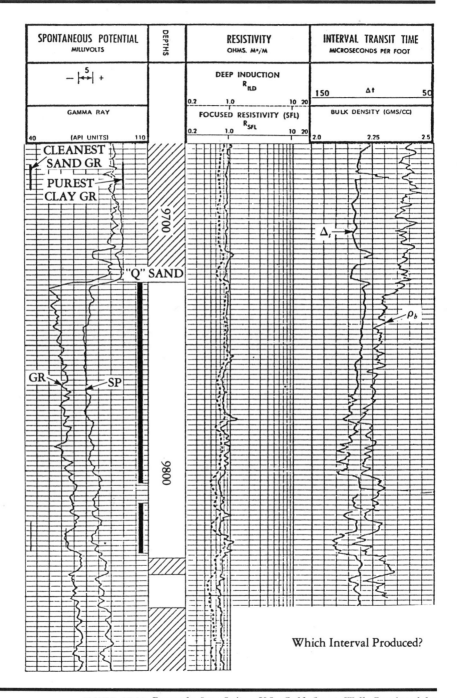

FIGURE 27.20 *Example Log Suite—U.S. Gulf Coast Well. Reprinted by permission of the SPWLA from Vajnar et al. 1977. SPWLA 18th Symposium.*

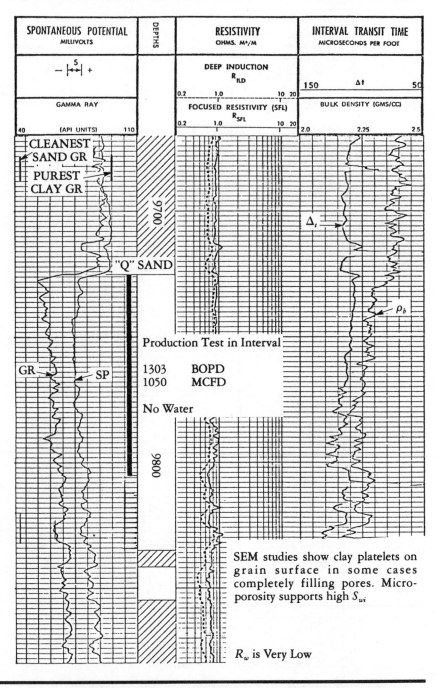

FIGURE 27.21 *Surprising Results in Low Resistivity Formation. Reprinted by permission of the SPWLA from Vajnar et al. 1977. SPWLA 18th Symposium.*

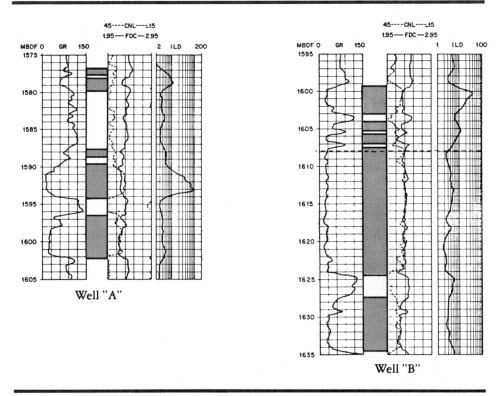

FIGURE 27.22 *Example Log Suite from North Sea Well. Reprinted by permission of the SPLWA from Bos 1982, Vol. 23, No. 5, The Log Analyst.*

If F_a is defined as C_w/C_o, and F as the simple porosity-dependent a/ϕ^m, then the expression F_a/F simplifies to

$$F_a/F = 1/(1 + \phi_0^2 C_{wB}/C_w),$$

where ϕ_0, the microporosity, is simply $(\phi_T - \phi_e)$. This function, when plotted, is very similar to the plot of figure 27.24, which was derived from published core analysis data. It would appear that Q_v can be equated to the product of a constant and the square of ϕ_0. If C_{wB} is set to 20 and the constant to 15, then the following equivalences are found:

Q_v (meq cm^{-3})	ϕ_0 %
0.001	0.8
0.023	3.9
0.17	10.0
1.47	31.3

that is, the least shaly formations have practically zero microporosity and the most shaly up to 30% or more.

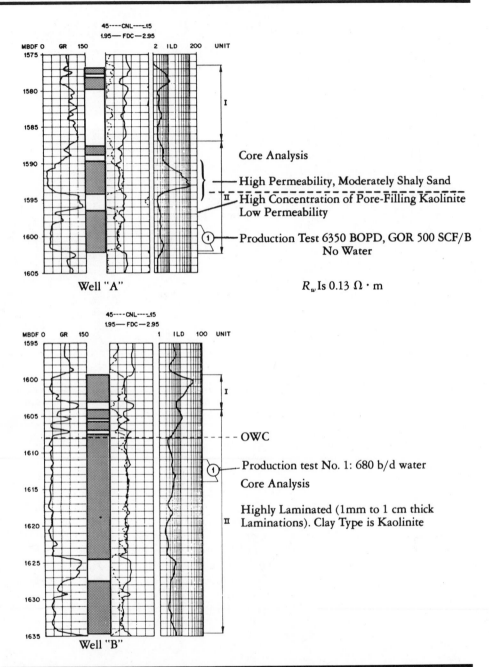

FIGURE 27.23 *Prolific Production from Apparent Wet Interval. Reprinted by permission of the SPLWA from Bos 1982, Vol. 23, No. 5, The Log Analyst.*

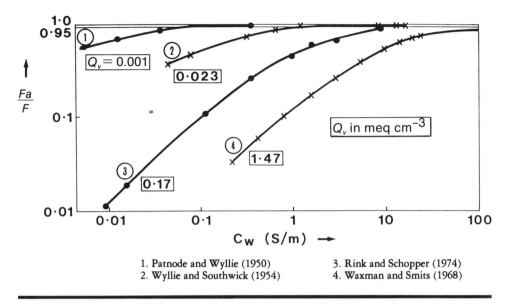

1. Patnode and Wyllie (1950)   3. Rink and Schopper (1974)
2. Wyllie and Southwick (1954)   4. Waxman and Smits (1968)

FIGURE 27.24 *Plot of C_w/C_o vs. C_w for Various Degrees of Shaliness. Reprinted by permission of the SPWLA from Worthington 1985, Vol. 26, No. 1, The Log Analyst.*

The question of transposing the wet clay model into one that includes hydrocarbons is tackled by the simple expedient of replacing total porosity in the Archie term with the product of total porosity and total water saturation, that is,

$$C_t = C_w (\phi_T \, S_{wT})^n + C_{wB} (\phi_T - \phi_e)^n \quad \text{(assuming } m = n\text{)}.$$

Note that the model proposes that no oil can enter the microporosity ($\phi_T - \phi_e$) because of capillary pressure effects, so the clay conductivity term remains unaltered.

The effective water saturation, S_{we}, is simply given by equating the hydrocarbon pore volume expressed in two different ways, that is,

$$\phi_e (1 - S_{we}) = \phi_T - \phi_T S_{wT},$$

hence,

$$S_{we} = 1 - \frac{\phi_T}{\phi_e} (1 - S_{wT}).$$

This proposed model may be compared with other models by algebraic, petrophysical, or practical means. It is certainly not the first such suggestion nor will it be the last.

BIBLIOGRAPHY

Almon, W.R.: "A Geologic Appreciation of Shaly Sands," paper WW, SPWLA Symposium, 1979.

Alquero, A.J., Fabris, A., Watt, H.B., and Wichmann, P.A.: "Effective Porosity and Shale Volume Determinations from Neutron Lifetime and Sidewall Neutron Logs," paper X, SPWLA Symposium, 1973.

Bos, M.R.E.: "Prolific Dry Oil Production from Sands with Water Saturations in Excess of 50%: A Study of a Dual Porosity System," *The Log Analyst* (September–October 1982) **23**.

Chilingar, G.V., Rieke, H.H., Sawabini, S.T., and Ershagi, I.: "Chemistry of Interstitial Solutions in Shales Versus That in Associated Sandstones," SPE paper 2527, SPE Symposium, October 1969.

Clavier, C., Coates, G., and Dumanoir, J.: "The Theoretical and Experimental Bases for the 'Dual Water' Model for the Interpretation of Shaly Sands," SPE paper 6859, AIME, 52nd Annual Fall Technical Conference and Exhibition, Denver, Oct. 9–12, 1977.

deWitte, A.J.: "Saturation and Porosity from Electric Logs in Shaly Sands," *Oil and Gas J.* (March 1957) **55**, 89.

deWitte, L.: "A Study of Electric Log Interpretation Methods in Shaly Formations," *Trans.,* AIME (1955) **204**, 103–110.

deWitte, L.: "Relations between Resistivities and Fluid Contents of Porous Rocks," *Oil and Gas J.* (August 1950) **49**, 120–132.

Doll, H.G.: "The SP Log in Shaly Sands," *Trans.,* AIME (1950) 189–205.

Fertl, W.H.: "Gamma-Ray Spectral Data Assists in Complex Formation Evaluation," *The Log Analyst* (1979) **20**, 3.

Heslop, A.: "Gamma-Ray Log Response of Shaly Sandstones," CWLS Symposium (1972) **5**, 29.

Heslop, A.: "Porosity in Shaly-Sands," paper F, SPWLA Symposium, 1975.

Hill, H.J., and Milburn, J.D.: "Effect of Clay and Water Salinity on Electrochemical Behavior or Reservoir Rocks," Petroleum Transactions, AIME (1950) **207**, 65–72.

Hill, H.J., Shirley, O.J., and Klein, G.E.: "Boundwater in Shaly Sands—Its Relation to Q_v and Other Formation Properties," edited by M.H. Waxman and E.C. Thomas, *The Log Analyst* (1979) **20**.

Hodson, G., Fertl, W.H., and Hammack, G.W.: "Formation Evaluation in Jurassic Sandstones in the Northern North Sea Area," *The Log Analyst* (1976) **17**.

Hossin, A., Delvaux P., Quint, M.A., and Gondouin, M.: "Application to the Hassi-Messaoud Cambrian Reservoir of a New Quantitative Interpretation Method for Shaly Sands," paper H, SPWLA Symposium, 1970.

Hoyer, W.A., and Rumble, R.C.: "Dielectric Constant of Rocks as a Petrophysical Parameter," paper 0, SPWLA Symposium, 1976.

Johnson, W.L., and Linke, W.A.: "Some Practical Applications to Improve Formation Evaluation of Sandstones in the Mackenzie Delta," paper R, 6th Formation Evaluation Symposium of CWLS, Calgary, Oct. 24–26, 1977.

Juhasz, I.: "The Central Role of Q_v and Formation-Water Salinity in the Evaluation of Shaly Formations," paper AA, SPWLA Symposium, 1979.

Kern, J.W., Hoyer, W.A., and Spann, M.M.: "High Temperature Electrical Conductivity of Shaly Sands," paper U, SPWLA Symposium, 1977.

Kern, J.W., Hoyer, W.A., and Spann, M.M.: "Low Porosity Gas Sand Analysis Using Cation Exchange and Dielectric Constant Data," paper PP, SPWLA Symposium, 1976.

Marett, G., Chevalier, P., Souhaite, P., and Suau, J.: "Shaly Sand Evaluation Using Gamma Ray Spectrometry Applied to the North Sea Jurassic," paper DD, SPWLA Symposium, 1976.

Moore, John E. "How to Combat Swelling Clays; Clay Mineralogy Problems in Oil Recovery," *The Petroleum Engineer* (February 1960) B 78–10; (March 1960) B 40–47.

Neasham, J.W.: "The Morphology of Dispersed Clay in Sandstones and its Effect on Sandstone Shaliness, Pore Space and Fluid Flow Properties," SPE Paper 6858, 52nd Annual Fall Technical Conference and Exhibition of the Society of Petroleum Engineers of AIME, Denver, Oct. 9–12, 1977.

Patchett, J.G.: "An Investigation of Shale Conductivity," paper U, SPWLA Symposium, 1975.

Patnode, H.W., and Wyllie, M.R.J.: "The Presence of Conductive Solids in Reservoir Rocks as a Factor in Log Interpretation," Petroleum Transactions, AIME (1950) **189**, 47–52.

Poupon, A., and Gaymard, R.: "The Evaluation of Clay Content from Logs," paper G, SPWLA Symposium, 1970.

Poupon, A., and Leveaux, J.: "Evaluation of Water Saturation in Shaly Formations," paper O, SPWLA Symposium, 1971.

Poupon, A., Loy, M.E., and Tixier, M.P.: "A Contribution to Electrical Log Interpretation in Shaly Sands," *Trans.,* AIME (1954) **201**, 138–145.

Poupon, A., Clavier, C., Dumanoir, J., Gaymard, R., and Misk, A.: "Log Analysis of Sand Shale Sequences—A Systematic Approach," *J. Pet. Tech.* (July 1970).

Ransom, R.C.: "Methods Based on Density and Neutron Well Logging Responses to Distinguish Characteristics of Shaly Sandstone Reservoir Rocks," *The Log Analyst* (1977) **18**, 47.

Rink, M., and Schopper, J.R.: "Interface Conductivity and Its Implications to Electric Logging," *Trans.,* SPWLA, 18th Annual Logging Symposium, 1974.

Ruhovets, N., and Fertl, W.H.: "Digital Shaly Sand Analysis Based on Waxman-Smits Model and Log-Derived Clay Typing," SPWLA Symposium, Paris, October, 1981.

Schlumberger: "Log Interpretation Volume I—Principles" (1972).

Schlumberger: "Well Evaluation Developments—Continental Europe" (1982).

Schlumberger: "VOLAN," Flyer SNP 5027.

Simandoux, P.: "Measures Dielectriques en Milieu Poreaux, Application a Mesure des Saturations en Eau, Etude du Comportement des Massifs Argileaux," (Dielectric Measurements in Porous Media and Application to Shaly Formations), *Revue de l'Institut Francaise du Petrole,* Supplementary Issue (1963) 193–215.

Smits, L.J.M.: "SP Log Interpretation in Shaly Sands," *Soc. Pet. Eng. J.* (1968) **8**, 123–136.

SPWLA: Reprint Volume—*Shaly Sand,* SPWLA (1982).

Thomas, E.C.: "The Determination of Q_v from Membrane Potential Measurements on Shaly Sands," *J. Pet. Tech.* (September 1976) 1087–1096.

Thomas, E.C., and Stieber, S.J.: "The Distribution of Shale in Sandstones and its Effect on Porosity," paper T, SPWLA Symposium, 1975.

Tixier, M.P., Morris, R.L., and Connell, J.G.: "Log Evaluation of Low Resistivity Pay Sands in the Gulf Coast," *The Log Analyst* (1968) **9**, 3.

Vajnar, E.A., Kidwell, C.M., and Haley, R.A.: "Surprising Productivity from Low-Resistivity Sands," *Trans.,* SPWLA, 18th Annual Logging Symposium, Houston, June 5–8, 1977.

Waxman, M.H., and Smits, L.J.M.: "Electrical Conductivities in Oil-Bearing Shaly Sands," *Soc. Pet. Eng. J.* (June 1968) 107–122. Presented as SPE Paper 1863-A at SPE 42nd Annual Fall Meeting, Houston, Oct. 1–4, 1967.

Waxman, M.W., and Thomas, E.C.: "Electrical Conductivities in Oil-Bearing Shaly Sands—I. The Relation between Hydrocarbon Saturation and Resistivity Index; II. The Temperature Coefficient of Electrical Conductivity," *SPE J.* (February 1974) **14**, 213–225.

Wilson, M.D.: "Origins of Clays Controlling Permeability in Tight Gas Sands," *J. Pet. Tech.* (December 1982) **35**, 2871–2876.

Wilson, M.D., and Pittman, E.D.: "Authigenic Clays in Sandstones: Recognition and Influence on Reservoir Properties and Paleoenvironmental Analysis," *J. Sed. Pet.* (March 1977) **47**, 3.

Winsauer, W.O., and McCardell, W.M.: "Ironic Double Layer Conductivity in Reservoir Rocks," *Trans.,* AIME (1953) **198**, 129–134.

Worthington, A.E.: "An Automated Method for the Measurement of Cation Exchange Capacity of Rocks," *Geophysics* (February 1973) **38**, 140–153.

Worthington, P.F.: "The Evolution of Shaly-Sand Concepts in Reservoir Evaluation," *The Log Analyst* (January–February 1985) **26**, 23–38.

Wyllie, M.R.J., and Southwick, P.F.: "An Experimental Investigation of the SP and Resistivity Phenomena in Dirty Sands," *J. Pet. Tech.* (February 1954).

Answers to Text Questions

QUESTION #27.1
$$\phi_e \;\; = \;\; 25\%$$
$$V_{\text{lam}} \;\; = \;\; 40\%$$

QUESTION #27.2
a. S_w = 30%
b. S_w = 119%
c. S_w = 66%

QUESTION #27.3
$$\phi_e \;\; = \;\; 15\%$$
$$V_{\text{dis}} \;\; = \;\; 10\%$$
$$S_{wT} \;\; = \;\; 67.7\%$$
$$S_{we} \;\; = \;\; 62.3\%$$

QUESTION #27.4
a. S_w = 25%
b. ρ_{by} = 0.4 g/cc
c. k_o = 30 md
d. ρ_{ma} = 2.65 g/cc

QUESTION #27.5
a. R_{wF} = 0.04 $\Omega\cdot\text{m}^2/\text{m}$
b. R_{wB} = 0.08 $\Omega\cdot\text{m}^2/\text{m}$
c. GR_{cl} = 26
d. GR_{sh} = 145
Yes, gas at 660–670

CHAPTER TWENTY-EIGHT

FORMATION EVALUATION IN COMPLEX LITHOLOGY

OBJECTIVES

The objectives of an analyst faced with the problem of formation evaluation in complex lithology may differ from the objectives of the analyst involved with evaluating shaly sands. Here, the problem is one of adequately accounting for an unknown mixture of rock types some of which may be reservoir rocks containing recoverable hydrocarbons. Conventionally, these reservoirs are classified as *hard rock* and characterized by low porosity, low permeability, and the existence of secondary porosity in the form of vugs and fractures.

Although most of the log analysis principles already covered (see chapter 23) are valid, the way in which they are applied to form a coherent method of foot-by-foot analysis is what is of most interest here. Some problem formations defy conventional analysis techniques and are attacked by local rules of thumb. Among these may be included conglomerates, which in some areas are prolific producers, and oolitic formations, where porosity evaluation is difficult. Some important types of analysis—such as geothermal projects; coal logging; and mineral logging for uranium, potassium, and sulfur—are unfortunately beyond the scope of this present work.

EXACTLY DETERMINED SYSTEMS

A classical method of solving multi-mineral log response problems is historically known as the *Chaveroo* method. It was developed when the Chaveroo field was being actively drilled and logged in the 1960s. The method reduces the problem to a simple inversion of a "response" matrix. It is most easily understood by reference to the Martini Problem:*

A log analyst, after a hard day's work, sought comfort and deserved repose in a bar. He ordered a martini and was immediately struck by the harmonious and mellow proportions of its ingredients. Wishing to learn the secret of the "perfect martini," he asked the bartender to tell him how much gin, dry vermouth, and sweet vermouth had been used. The barman (a geologist perhaps?) could only reply that it was mostly made of gin with occasional vermouthian tendencies. The analyst thereupon set out to back-calculate the relative proportions of the mix using the alcohol and sugar contents of the ingredients and of the mixture. He "logged" his martini to find it

*I am indebted to John Doveton of the Kansas Geological Survey for the original Martini Problem, which I have slightly adapted for the present purpose.

had 35% alcohol and 3.4% sugar. He consulted a reference book to find the amount of alcohol and sugar in each ingredient. Here is what he found:

	Alcohol	Sugar
Martini	35%	3.4%
Gin	47%	0%
Dry vermouth	18%	3%
Sweet vermouth	16%	16%

He solved the problem by setting up three simultaneous equations:

Alcohol: $35 = V_G \times 47 + V_{DV} \times 18 + V_{SV} \times 16,$

Sugar: $3.4 = V_G \times 0 + V_{DV} \times 3 + V_{SV} \times 14,$ and

Material balance: $1 = V_G + V_{DV} + V_{SV},$

where V_G is the fraction of gin, V_{DV} the fraction of dry vermouth, and so on. He then solved the three equations to find V_G, V_{DV}, and V_{SV}, the respective fractions of gin, dry vermouth, and sweet vermouth.

QUESTION #28.1
Compute the values of V_G, V_{DV}, and V_{SV}.

This is not a trivial example. Each logging tool can be used to set up a response equation describing a lithology mix. If V_L, V_D, and V_S are the bulk volumes of limestone, dolomite, and sandstone in some lithology mixture, then, for example, the density of the mixture can be expressed in terms of the relative quantities of each, and the density-tool response to each:

$$\rho_b = (V_L \times 2.71) + (V_D \times 2.87) + (V_S \times 2.65) + (\phi \times 1.0).$$

In a similar fashion, the sonic and the neutron logs can also define a response equation:

$$\Delta t = (V_L \times 47.5) + (V_D \times 43.5) + (V_S \times 55.5) + (\phi \times 189).$$

$$\phi_N = (V_L \times 0) + (V_D \times 0.07) + [V_S \times (-0.03)] + (\phi \times 1.0).$$

Finally, the material-balance equation reveals that

$$1 = V_L + V_D + V_S + \phi.$$

QUESTION #28.2

Given $\Delta t = 69 \ \mu s/ft,$

$\rho_b = 2.49 \ g/cc,$ and

$\phi_N = 16.5\%.$

Find the porosity and the fractions of limestone, dolomite, and sandstone in the mixture.

Elaboration of examples such as these should convince even the most mathematically inclined analyst that level-by-level solutions to matrix inversion problems using only a hand-held calculator are not to be recommended. Clearly this type of processing is best left to a computer that has been fed log data in digital form.

In the example of question 28.2, three independent formation measurements were combined to solve for four unknowns—porosity and the volume fractions of limestone, dolomite, and sandstone. Since it requires four equations to solve for four unknowns, the material-balance equation stating that the sum of the component fractions is equal to one provided the missing equality. Thus, in general, given n independent logging measurements, a solution may be found for $n + 1$ components. For example if, in addition to neutron, density, and sonic measurements, there were also available measurements of P_e and gamma ray, then a solution could be sought for two additional components, anhydrite and salt, for example.

The mathematics of this, the Chaveroo, method are thus straightforward. The weakness of the method lies in the analyst's choice of components that are to be solved for. What happens, for example, if a heavy mineral (e.g., pyrite with $\rho_{ma} = 4.99 \ g/cc$) is present but not specified as a component to be solved for? There is a good chance that in such a case the point in question will fall outside the solution polygon and result in an apparent *negative* amount of one of the other components that was specified to be solved for.

Many computer programs have been written to handle the Chaveroo approach to log analysis in multimineral environments. Some have elegant logic to handle cases where an unspecified component appears or a negative amount of a specified component results. Such programs have their place and their application in situations where the lithology is reasonably well known (from experience, core analysis, etc.) and a knowledgeable analyst is employed to select both the components and their correct endpoints. An example of the result obtained from this kind of processing is shown in figure 28.1. The column logged is predominantly anhydrite and dolomite with occasional appearances of sand and gypsum. The well was completed in the intervals shown,

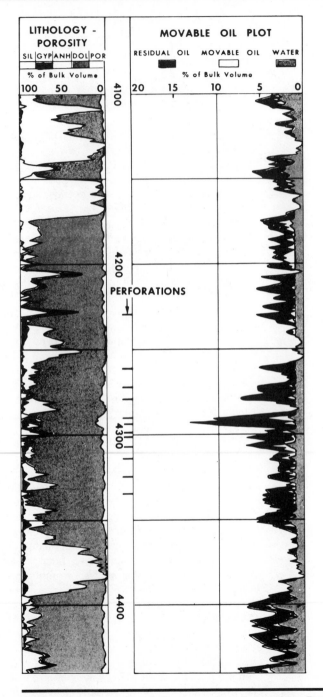

FIGURE 28.1 *Example of Chaveroo Log Processing. Reprinted by permission of the SPE-AIME from Burke et al. 1966, fig. 4, p. 4.* © *1966 SPE-AIME.*

acidized and fractured and put on production at 609 BOPD and 6 BWPD. The Chaveroo method can only generate one answer set for a given input set, since the system is exactly determined. Once the response equations are written and the endpoints chosen for each component then, for any given set of log data, one and only one solution can be found. What happens if the log data are subject to statistical variations (neutron, density, gamma ray) or the response model is inaccurate? Logically, the answers produced will also suffer from such variations and inaccuracies and there will be no way to assess the magnitude of the error. In general it can be stated that the most accurate assessment made by Chaveroo processing will be that of porosity. However, if the inaccuracies are to be properly gauged, another type of processing is called for.

OVERDETERMINED SYSTEMS

Overdetermined systems, unlike exactly determined systems, have more equations than unknowns. For example, with five log measurements (ρ_b, ϕ_N, Δt, P_e, and GR), finding the relative proportions of three components (sand, lime, and dolomite) is a relatively simple problem. Instead of there being only one possible solution there are many possible solutions. Any three log measurements suffice to form an exact solution subset, and in this case there are ten different ways to choose three logs out of five. (The reader may satisfy himself on this score by writing down all possible combinations of any two log measurements to be omitted.) If the log measurements are perfect and the model perfect, all ten possible solutions will coincide. In practice this never happens, and the result is a set of ten possible answers for each component. Which one is correct? For each possible answer a value for the original log measurements may be back-calculated. For example, the density-tool response equation may be written

$$\rho_b = \rho_1 V_1 + \rho_2 V_2 + \cdots \rho_n V_n,$$

where ρ_1, ρ_2, . . . are the log responses to components 1, 2, . . . and V_1, V_2, . . . are the volume fractions of components 1, 2, ... If $V_{1,1}$ is the estimated value of V_1 by the first of our ten possible answer sets, then the back-calculated value of ρ_b will be

$$\rho_{b_1} = \rho_1 V_{1,1} + \rho_2 V_{2,1} + \cdots + \rho_n V_{n,1}.$$

This may then be compared with the actual log reading of ρ_b and an error function defined as

$$E_{\rho_b} = \rho_b - \rho_{b_1}.$$

If the process is repeated for all the possible solutions, then the individual error functions may be summed. The same procedure may then be repeated for each of the logging measurements in turn.

The problem now transforms itself into one of minimizing the error function. That is, a solution set is sought such that the differences between the observed log readings and the back-calculated log values is a minimum. This will result in the most probable solution set. This type of processing is generically referred to as *global* logic and may equally be applied to any log-response equation including resistivity log responses to both the invasion process and to variations in water saturation. Figure 28.2 shows a generalized flow diagram of the logical steps involved in such a processing chain, and figure 28.3 illustrates the results of computer processing using such steps.

Of particular interest on figure 28.3 is that Track 1 displays a continuous plot of "reduced incoherence." This is a measure of the reliability of the results obtained, which, in this example, is good. The remainder of the plot is fairly conventional with S_w, $\phi(1 - S_{xo})$, and $\phi(1 - S_{xo}) \rho_{by}$ in Track 2 (indicating gas pay), a movable oil plot in Track 3, and a lithology breakdown in Track 4 solved for porosity, siderite, quartz, silt, and clay.

The advantage of global processing is that places where the raw log data are unreliable will become evident, and in zones of special interest the analyst will have some rigidly mathematical estimate of the probability of the analysis being correct. The disadvantages include the complexity of the processing and the need for specialists to apply it.

CROSSPLOT-BASED SYSTEMS

Straddling the Chaveroo and global methods are more generalized methods based on crossplot analysis. These methods rely on a neutron–density crossplot to resolve the question of porosity and lithology in a more general way. No special effort is made to determine the exact fractions of the different lithologic components; rather an apparent matrix density is computed and the analyst is left to his own devices to interpret its significance. The key steps in such a logical chain include:

Determination of adverse hole conditions
Identification of special minerals (e.g., coal, salt, etc.)
Initial estimates of porosity and saturations
Estimation of hydrocarbon effects on porosity logs
Reiteration to satisfy constraints
Final computations including S_w

Because of the nature of the problem, (i.e., the number of unknowns), the conventional wisdom calls for the analyst to assume the most probable minerals present and to assume a value for the hydrocarbon density, ρ_{by}. Normally, available log data are insufficient to resolve exactly enough unknowns to avoid the need for

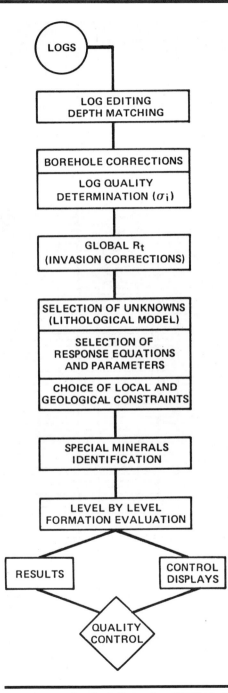

FIGURE 28.2 *Logical Steps for GLOBAL Log Processing. Courtesy Schlumberger Well Services.*

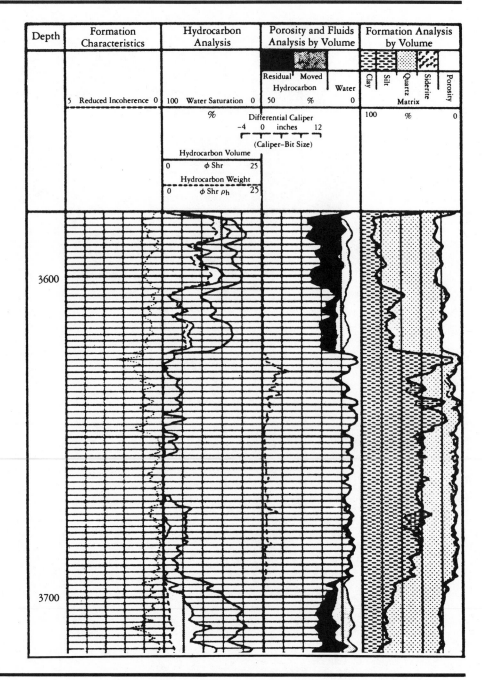

FIGURE 28.3 *Example of Log Analysis Using GLOBAL Method. Courtesy Schlumberger Well Services.*

these assumptions. There are many logical ways to perform such analysis, the choice of which depends on the whim of the analyst whose duty it is to supply the required input parameters.

Service companies offer crossplot log processing under a number of trade names such as *Coriband* (Schlumberger) and *Epilog-Complex Reservoir Analysis* (Dresser Atlas). The advantages of these services include continuous processing over the entire section of interest and the speed with they can be produced. The disadvantages of crossplot systems include the lack of control an operator may have over the computation parameters used and the options selected in the processing chain. Figure 28.4 gives a general outline of the logic flow in the Coriband processing chain.

An example of the output of this sort of crossplot processing is given in figure 28.5, which shows a Coriband processing of a shaly carbonate section with some secondary porosity. The format of the presentation is:

Track	Parameter
1	ρ_{maa} and secondary porosity, ϕ_2
2	S_w
3	ϕ, ϕS_w, and ϕS_{xo} (movable oil plot)
4	V_{sh} and ϕ (coded lithology track)

On checking these computer-generated products, look on the headings or listings for reports on the model used and the options for minerals that were input. They all require that the analyst input a fixed value for the hydrocarbon density; check that value too.

PRACTICAL QUICK-LOOK METHODS

Although analysis of complex lithology is a natural candidate for computer processing, there are conditions when the analyst will need to make his own evaluation with recourse only to a set of analysis charts and a calculator. In that case, some logical guidelines are in order. The steps might be ordered thus:

1. Hand pick levels of obvious interest using such criteria as GR and resistivity response.
2. Tabulate raw log readings.
3. Compute shale volume from available indicators (GR, SP, etc.)
4. Correct ρ_b, ϕ_N, and Δt readings to equivalent clean-formation values.
5. Calculate crossplot porosity and ρ_{maa}.
6. Calculate S_w and S_{xo}.
7. If the bulk volume of hydrocarbon in the invaded zone, $\phi(1 - S_{xo})$, exceeds 0.1, consider an iteration to further correct ρ_b and ϕ_N for the effects of light hydrocarbons.
8. Calculate Δt_{maa} and ϕ_S as a check for secondary porosity.

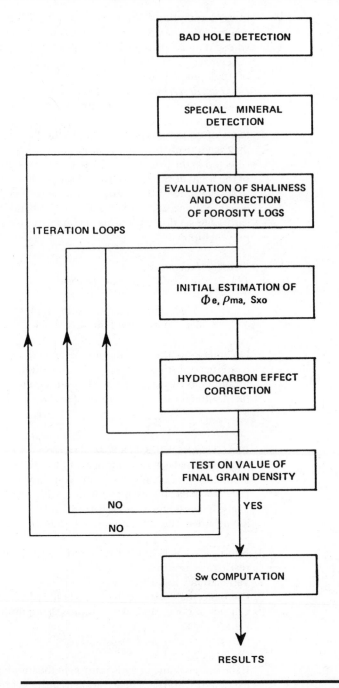

FIGURE 28.4 *Coriband Flow Diagram. Courtesy Schlumberger Well Services.*

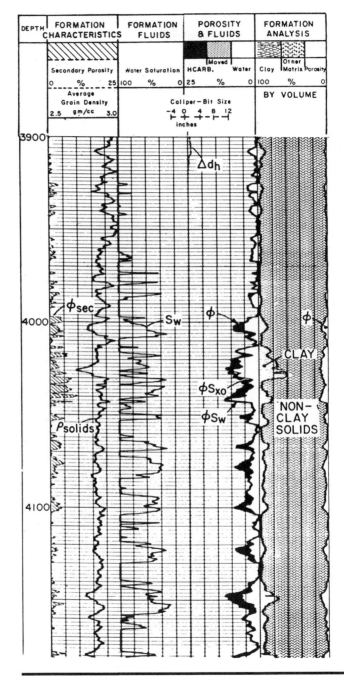

FIGURE 28.5 *Example of Logs Processed by Coriband. Reprinted by permission of the SPE-AIME from Poupon et al. 1971, fig. 8, p. 1003,* © *1971 SPE-AIME.*

Most of these steps have already been covered elsewhere in the text. The relevant relationships are restated in the form of a flow chart in figure 28.6.

The computation of Δt_{maa} for the purpose of estimating secondary porosity deserves some further explanation. The procedure is as follows:

a. Use ρ_b and ϕ_N to find ϕ_X.
b. At the plotted point, estimate the matrix density.
c. Use this value of matrix density to estimate the most probable matrix travel time, Δt_{maa}.
d. Use this matrix travel time to compute a value for ϕ_S.
e. Compare ϕ_X with ϕ_S.

The estimation of ρ_{ma} in step (b) can be made by using the MID plot (see chapter 23) or by calculation using the relationship

$$\rho_{ma} = \frac{\rho_b - \phi_X \rho_f}{1 - \phi_X}.$$

Estimating the value of Δt_{maa} requires the use of figure 28.7. Note that a line may be drawn between any pair of matrix materials to define a Δt_{ma} as a function of ρ_{ma}. For example, in a lime-dolomite matrix, a value of 2.8 g/cc for ρ_{ma} implies a Δt_{ma} of 45.5 μs/ft.

QUESTION #28.3
Apply the above method to the following data:

ρ_b = 2.575 g/cc,
$\phi_{N(lime)}$ = 11%, and
Δt = 55.5 μs/ft.

Is there any secondary porosity?

LOGGING-TOOL RESPONSES TO RESERVOIR ROCKS AND MINERALS

Whatever the method selected to perform log analysis in complex lithology, a detailed knowledge of logging-tool response is required. The work of Edmundson and Raymer (1979) is particularly useful here. See also the appropriate tables in the chapter appendix.

SUMMARY

Analysis of complex lithology need not be complex if the analyst has a clear understanding of log-response equations and logging-tool response to reservoir rocks and minerals. Problems of clay conduc-

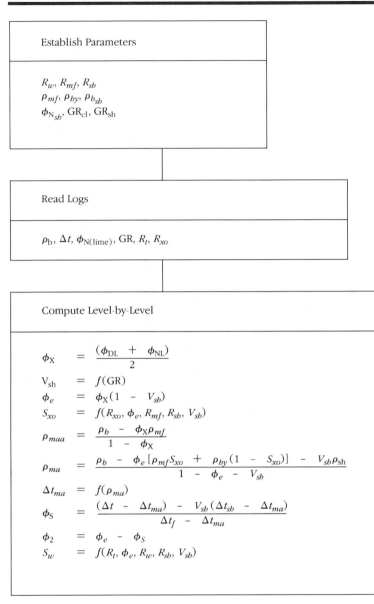

Establish Parameters

R_w, R_{mf}, R_{sb}
ρ_{mf}, ρ_{by}, $\rho_{b_{sb}}$
$\phi_{N_{sb}}$, GR_{cl}, GR_{sh}

Read Logs

ρ_b, Δt, $\phi_{N(\text{lime})}$, GR, R_t, R_{xo}

Compute Level-by-Level

$$\phi_X \quad = \quad \frac{(\phi_{DL} + \phi_{NL})}{2}$$

$$V_{sh} \quad = \quad f(GR)$$

$$\phi_e \quad = \quad \phi_X(1 - V_{sb})$$

$$S_{xo} \quad = \quad f(R_{xo}, \phi_e, R_{mf}, R_{sb}, V_{sb})$$

$$\rho_{maa} \quad = \quad \frac{\rho_b - \phi_X \rho_{mf}}{1 - \phi_X}$$

$$\rho_{ma} \quad = \quad \frac{\rho_b - \phi_e[\rho_{mf}S_{xo} + \rho_{by}(1 - S_{xo})] - V_{sb}\rho_{sh}}{1 - \phi_e - V_{sb}}$$

$$\Delta t_{ma} \quad = \quad f(\rho_{ma})$$

$$\phi_S \quad = \quad \frac{(\Delta t - \Delta t_{ma}) - V_{sb}(\Delta t_{sb} - \Delta t_{ma})}{\Delta t_f - \Delta t_{ma}}$$

$$\phi_2 \quad = \quad \phi_e - \phi_S$$

$$S_w \quad = \quad f(R_t, \phi_e, R_w, R_{sb}, V_{sb})$$

FIGURE 28.6 *Simplified Flow Chart for Analysis of Complex Lithology.*

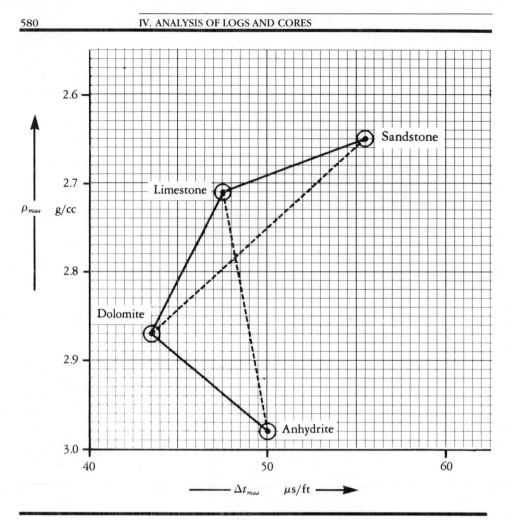

FIGURE 28.7 *Complex Lithology—ρ_{ma} vs. Δt_{ma}.*

tivity can, as a rule, be ignored without prejudice. The analyst's main concern need be only about the degree of mathematical complexity the data are to be subjected to. If in doubt, the simplest approach often gives the most reliable result since the impact of gross computational error is immediately obvious and simply corrected. The sensitivity of the end result to changes in input parameters is not always so easy to see or anticipate when using the more complex methods.

APPENDIX 28A. *Properties of Evaporites and Coals*

Name	Formula	M	ρ (g/cc)	ρ_{FDC}	P_{eff}	Z	GR_k 10^{15} atoms/g	Σ cu	ϕ_{SNP} pu	ϕ_{CNL} pu
Evaporites:										
Halite	$NaCl$	58.44	2.17	2.04	4.65	15.33	—	754.2	-2.	-3.
Anhydrite	$CaSO_4$	136.14	2.96	2.98	5.05	15.68	—	12.45	-1.	-2.
Gypsum	$CaSO_4(H_2O)_2$	172.17	2.32	2.35	3.99	14.68	—	18.5	50+	60+
Trona	$Na_2CO_3NaHCO_3H_2O$	208.02	2.12	2.08	0.71	9.08	—	15.92	24.	35.
Tachydrite	$CaCl_2(MgCl_2)_2(H_2O)_{12}$	517.60	1.68	1.66	3.84	14.53	—	406.02	50+	60+
Sylvite	KCl	74.55	1.98	1.86	8.51	18.13	953.15	564.57	-2.	-3.
Carnalite	$KClMgCl_2(H_2O)_6$	277.86	1.61	1.57	4.09	14.79	255.51	368.99	41.	60+
Langbeinite	$K_2SO_4(MgSO_4)_2$	414.99	2.83	2.82	3.56	14.23	342.46	24.19	-1.	-2.
Polyhalite	$K_2SO_4MgSO_4(CaSO_4)_2(H_2O)_2$	602.94	2.78	2.79	4.32	15.01	235.72	23.70	14.	25.
Kainite	$MgSO_4KCl(H_2O)_3$	248.97	2.13	2.12	3.50	14.17	285.42	195.14	40.	60+
Kieserite	$MgSO_4H_2O$	138.38	2.57	2.59	1.83	11.82	—	13.96	38.	43.
Barite	$BaSO_4$	233.40	4.48	4.09	266.82	47.20	—	6.77	-1.	-2.
Celestite	$SrSO_4$	183.68	3.97	3.79	55.19	30.47	—	22.53	-1.	-1.
Sulfur	S	32.06	2.07	2.02	5.43	16.00	—	20.22	-2.	-3.
Pyrite	FeS_2	119.97	5.0	4.99	16.97	21.96	—	90.10	-2.	-3.
Marcasite	FeS_2	119.97	4.89	4.87	16.97	21.96	—	88.12	-2.	-3.
Pyrrhotite	Fe_7S_8	647.41	4.6	4.53	20.55	23.16	—	94.18	-2.	-3.
Sphaterite	ZnS	97.44	4.0	3.85	35.93	27.04	—	25.34	-3.	-3.
Chalcopyrite	$CuFeS_2$	183.51	4.2	4.07	26.72	24.91	—	102.13	-2.	-3.
Galena	PbS	239.26	7.5	6.39	1631.37	78.05	—	13.36	-3.	-3.
Coals:										
Anthracite	$CH_{.358}N_{.009}O_{.022}$	—	1.51	1.47	0.16	6.0	—	8.65	37.	38.
Bituminous	$CH_{.793}N_{.015}O_{.078}$	—	1.27	1.24	0.17	6.09	—	14.30	50+	60+
Lignite	$CH_{.849}N_{.015}O_{.211}$	—	1.23	1.19	0.20	6.41	—	12.79	47.	52.

Source: Schlumberger Well Services.

APPENDIX 28B. *Properties of Silicates, Carbonates, Oxidates, Phosphates, and Feldspars*

Name	Formula	M	ρ (g/cc)	ρ_{FDC}	P_{eff}	Z	GR_k 10^{15} atoms/g	Σ cu	ϕ_{SNP} pu	ϕ_{CNL} pu
Silicates:										
Quartz	SiO_2	60.09	2.65	2.64	1.81	11.78	—	4.26	-1.	-2.
β-Cristobalite	SiO_2	60.09	2.19	2.15	1.81	11.78	—	3.52	-2.	-3.
Opal 3.5% H_2O	$SiO_2(H_2O)_{.1209}$	62.26	2.16	2.13	1.75	11.68	—	5.03	4.	2.
Opal 6.33% H_2O	$SiO_2(H_2O)_{.2253}$	64.15	2.10	2.07	1.70	11.60	—	6.12	7.	6.
Opal 8.97% H_2O	$SiO_2(H_2O)_{.3285}$	66.0	2.04	2.01	1.66	11.52	—	7.05	11.	11.
Garnet	$Fe_3Al_2(SiO_4)_3$	497.76	4.32	4.31	11.09	19.51	—	44.91	3.	7.
Hornblende	$Ca_2NaMg_2Fe_2AlSi_8O_{22}(O,OH)_2$	900.13	3.2	3.2	5.99	16.44	—	18.12	4.	8.
Tourmaline	$NaMg_3Al_6B_3Si_6O_2(OH)_4$	558.78	3.03	3.02	2.14	12.35	—	7449.82	16.	22.
Zircon	$ZrSiO_4$	183.31	4.67	4.5	69.10	32.43	—	6.92	-1.	-3.
Carbonates:										
Calcite	$CaCO_3$	100.09	2.71	2.71	5.08	15.71	—	7.08	0.	-1.
Dolomite	$CaCO_3MgCO_3$	184.41	2.87	2.88	3.14	13.74	—	4.70	2.	1.
Ankerite	$Ca(Mg,Fe)(CO_3)_2$	240.25	2.9	2.86	9.32	18.59	—	22.18	0.	1.
Siderite	$FeCO_3$	115.86	3.94	3.89	14.69	21.09	—	52.31	5.	12.
Oxidates:										
Gibbsite	$Al(OH)_3$	78.01	2.44	2.49	1.10	10.28	—	23.11	50+	60+
Hematite	Fe_2O_3	159.69	5.27	5.18	21.48	23.44	—	101.37	4.	11.
Magnetite	Fe_3O_4	231.54	5.18	5.08	22.24	23.67	—	103.08	3.	9.
Geothite	$FeO(OH)$	88.86	4.37	4.34	19.02	22.66	—	85.37	50+	60+
Limonite	$FeO(OH)(H_2O)_{2.05}$	125.79	3.5	3.59	13.00	20.39	—	71.12	50+	60+

Phosphates:										
Hydroxyapatite	$Ca_5(PO_4)_3OH$	502.33	3.15	3.17	5.81	16.30	—	9.60	5.	8.
Chlorapatite	$Ca_5(PO_4)_3Cl$	520.78	3.18	3.18	6.06	16.50	—	130.21	-1.	-1.
Fluorapatite	$Ca_5(PO_4)_3F$	504.32	3.20	3.21	5.82	16.31	—	8.48	-1.	-2.
Carbonapatite	$(Ca_5(PO_4)_3)_2CO_3H_2O$	1048.67	3.11	3.13	5.58	16.20	—	9.09	5.	8.
Feldspars — Alkali:										
Orthoclase	$KAlSi_3O_8$	278.34	2.55	2.52	2.86	13.39	255.29	15.51	-2.	-3.
Anorthoclase	$KAlSi_3O_8$	278.34	2.62	2.59	2.86	13.39	255.29	15.91	-2.	-2.
Microcline	$KAlSi_3O_8$	278.34	2.56	2.53	2.86	13.39	255.29	15.58	-2.	-3.
Feldspars — Plagioclase:										
Albite	$NaAlSi_3O_8$	262.23	2.62	2.59	1.68	11.55	—	7.47	-1.	-2.
Oligoclase	10–30% Anorthite									
Andesine	30–50% Anorthite	Interpolate linearly								
Labradorite	50–70% Anorthite	between Albite and								
Bytownite	70–90% Anorthite	Anorthite.								
Anorthite	$CaAl_2Si_2O_8$	278.22	2.76	2.74	3.13	13.73	—	7.24	-1.	-2.

Source: Schlumberger Well Services.

APPENDIX 28C. *Properties of Clays, Zeolites, and Micas*

Name	M	Formula	ρ (g/cc)	ρ_{FDC}	P_{eff}	Z	GR_k 10^{15} atoms/g	Σ cu	ϕ_{SNP} pu	ϕ_{CNL} pu
Clays:										
Kaolinite	W	$Al_4Si_4O_{10}(OH)_8$	2.42	2.41	1.83	11.83	7.79	14.12	34.	37.
Kaolinite	D		2.44	2.44	1.84	11.85	7.85	13.98	33.	36.
Chlorite	W	$(Mg, Fe, Al)_6(Si, Al)_4O_{10}(OH)_8$	2.77	2.76	6.30	16.67	1.06	24.87	37.	52.
Chlorite	D		2.79	2.79	6.33	16.70	1.07	24.91	36.	51.
Illite	W	$K_{1-1.5}Al_4(Si_{7-6.5}, Al_{1-1.5})O_{20}(OH)_4$	2.53	2.52	3.45	14.10	86.68	17.58	20.	30.
Illite	D		2.65	2.63	3.55	14.22	89.26	17.22	15.	24.
Mixed Layer Aggregate 1	W		2.37	2.37	3.96	14.66	25.91	18.80	35.	44.
Mixed Layer Aggregate 1	D		2.70	2.70	4.32	15.02	28.23	17.98	20.	29.
Mixed Layer Aggregate 2	W		2.36	2.35	2.23	12.50	105.07	15.22	25.	33.
Mixed Layer Aggregate 2	D		2.57	2.56	2.36	12.69	111.38	14.14	16.	22.
Mixed Layer Aggregate 3	W		2.62	2.61	3.01	13.58	41.84	13.60	13.	19.
Mixed Layer Aggregate 3	D		2.67	2.67	3.05	13.63	42.42	13.29	10.	15.
Montmorillonite	W	$(Ca, Na)_7(Al, Mg, Fe)_4(Si, Al)_8$ $O_{20}(OH)_4(H_2O)_8$	2.12	2.12	2.04	12.19	3.89	14.12	40.	44.
Montmorillonite	D		2.53	2.52	2.30	12.61	4.45	11.23	20.	24.

Zeolites:

Mineral		Formula									
Heulandite	W	(Ca, Na$_2$)Al$_2$Si$_7$O$_{18}$(H$_2$O)$_6$	—	2.23	2.23	2.86	13.39	18.44	11.77	25.	30.
Heulandite	D		—	2.34	2.34	2.96	13.52	17.78	12.61	31.	36.
Laumontite	W	CaAl$_2$Si$_4$O$_{12}$(H$_2$O)$_{3.5-4}$	—	2.53	2.53	2.64	13.10	3.94	11.23	22.	26.
Laumontite	D		—	2.56	2.56	2.67	13.13	3.99	11.00	20.	24.
Mordenite	W	(Na$_2$, K$_2$, Ca)Al$_2$Si$_{10}$O$_{24}$(H$_2$O)$_7$	—	2.24	2.24	1.91	11.97	4.39	11.55	27.	31.
Mordenite	D		—	2.39	2.38	1.99	12.11	4.60	10.30		23.
Analcine	W	NaAlSi$_2$O$_6$H$_2$O	—	2.45	2.43	1.56	11.31	4.52	11.79	17.	23.
Analcine	D		—	2.45	2.44	1.56		4.53	11.77	17.	23.

Micas:

Mineral		Formula									
Muscovite		KAl$_2$(Si$_3$AlO$_{10}$)(OH)$_2$	398.32	2.83	2.82	2.40	12.76	178.40	16.85	12.	20.
Glauconite Sample 1	W	K$_2$(Mg, Fe)$_2$Al$_6$(Si$_4$O$_{10}$)$_3$(OH)$_{12}$	—	2.51	2.49	6.77	17.01	90.51	25.31	26.	42.
Glauconite Sample 1	D		—	2.71	2.69	7.11	17.25	94.86	25.71	18.	33.
Glauconite Sample 2	W	K$_2$(Mg, Fe)$_2$Al$_6$(Si$_4$O$_{10}$)$_3$(OH)$_{12}$	—	2.53	2.52	5.32	15.91	113.22	22.37	23.	37.
Glauconite Sample 2	D		—	2.69	2.67	5.53	16.08	117.53	22.39	16.	29.
Glauconite Sample 3	W	K$_2$(Mg, Fe)$_2$Al$_6$(Si$_4$O$_{10}$)$_3$(OH)$_{12}$	—	2.65	2.63	7.04	17.2	112.44	26.70	21.	37.
Glauconite Sample 3	W		—	2.80	2.77	7.27	17.35	115.92	27.09	16.	30.
Biotite Sample 1	W	K(Mg, Fe)$_3$(AlSi$_3$O$_{10}$)(OH)$_2$	—	3.08	3.06	6.22	16.61	119.81	29.63	8.	16.
Biotite Sample 1	D		—	3.12	3.10	6.26	16.64	120.54	29.77	6.	13.
Biotite Sample 2	W	K(Mg, Fe)$_3$(AlSi$_3$O$_{10}$)(OH)$_2$	—	2.95	2.93	6.32	16.69	136.25	30.04	14.	26.
Biotite Sample 2	D		—	3.00	2.97	6.37	16.73	137.29	30.22	12.	23.

Source: Schlumberger Well Services.

BIBLIOGRAPHY

Alberty, M., and Hashmy, K.H.: "Application of ULTRA to Log Analysis," paper Z, *Trans.,* SPWLA, 25th Annual Logging Symposium, New Orleans, June 10–13, 1984.

Burke, J.A., Campbell, R.L., and Schmidt, A.W.: "The Litho-Porosity Crossplot," SPWLA, 10th Annual Logging Symposium, Tulsa, May, 1969.

Burke, J.A., Curtis, M.R., and Cox, J.T.: "Computer Processing of Log Data Enables Better Production in Chaveroo Field," SPE paper 1576, presented at the 41st Annual Meeting, Dallas, Oct. 2–5, 1966.

Edmundson, H., and Raymer, L.L.: "Radioactive Logging Parameters for Common Minerals," *Trans.,* SPWLA, 20th Annual Logging Symposium, Tulsa, June 3–6, 1979.

Mayer, C., and Sibbit, A.: "GLOBAL, A New Approach to Computer-Processed Log Interpretation," SPE paper 9341, presented at the 55th Annual Meeting, Dallas, Sept. 21–24, 1980.

Nugent, W.H., Coates, G.R., and Peebler, R.P.: "A New Approach to Carbonate Analysis," *Trans.,* SPWLA, 19th Annual Logging Symposium, El Paso, June 13–16, 1978.

Poupon, A., Hoyle, and Schmidt: "Log Analysis in Formations with Complex Lithologies," *J. Pet. Tech.* (August 1971) **24**.

Savre, W.C.: "Determination of a More Accurate Porosity and Mineral Composition in Complex Lithologies with the Use of Sonic, Neutron and Density Surveys," *J. Pet. Tech.* (September 1963) **16**, 945–959.

Raymer, L.L., and Biggs, W.P.: "Matrix Characteristics Defined by Porosity Computations," SPWLA 4th Annual Logging Symposium, May 1963.

Schlumberger: "Well Evaluation Conference South East Asia" (October 1981).

Schlumberger: "Litho Density Tool Interpretation," M 086108 (1982).

Schlumberger: "Well Evaluation Developments, Continental Europe" (1982).

Schmidt, A.W., Land, A.G., Yunker, J.D., and Kilgore, E.C.: "Applications of the Coriband Technique to Complex Lithologies," *Trans.,* paper Z, SPWLA, 12th Annual Logging Symposium, Dallas, May 2–5, 1971.

Tilly, H.P., Gallagher, B.J., and Taylor, T.D., "Methods for Correcting Porosity Data in a Gypsum-Bearing Carbonate Reservoir," *J. Pet. Tech.* (October 1982) **35**, 2449–2454.

Answers to Text Questions

QUESTION #28.1
Gin 60%
Sweet vermouth 20%
Dry vermouth 20%

QUESTION #28.2

V_{lime}	=	40%
V_{dol}	=	30%
V_{sand}	=	15%
ϕ	=	15%
Total	=	100% bulk volume

Hence fractions of matrix materials are:

Limestone: 40/85 = 47%

Dolomite: 30/85 = 35%

Sandstone: 15/85 = 18%

Total: = 100% solids

QUESTION #28.3

ϕ_X	=	9.45%
ρ_{maa}	=	2.74 g/cc
Δt_{maa}	=	46.75 μs/ft
ϕ_S	=	6.15%
Secondary Porosity	=	9.45 – 6.15 = 3.3%

PART V

FORMATION TESTING

CHAPTER TWENTY-NINE

WIRELINE FORMATION TESTING

Wireline formation testers serve a number of useful purposes including:

Obtaining a sample of formation fluid
Gauging formation permeability
Measuring formation pressure
Determination of formation pressure gradients

Wireline formation testers have been used for many years to recover samples of formation fluid in both open and cased hole. Nevertheless, the traditional tools suffered from a number of drawbacks. One was the lack of resolution and accuracy of their pressure gauges and the other was the inability of the instrumentation to tell the operator whether or not a good packer seal was obtained until it was too late to rectify a bad situation.

These inadequacies have now largely been overcome by the introduction of two key features of modern *repeat formation tester* (RFT) tools, namely quartz crystal pressure gauges and pretest capabilities that allow the operator to rectify a bad seal before it leads to undesirable results. An added bonus is the ability of these tools to make pressure tests independently of actual sample taking. Indeed, in practice nowadays it is quite common to use these tools solely to make pressure tests.

TOOL CHARACTERISTICS AND APPLICATIONS

Most service companies now offer a repeat formation tester with the following features:

Pretest chambers
Sample chambers
High-resolution pressure gauge

Table 29.1 summarizes the commonly available tools. Wireline formation testers are particularly useful when:

1. investigating a zone of interest where conventional tests are not feasible, such as too far above TD, lack of good intervals to set straddle packers, or a very short interval where depth control is critical.
2. to pin down water/oil, gas/oil, or gas/water contacts.
3. where rig time is critical.
4. where pressure control is critical due to time of day or rig locations.

TABLE 29.1 *Repeat Formation Tester Specifications*

	Gearhart	Schlumberger	Dresser Atlas	Welex
Pressure rating (psi)	15,000	20,000	15,000	20,000
Temperature rating (°F)	350	350	350	350
Minimum hole size (in.)	6¼	6	6¼	7
Maximum hole size (in.)	12¾	14¾	12¼	13
Basic make-up length (ft)	16	33	35½	?
Formation pressure readings per trip in hole	Any number	Any number	Any number	Any number
Sample chamber sizes (gal)	2¼, 5	1, 2¾, 6, & 12	2¼, 5	1, 2½, 5, & 7

When ordering this service, give plenty of notice to the service company. The desired choice of variables such as sample chamber size, packer hardness, choke size, pressure gauges, and water cushions may not be universally available. If a sample of recovered hydrocarbons is needed for PVT lab analysis, a special pressure cylinder should be requested.

When running the tool, a valid test may be considered one that recovers significant quantities of fluid and/or records formation and hydrostatic pressure. A dry test is indeterminate and the tool should be repositioned several times to determine if (1) the formation is impermeable—in which case all tests will be dry—or (2) the tool was set in a shale or tight streak—in which case repositioning should result in a valid test. A lost packer seal is also indeterminate—the tool should be repositioned. Open-hole logs are particularly helpful in resolving dry tests and lost packer seals. The microlog, if available, is useful as an indicator of tight streaks; and caliper logs, particularly the 4-arm type, are useful for avoiding hole conditions leading to lost packer seals.

OPERATING PRINCIPLES

Mechanics

Figure 29.1 shows the RFT tool in the closed position (a), for descent into the well and in the open (set) position (b) for pressure measurement and sample taking.

Hydraulics

Communication between the formation and the tool interior is established through the probe. Figure 29.2 shows a schematic of the tool's sampling system. Note the details of the actuation of the filter probe, which in the setting cycle is forced to cut through any mudcake and during the sampling cycle is retracted to give an open path to the formation fluids.

Note also the pretest chambers and the position of the sample chambers. The two pretest chambers, automatically activated every time the tool is set, withdraw 10 cc of formation fluid. Chamber 2

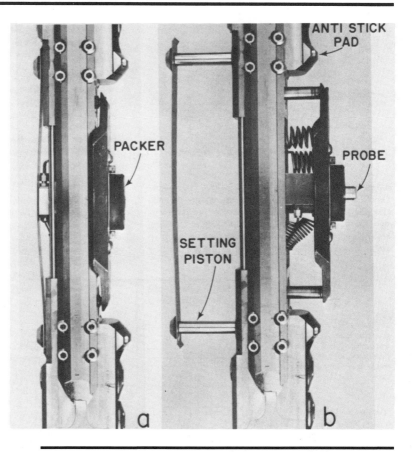

FIGURE 29.1 *Repeat Formation Tester Tool. (a) Closed. (b) Open. Reprinted by permission of the SPE-AIME from Smolen and Litsey 1977, fig. 1, p. 3.* © *1977 SPE-AIME.*

has a higher flowrate than chamber 1. The actual rates of fluid withdrawal vary with the tool and the downhole conditions but are in the neighborhood of 50 cc/min for chamber 1 and 125 cc/min for chamber 2, resulting in pretest times of roughly 12 and 5 seconds. The pretest samples are expelled back into the mud column and are not saved.

Figure 29.3 shows a typical log produced during a test. Since the tool is stationary in the hole during the test, the recording is made on a time scale with increasing time in the downhole direction on the log. Notice that in Track 1 pressure is recorded in analog form. Four subtracks record the units, tens, hundreds, and thousands of pounds.

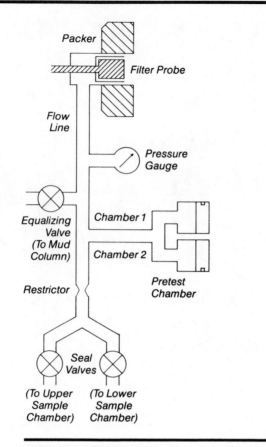

FIGURE 29.2 *RFT Sampling System Schematic. Courtesy Schlumberger Well Services.*

QUESTION #29.1
Read the pressure from the log at time *T*

a. from the analog curve;
b. from the digital curve.

Each record shows the following pressures:

1. Hydrostatic pressure—before tool is set
2. Drawdown pressure—during pretest
3. Buildup pressure—after pretest
4. Formation pressure—after buildup

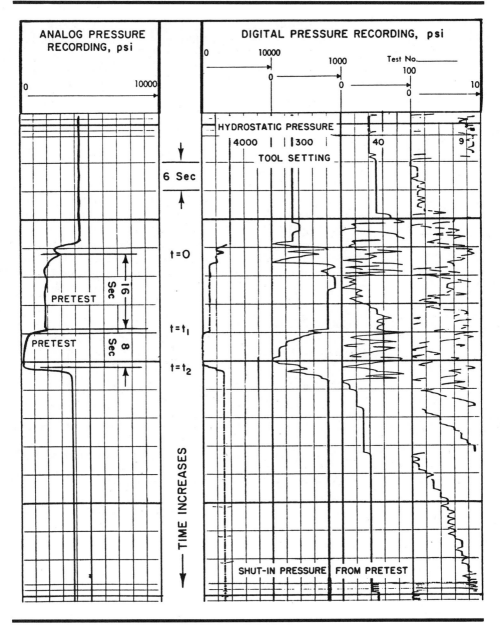

FIGURE 29.3 *RFT Pressure Recording. Reprinted by permission of the SPE-AIME from Smolen and Litsey 1977, fig. 4, p. 4. © 1977 SPE-AIME.*

Pressure Gauges

The standard gauge used in the RFT is a strain gauge calibrated by a dead-weight tester. The accuracy of this system, after applying temperature corrections, is 0.41% of full scale (i.e., 41 psi for a 10,000-psi gauge). The resolution of the gauge is about 1 psi with a repeatability of 3 psi. The accuracy may be improved to 0.13% full scale with a special calibration technique in which the gauge and the downhole electronics are placed in a temperature-controlled oven.

Where greater accuracy is required, a high-precision quartz gauge may be used. Accuracy is then 0.5 psi, provided that the temperature is known to 1°C. Resolution is of the order 0.01 psi. Table 29.2 summarizes these gauge specs. Note (see fig. 29.4) that the quartz gauge is physically located lower in the tool than the reference measurement point (which is the strain gauge) and that, therefore, the pressure recorded by the two gauges is different as a result of the hydrostatic head of a column of silicone oil. In some cases, a further pressure difference may be noted between the two gauges since the strain gauge is calibrated in psig and the quartz gauge is psia.

INTERPRETATION

In order to make the greatest use of RFT data, the analyst should be able to interpret the following types of RFT records:

Pretest records for formation permeability
Post pretest buildup for formation permeability
Large-sample fill-up time for formation permeability
Sequential pressure readings versus depth for pore-pressure gradients
Large-sample collection data for expected formation production

Pretest Records for Formation Permeability

Figure 29.5 shows a typical pretest record. In reality, only one pretest is required to estimate formation permeability. The magnitude of the pressure differential Δp between pretest sampling pressure and formation pressure, coupled with the flow rate during pretest, is sufficient to define permeability. In general, this may be found by a relation of the form

$$k = A \times C \times q \times \mu/\Delta p,$$

TABLE 29.2 *Comparisons of RFT Pressure Gauges*

	Strain Gauge	HP Quartz Gauge
Resolution, psi	1	0.01
Repeatability, psi	3	
Accuracy:		
Normal	0.41% full scale	0.5 psi
Special	0.31% full scale	

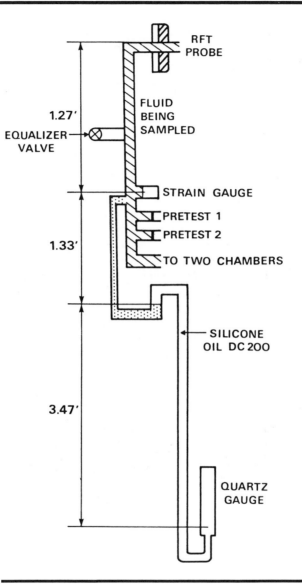

FIGURE 29.4 *Placement of the Strain and Quartz Gauges. Courtesy Schlumberger Well Services.*

where

k is permeability in md,
A is constant to take care of units,
C is the flow-shape factor,
q is the flow rate in cc/s,
μ is the viscosity of the fluid in cp, and
Δp is the drawdown in psi.

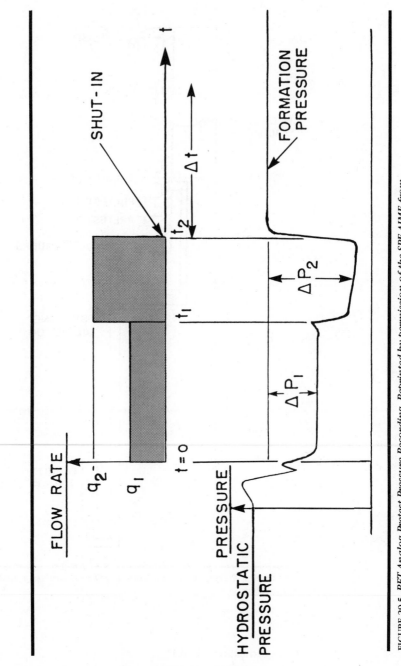

FIGURE 29.5 *RFT Analog Pretest Pressure Recording. Reprinted by permission of the SPE-AIME from Smolen and Litsey 1977, fig. 3, p. 4. © 1977 SPE-AIME.*

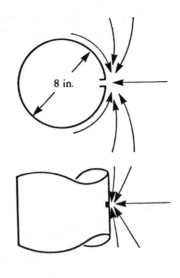

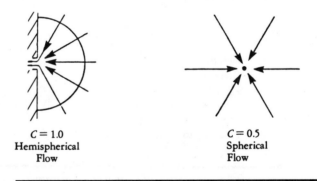

FIGURE 29.6 *Flow-Shape Factor C. Courtesy Schlumberger Well Services.*

Figure 29.6 shows a number of possible flow regimes around an RFT tool and the borehole. It is generally agreed that the flow shape is somewhere between hemispherical and spherical. Computer modeling of the probe/formation system shows that the combination of constants *A* and *C* to be used should be such that

$$k = \frac{5660\, q\mu}{\Delta p}.$$

The flow rate is derived by dividing the 10-cc volume of the pretest chamber by the sampling time read from the pressure record. The viscosity μ is considered to be that of the mud filtrate and may be estimated, using a knowledge of temperature and filtrate salinity, from the standard water viscosity chart shown in figure

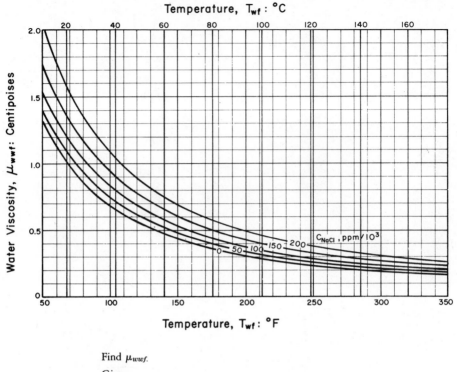

Find μ_{wwf}.

Given:

$C_{NaCl} = 150,000$ ppm.
$T_{wf} = 200°F$.

1. Enter abscissa at $T_{wf} = 200$.
2. Go up to $C_{NaCl} = 150,000$.
3. $\mu_{wwf} = 0.43$ cp.

FIGURE 29.7 *Estimation of Water Viscosity. Courtesy Schlumberger Well Services.*

29.7. Δp is read from the pressure recording as the difference between pretest sampling pressure and formation pressure.

There are variations on the standard probe and packer and these introduce changes in the value of the flow-shape factor. To compensate, the numbers to use are

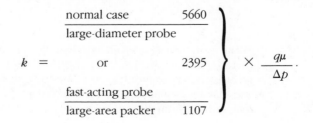

$$k = \left. \begin{array}{c} \dfrac{\text{normal case} \quad 5660}{\text{large-diameter probe}} \\[2ex] \text{or} \quad 2395 \\[2ex] \dfrac{\text{fast-acting probe}}{\text{large-area packer} \quad 1107} \end{array} \right\} \times \dfrac{q\mu}{\Delta p}.$$

QUESTION #29.2
If a normal probe and packer are used and

shut-in formation pressure	=	3520 psi,
first pretest pressure	=	3257 psi,
time to fill first pretest	=	10 s,
second pretest pressure	=	2993 psi,
time to fill second pretest	=	5 s,
filtrate salinity	=	50,000 ppm, and
formation temperature	=	175°F,

find the formation permeability.

The limitations of the pretest method for determination of permeability are:

a. If the permeability is very high, the drawdown is very small and cannot be measured accurately.
b. If the permeability is very low, the sampling pressure may drop below the bubble point, in which case gas or water vapor is liberated and the flow rate of the liquid withdrawn is less than the volumetric displacement of the pretest pistons.
c. The volume of formation investigated is small and hence the permeability measured may be that of the damaged zone, if present, and thus not representative of the formation as a whole.

In general, a good estimate of formation permeability may be obtained from a visual inspection of the pretest record. Figure 29.8 shows a number of pretest records and their corresponding permeabilities.

Post Pretest Buildup for Formation Permeability

Permeabilities obtained from pretest may be subject not only to the errors mentioned above but in addition may be measuring instead of absolute permeability, relative permeability to the water in the flushed zone. Figure 29.9 marks the pretest region on a set of relative permeability curves from which it can be deduced that pretest permeabilities are less than half absolute permeability when measured in an invaded oil zone.

A preferred method of calculating permeability is the analysis of the late-time portion of the pressure buildup record after the pretest disturbance has been made. A much larger rock volume can be investigated with this method. Effectively, the method will measure k_{ro} at a saturation close to S_{w_i} which will be very close to $k_{absolute}$ (see fig. 29.9), provided the measurement is made above the transition zone.

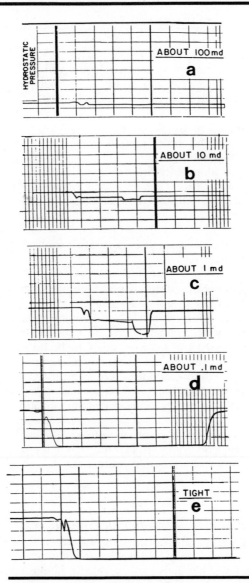

FIGURE 29.8 *Permeability Estimates from Pretest Pressure Records. Courtesy Schlumberger Well Services.*

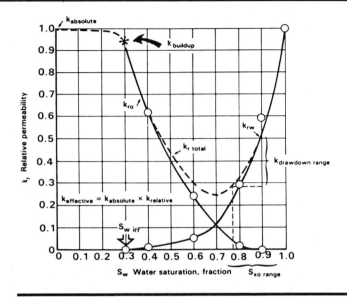

FIGURE 29.9 *Relative Permeability Curves. Courtesy Schlumberger Well Services.*

Figure 29.10 illustrates two modes of propagation of a pressure disturbance—spherical and cylindrical. If a test is conducted in a thin bed, the cylindrical mode will predominate; but, in a thick bed, the spherical mode will predominate.

SPHERICAL BUILDUP. The probe pressure response during buildup is obtained by superposition of the two single drawdown responses, and calculations based on spherical flow in an infinite homogeneous medium lead to the expression

$$P_i - P_s = \frac{(8 \times 10^4)\, q_1\, \mu (\phi \mu C_t)^{\frac{1}{2}}}{(k_s)^{3/2}} \times f_s(\Delta t),$$

where

$$f_s(\Delta t) = \frac{\dfrac{q_2}{q_1}}{\sqrt{\Delta t}} - \frac{\dfrac{q_2}{q_1} - 1}{\sqrt{T_2 + \Delta t}} - \frac{1}{\sqrt{T_1 + T_2 + \Delta t}},$$

p_i = initial formation pressure, in psi;
p_s = probe pressure (spherical buildup), in psi;
q_1 = flow rate during first sampling period, in cc/s;
q_2 = flow rate during second sampling period, in cc/s;
μ = viscosity of fluid in uncontaminated formation, in cp;
ϕ = formation porosity, fraction;

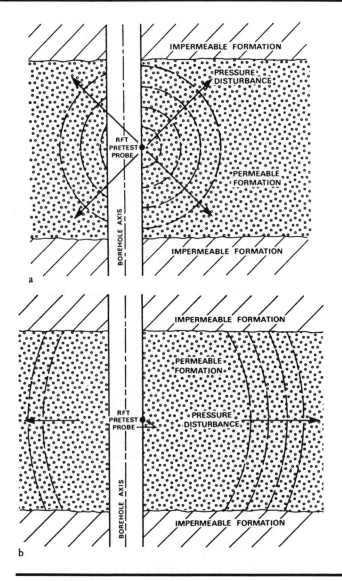

FIGURE 29.10 *Modes of Propagation.* (a) *Spherical.* (b) *Cylindrical. Courtesy Schlumberger Well Services.*

C_t =   total compressibility of fluid in uncontaminated formation, psi^{-1};

k_s =   isotropic spherical buildup permeability, in md;

T_1 =   sampling time related to q_1, in seconds or minutes;

T_2 =   sampling time related to q_2, in seconds or minutes; and

Δt =   elapsed time after shut-in, in seconds or minutes.

A plot of p_s, the observed pressure during buildup, versus f_s, the spherical time function, on linear graph paper will ideally result in a straight line of slope m. Extrapolation of this line to $f_s (\Delta t) = 0$ gives the undisturbed formation pressure, p_i. The straight line may be described by

$$p_s = m \times f_s (\Delta t) + p_i,$$

where

$$m = \frac{(8 \times 10^4)\, q_1\, (\phi \mu C_t)^{1/2}}{(k_s)^{3/2}}.$$

Then, from the slope m, the isotropic permeability k_s can be found from

$$k_s = 1856\, \mu\, (q_1/m)^{2/3}\, (\phi C_t)^{1/3}.$$

CYLINDRICAL BUILDUP. In relatively thin beds, a radial—cylindrical—flow pattern predominates and the buildup is affected only by the horizontal permeability. The buildup equation is then given by

$$p_i - p_c = 2687\mu\,(q_1\, \mu/k_r h) \times f_c\, (\Delta t),$$

where

$$f_c(\Delta t) = \log \left[\frac{T_1 + T_2 + \Delta t}{T_2 + \Delta t} \right] + \frac{q_2}{q_1} \log \left[\frac{T_2 + \Delta t}{\Delta t} \right],$$

and

p_i = initial formation pressure, in psi;
p_c = probe pressure (cylindrical buildup), in psi;
q_1 = flow rate during first sampling period, in cc/s;
q_2 = flow rate during second sampling period, in cc/s;
μ = viscosity of fluid in uncontaminated formation, in cp;
k_r = cylindrical buildup permeability, in md;
h = distance between the two impermeable boundaries, in cm;
T_1 = sampling time related to q_1, in seconds or minutes;
T_2 = sampling time related to q_2, in seconds or minutes; and
Δt = elapsed time after shut-in, in seconds or minutes.

Again, a plot of p_c versus $f_c(\Delta t)$ will produce a straight line of slope m such that

$$p_c = m \times f_s (\Delta t) + P_i,$$

where

$$m = \frac{2687\, q_1 \mu}{k_r h}.$$

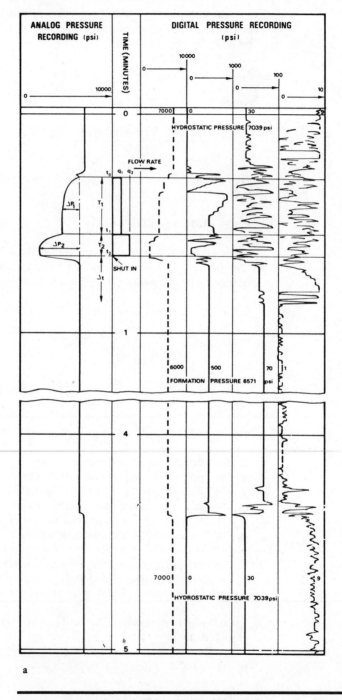

FIGURE 29.11 *Pressure Buildup Permeability.* (a) *Pressure Record.* (b) *Buildup Plot. Courtesy Schlumberger Well Services.*

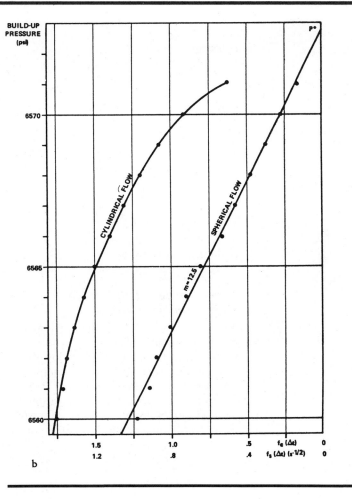

Whence

$$k = \frac{2687 \; q_1 \mu}{m \times b}.$$

Figure 29.11 shows a pressure buildup record together with a corresponding plot of probe pressure against both the spherical and cylindrical time functions.

QUESTION #29.3
Inspect figure 29.11.
Which type of buildup is relevant to this test, cylindrical or spherical?

QUESTION #29.4
Figure 29.12 shows a recording of pressure buildup. Use the data given to determine formation pressure and permeability.

T(seconds)	p	T(seconds)	p
105	3990	217	4006
113	3992	249	4008
121	3994	290	4010
131	3996	347	4012
143	3998	431	4014
157	4000	565	4016
173	4002	816	4018
193	4004		

Bed thickness is 0.5 meters.

Large-Sample Fill-up Time for Formation Permeability

When a large sample of formation fluid is recovered, the time taken to fill the sample chamber can be used as an indicator of permeability. In this case, the drawdown is considered to be the formation pressure itself since the sample chamber is for all practical purposes at atmospheric pressure. This may not hold true if the fill-up time is limited by a water cushion and a choke, so use this method with discretion and take it for what it is, a "quick and dirty" method of finding permeability. The equation is

$$k = \frac{4388 \, Cq}{p},$$

where:

k = fill-up permeability, in md;
C = flow-shape factor;
q = flow rate, in cc/s;
μ = fluid viscosity, in cp; and
p = drawdown pressure, in psi.

QUESTION #29.5
A large sample is collected in the 2¾-gal sample chamber, which took 5 seconds to fill. Given formation pressure is 3542 psi, $\mu = 0.5$ cp, $C = 0.75$, find k.

Sequential Pressure Readings versus Depth for Pore-Pressure Gradients

Since many formation pressure measurements may be made on one trip in the hole, pressure gradients may be calculated and plotted.

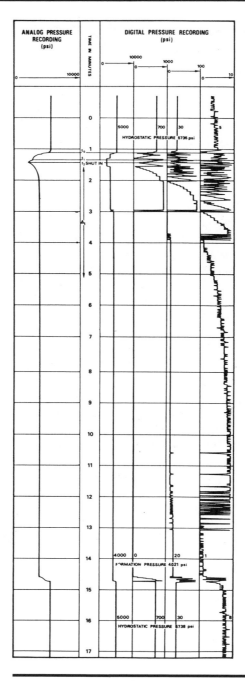

FIGURE 29.12 *Pressure Buildup Recording for Question #29.4. Courtesy Schlumberger Well Services.*

The easiest way to do this is to plot formation pressure against depth. Figure 29.13 is an example of this kind of plot. It is also useful to plot hydrostatic pressure on the same plot. Gas/oil and oil/water contacts are evident on a plot of this nature. The fluid density can be deduced from the pressure gradient using

$$\text{fluid density (g/cc)} = \text{pressure gradient (psi/ft)} \times 2.3072.$$

Care should be taken in low porosity transition zones where capillary pressure effects are pronounced. Log-derived OWC's, for example, may appear somewhat shallower in the well than the free water level indicated from plots of formation pressure versus depth (see fig. 29.14).

FORMATION PRODUCTION ESTIMATES

When a large sample is recovered, it is possible to predict formation productivity by analysis of the recovered material. A miniseparator (fig. 29.15) is used at the surface to measure the recovered volumes of oil, water, and gas. The water recovered will be a mixture of mud filtrate and formation water. The amount of formation water is calculated from figure 29.16 or from the relation

$$\% \text{ formation water} = \frac{\text{ppm recovered water} - \text{ppm filtrate}}{\text{ppm formation water} - \text{ppm filtrate}}.$$

Empirical charts then link recovered volumes to predicted production. Figure 29.17 shows a plot of oil recovery (in cc) versus gas recovery (in scf). Three areas are delineated on the chart indicating formations that are gas, oil, and water productive. An estimate of water cut can also be made using

$$\text{watercut \%} = \frac{\text{volume of formation water}}{\text{volume of formation water} + \text{volume of oil}} \times 100$$

QUESTION #29.6

On an RFT test the following recoveries were measured at the surface:

Gas: 23 scf
Oil: 7250 cc
Water: 1350 cc

The resistivity of the recovered fluid was 0.88 at 78°F.
The resistivity of the mud filtrate was 1.9 at 78°F.
The resistivity of the formation water was 0.32 at 78°F.
Use figure 29.17 to predict oil and gas production.
Use figure 29.16 to help predict water cut.

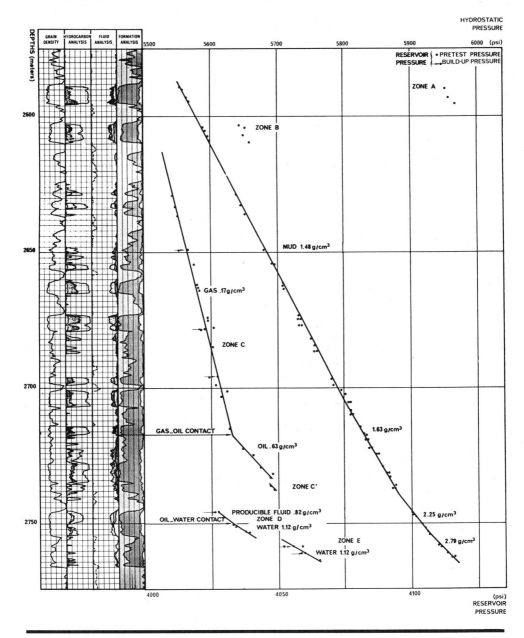

FIGURE 29.13 *Formation Pressure vs. Depth Plot. Courtesy Schlumberger Well Services.*

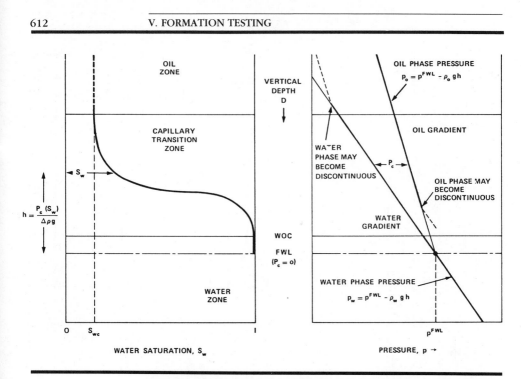

FIGURE 29.14 *Pressure Gradients around the Oil/Water Contact. Courtesy Schlumberger Well Services.*

BIBLIOGRAPHY

Dresser Atlas: "Formation Multi-Tester Interpretation Manual" (reference 06/80 9404).

Schlumberger: "Formation Tester Interpretation—Methods and Charts" (1966).

Schlumberger: "Fluid Conversions in Production Logging" (1974).

Schlumberger: "The Essentials of Wireline Formation Tester" (1981).

Schlumberger: "RFT Essentials of Pressure Test Interpretation" (1982) (available from ATL-Marketing, publication M-081022).

Smolen, J., and Litsey, L.: " Formation Evaluation Using Wireline Formation Tester Pressure Data," SPE paper 6822, presented at the 18th Annual Technical Conference and Exhibition, Denver, October 1977.

Stewart, G., and Wittmann, M. J.: " Interpretation of the Pressure Response of the Repeat Formation Tester," SPE paper 8362, presented at the 54th Annual Technical Conference and Exhibition, Las Vegas, October 1979.

Stewart, G., and Wittmann, M. J.: " The Application of the Repeat Formation Tester to the Analysis of Naturally Fractured Reservoirs," SPE paper 10181, presented at the 22nd Annual Technical Conference and Exhibition, San Antonio, Oct. 5–7, 1981.

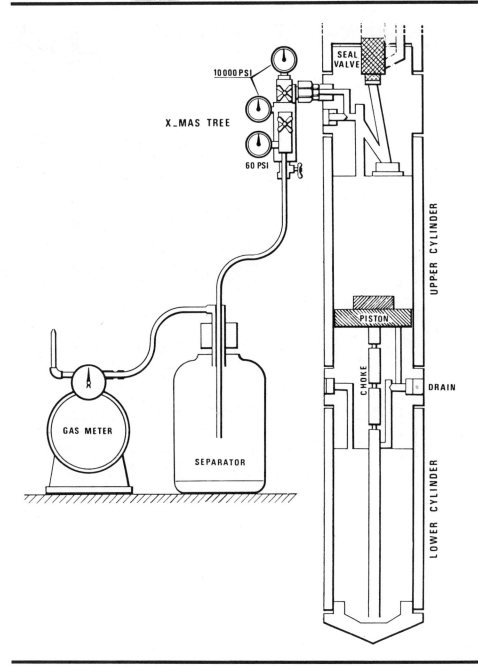

FIGURE 29.15 *Formation Tester Sample Recovery Measurement. Courtesy Schlumberger Well Services.*

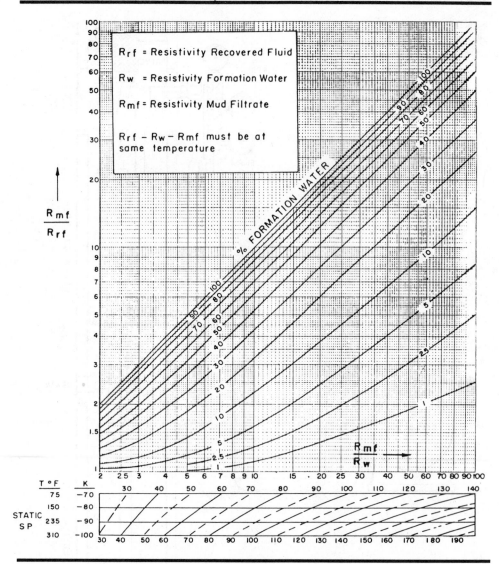

FIGURE 29.16 *Determination of Formation Water Percentage. Courtesy Schlumberger Well Services.*

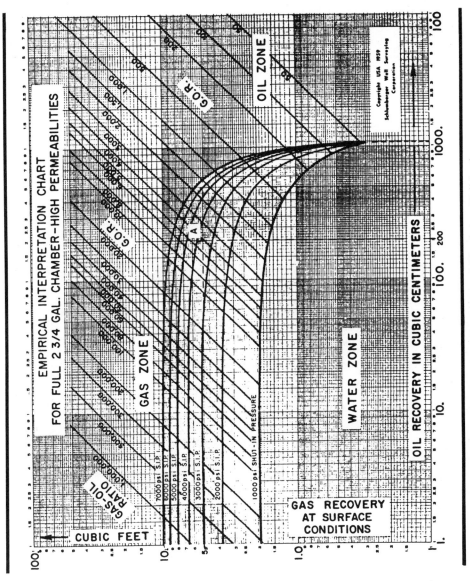

FIGURE 29.17 *Empirical Estimation of Productivity. Courtesy Schlumberger Well Services.*

Answers to Text Questions

QUESTION #29.1
3845 psi

QUESTION #29.2
Pretest 1: $k = 8.61$ md
Pretest 2: $k = 8.59$ md

QUESTION #29.3
Spherical

QUESTION #29.4
Formation pressure $= 4022.6$ psi
$k = 0.06$ md

QUESTION #29.5
$k = 967$ md

QUESTION #29.6
$R_{mf}/R_w = 5.94$
$R_{mf}/R_{rf} = 2.16$
Formation water % $= 21\%$
Formation water volume $= 284$ cc
Watercut $= \dfrac{284}{7250 + 284} = 4\%$

Gas and oil recoveries plot in oil zone with GOR $= 500$
Well actually produced 120 BOPD, no water, and GOR 700 scf/B

CHAPTER THIRTY

DRILLSTEM TESTING

OBJECTIVES

The purposes in running a drillstem test are:

To ascertain if the formation can achieve sustained production
To establish the type of production (oil or gas)
To collect formation fluids for laboratory analysis (PVT, water salinity, etc.)
To establish formation pressure
To estimate formation permeability

Although a wireline repeat formation tester (RFT) may give some of this information, the drillstem test (DST) is the only proof positive that recoverable hydrocarbons are present in the formation and will flow once the well is put on production. For this reason the DST is a popular formation evaluation technique.

MECHANICS OF DST

Essentially the mechanics of running a DST are quite straightforward. The drillpipe is lowered into the open hole with a testing assembly on the bottom of the string. This assembly comprises a packer and a system of valves. The drillpipe itself is maintained either fully or partially empty. When the assembly reaches the bottom of the hole, the packer is set in order to seal the bottom of the test assembly from the mud column. The flow valve is then opened and the formation is placed in communication with the inside of the drillpipe, which is at a much lower pressure than the formation. Fluids then flow into the drillpipe. If formation pressure and permeability are sufficiently high, formation fluids will flow to the surface. Figure 30.1 illustrates the principle.

When testing is completed, the test valve is closed and a pressure-equalizing valve is opened so that the mud in the formation/drillpipe annulus once again comes into contact with the formation and kills further flow. The packer may then be unseated and the formation fluid still in the drillpipe circulated out of the hole through a reverse circulating valve.

PRESSURE RECORDING

The test assembly normally carries two pressure recorders; one is placed in the flow stream and one outside the tool. If any dif-

617

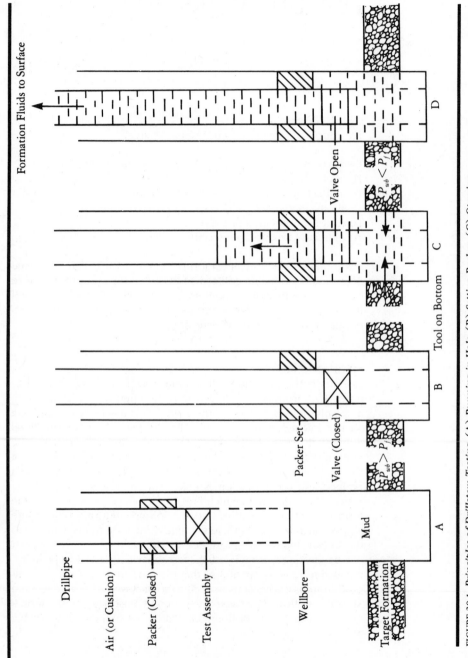

FIGURE 30.1 *Principles of Drillstem Testing.* (A) *Running in Hole.* (B) *Setting Packer.* (C) *Opening Flow Valve.* (D) *Fluids to Surface.*

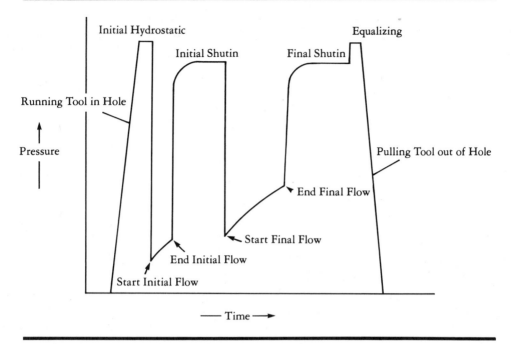

FIGURE 30.2 *Typical DST Plot of Pressure vs. Time.*

ferences are noted between the readings of the two pressure recorders, tool plugging is indicated. A typical pressure recording is shown in figure 30.2.

By plotting the pressure observed as a function of time it is possible to determine formation permeability; reservoir limits; formation damage (reduced permeability adjacent the wellbore), if present; and formation pressure. A number of plotting techniques are used including the *Horner plot,* where pressure is plotted versus the log of $(T + \Delta T)/\Delta T$, where T is the time during which the formation was open to flow and ΔT is the cumulative time since flow started (see fig. 30.3). If both the first and second buildup extrapolate to the same formation pressure, it may be assumed that the reservoir is homogeneous and that the flow periods were of the correct duration.

PLANNING THE DST

When planning the DST, a number of points are worth remembering:

1. In deep wells, a water or other cushion will be needed in order to prevent collapse of the drillpipe. Normally a water or pressurized nitrogen gas cushion will be used to limit the initial drawdown on the formation to 4000 psi or less.

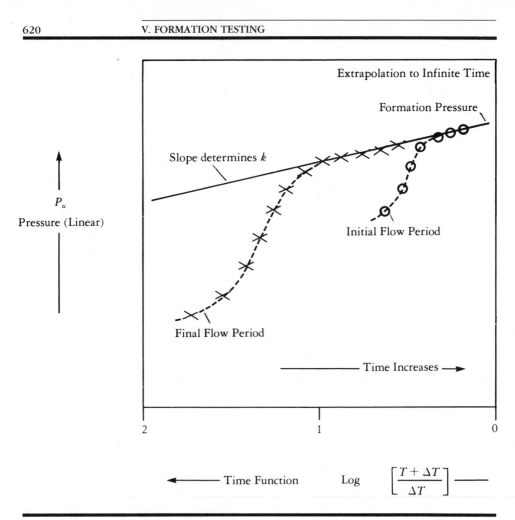

FIGURE 30.3 *Horner Plot for DST Data.*

2. Consult the open-hole logs for a choice of depth at which to set the packer. Avoid unconsolidated or fractured formations and washouts.
3. Plan to adjust the flow and buildup periods according to the following guidelines:

Initial Flow:	3-5 minutes
Initial Shut-in:	75 to 90 minutes depending on blow during initial flow period. Less time is required if blow was strong and vice versa.
Final Flow:	60 to 120 minutes or longer if flow rates are to be gauged. If well dies shut it in.
Final Shut-in:	1½ to 3 times final flow period depending on blow. Strong blow requires less final shut-in and vice versa. If well has been produced for any length of time, shut-in should equal final flow time.

PERSPECTIVE

The DST should be viewed in perspective along with all the other formation evaluation techniques. By its nature, it measures macro formation properties. Because of the large pressure drawdown created, the pressure disturbance travels much farther in the formation than any other formation evaluation tool can reach. A skilled interpreter of DST records can pick neighboring faults or other boundaries to the reservoir and estimate reservoir size and producibility. That kind of information is hard to beat and is of inestimable value. A DST is the acid test of all the other formation evaluation techniques.

However, a DST cannot evaluate rock type or identify the precise porosity and water saturation, so it is not the be-all and end-all. It is a powerful tool, and the analyst should combine all the available data to help in deciding where and when to use it and then use the additional data gathered to improve the overall picture of the reservoir.

BIBLIOGRAPHY

Amyx, James W., et al.: *Petroleum Reservoir Engineering Physical Properties,* McGraw-Hill, New York, 1960.

Horner, D.R.: "Pressure Buildup in Wells," Proc. of the Third World Petroleum Congress, Leyden, Holland, Section II, p. 503, 1951.

Lee, J.: "Well Testing," Society of Petroleum Engineers of AIME, New York/Dallas, SPE Textbook Series (1982).

Miller, C. C., Dyes, A. B., and Hutchinson, C. A., Jr.: "The Estimation of Permeability and Reservoir Pressure from Bottom-Hole Pressure Buildup Characteristics," *Trans.,* AIME (1953) **198**, 125.

PART VI

INTEGRATED
FORMATION EVALUATION

CHAPTER THIRTY-ONE

INTEGRATED FORMATION EVALUATION PLAN

OBJECTIVES

When a serious formation evaluation program is considered, objectives must be set and plans made to meet them. In this chapter the goals and strategies involved will be laid out.

In its broadest sense, the objective of a formation evaluation program should be to gather sufficient data before, during, and after drilling the well so that all subsurface formations and their contents can be completely evaluated. Thus, many disciplines have to be involved.

The *geophysicist* must plan where to run seismic lines, quality check the actual shooting, and then interpret the data in order to assist in choosing a drilling location. The *geologist* and the *geochemist* must work together to map the locations of traps, reservoir rocks, and source rocks. The *drilling engineer* must plan how the well is to be drilled and supervise the drilling process and any testing and/or completion work. The *reservoir engineer* must work with the *petrophysicist* and the *log analyst* in order to provide the *manager* with reserve figures and production decline estimates so that logical economic decisions can be made regarding field development.

Seldom do all of these professionals work in the same group, in the same office building, or even in the same city; and seldom is there one line manager with enough authority over these diverse groups to command an integrated formation evaluation plan. Too often, each group optimizes its own plan to suit its immediate objective without regard to the overall plan—with the result that much of the data required to make a complete formation evaluation is lost or cannot be gathered. In the ultimate analysis, the loss is an economic one that must eventually return to haunt the captains of the industry. Which of us has not heard of leases abandoned by one company only to be proven productive and profitable by another, more willing or able to evaluate fully the formations drilled through.

Who, then, can coordinate all the individual efforts so that the objectives are met? Surely it must be the formation evaluator. In his hands lie the keys to economic success. All other efforts are directed toward providing the data required to evaluate fully the formations drilled. He must, therefore, be somewhat of a politician if he is to succeed and to acquire the necessary authority to ensure that all phases of the operation dove-tail logically and harmoniously. It is not an easy task. Good luck to the intrepid who attempt it.

PLANNING

Before a well is drilled, it is customary to make detailed plans to include:

Bit size
Mud type
Logging points
Casing points
Intervals to be cored
Intervals to be tested

Many of the decisions involved interact with each other. In particular:

1. The *hole size* should be planned so that where wireline logs are required the tools can reasonably be expected to give valid measurements. Many logging tools will not give representative readings in overly large holes; and in these cases it is best to plan on drilling a modest hole that, after logging, can be under-reamed to an appropriate diameter to accommodate the casing string.
2. The *mud type* can seriously affect the response of logging tools and also the validity of saturation measurements from cores. Situations can arise where no amount of tinkering with the mud chemistry can satisfy all the requirements of all the interested parties. At least, the attempt should be made to ensure:

Control of problem shales
Low water loss (to reduce invasion effects)
Compatibility with core-analysis requirements

3. *Logging points* are usually picked to coincide with bit changes, casing points, or other naturally occurring interruptions to the drilling of the well. On occasion wells have been logged on such festive days as Christmas Day or Superbowl Sunday, or even just plain old golfing weekends. Understandable though such practices are, they usually lead to situations where data quality is compromised.
4. *Logging suites* should be planned to complement the type of mud and the objectives of the formation evaluation. With modern multiple-tool strings, the practice of foregoing a particular survey because of time considerations is, in most cases, no longer valid. Logging today is probably three times faster than it was 20 years ago.

MONITORING

While the well is being drilled, continuous monitoring of the mud log is vital for several reasons:

Detection of hydrocarbon shows
Detection of adverse drilling conditions
Safety

In particular, a drilling break combined with mud-log shows may call for either a core to be cut or a drillstem test. Such points can never be exactly anticipated in advance, and if the opportunity is lost the data that could have been gathered from core analysis or the DST can never be collected.

Equally important is the monitoring of wireline log quality. Once pipe is set in the well, no open-hole log can be run.

DATA ANALYSIS

Normalization

Before any kind of data analysis is attempted, all types of recording should be normalized to a common system of depth measurement. This calls for a reconciliation between:

Driller's depths
Mud logger's depths
Wireline log depths
Cored intervals
DST points
VSP and seismic surveys

Often this becomes a monumental task. Many horror stories are told of wells drilled in meters and logged in feet; of holes with unintentional side-tracks with the mud log in one bifurcation and the wireline log in the other; of core barrels turned top to bottom, etc., etc. Only when all records refer to the same depth can any serious data analysis begin.

Environmental Effects

All *wireline logs,* to some extent or another, suffer from wellbore environmental effects due to temperature, pressure, salinity, hole size, mud weight, and so on, and require appropriate corrections before quantitative analysis is attempted. *Mud log* gas shows likewise need adjustment for background gas before any calculations are made. When brought to the surface, *core samples* are relieved of overburden stress, and their measured porosities and permeabilities do not, therefore, truly reflect the formation properties in situ. All of these effects must be accounted for before data from different sources can be combined in a total formation evaluation package.

Rock Typing

Once all the data, be it from mud logs, wireline logs, cores, or drillstem tests, have been gathered, normalized, and corrected, the next step will be to establish rock types. In simple reservoirs, only one rock type may be present, in which case the task is simplified. However, the norm is to find several different rock types, each of which has its own petrophysical characteristics such as the relationship between, for example,

F and ϕ
ϕ and k.
S_w and R_0/R_t

These relationships must be properly established through core analysis or other means. Usually, simple crossplots will suffice to establish the existance of different rock types; and, with the help of logs, appropriate core sections can be selected for analysis.

Log Calibration

The next step in the chain is to calibrate logs to some known standard. Like it or not, even the most carefully planned and quality checked logging suites can still contain records that are systematically shifted or skewed with respect to true response. Of help in this respect are sundry techniques such as comparison of log response in one well with the cumulative average response of logs in other wells in the same field at a particular horizon. Core-analysis measurements are also of great value in this respect.

At this time it is also of value to attempt to establish the types of clay minerals associated with the reservoir rock, particularly if it is a shaly sand. X-ray diffraction and SEM photos are of particular help in this task.

Numerical Computations

Given that a well-calibrated data set is at hand, the analyst may then embark on numerical computations of the log data. A number of choices exist including:

Selective level-by-level analysis with a calculator or chart book
Interactive computer analysis over limited intervals of digitized data
Batch processing of the entire logged column from the original data
    tape

Each method has its advantages and disadvantages. Although, usually, desk-top-calculator analysis does not include iterative processing (because it is complex and time consuming), it at least gives the analyst some immediate feedback regarding his choice of parameters and the sensitivity of the computed results to those parameters.

The interactive mode allows far more data to be processed in a much shorter time using more complex logic, but it opens the door to endless recomputations when the analyst decides to play "what-if" games and attempts to vary his method of analysis to suit some preconceived notion of what the final answer should be.

Batch processing of large data bases is a more time-consuming business, but it allows the most modern and complex algorithms to be employed. Some loss of feedback between the analyst and the processing is to be expected. The output from batch processing can at times be overwhelming, and valuable time can be lost sifting through mounds of computer print outs in the search for a simple item such as the integrated hydrocarbon pore feet.

SUMMARY

Formation evaluation will only be effective when it is planned to be so. If integrated formation evaluation is a serious objective, as indeed it should be, then plans should be made with the active participation of many different groups. If the plan is to succeed, one party must be responsible, and that party must be given authority appropriate to the responsibility. As a guide to formation evaluators, a flow chart is offered as a starting point.

Integrated Formation Evaluation Plan Flow Chart

Formulate Plan	Geophysicist Geologist Drilling engineer Log analyst/petrophysicist
Select	Well geometry Mud type Logging suites Coring intervals Testing intervals
Monitor	Mud logging Wireline logging Coring Testing
Normalize	Driller's depths Mud log depths Wireline depths Core depths DST depths Sonic and Seismic
Correct data	Environmental effects
Rock typing	Cuttings Cores Logs
Log Calibration	Core Analysis Logs
Computations	

BIBLIOGRAPHY

Crain, E.R.: "Economics of Log Evaluation for Large Projects," paper A, *Trans.,* SPWLA, 21st Annual Logging Symposium, Lafayette, LA, July 8–11, 1980.

Johnson, H.: "The Role of the Log Analyst," *The Log Analyst* (April–June 1966) 7.

Pirson, S.J.: "The Education of a Log Analyst," *Trans.,* SPWLA, 6th Annual Logging Symposium, Dallas, May 1965.

Threadgold, P.: "The Function and Training of a Log Analyst," *The Log Analyst* (September–October 1975) **16**, 11, 13.

Towle, G.: "The Log Analyst with a Service Company," *Trans.,* SPWLA, 6th Annual Logging Symposium, Dallas, May 1965.

APPENDIX: LOG ANALYSIS EQUATIONS SUMMARY

SATURATION

$$S_w^n = R_0/R_t$$

Archie

$$R_0 = F \times R_w$$

$$F = a/\phi^m$$

$$\left.\begin{array}{l} S_w^n = a \times R_w/\phi^m \times R_t \\[1em] S_{xo}^n = a \times R_{mf}/\phi^m \times R_{xo} \end{array}\right.$$

Ratio

$$S_w = \left[\frac{R_{xo}}{R_t} \times 10^{(-SP/k)}\right]^{5/8}$$

Laminated Shaly Sand

$$S_w = \left[\left(\frac{1}{R_t} - \frac{V_{lam}}{R_{sh}}\right) \frac{a \times R_w}{\phi_e^m(1 - V_{lam})}\right]^{1/n}$$

Modified Total Shale

$$\frac{1}{R_t} = \frac{\phi_e^m \times S_w^n}{a \times R_w \times (1 - V)} + \frac{V_{sh} \times S_w}{R_{sh}}$$

POROSITY

$$\phi_D = \frac{\rho_{ma} - \rho_b}{\rho_{ma} - \rho_f}$$

$$\phi_S = \frac{\Delta t - \Delta t_{ma}}{\Delta t_f - \Delta t_{ma}}$$

$$\left.\begin{array}{l} \\[2em] \end{array}\right\} \text{ Clean formations}$$

or $\quad \phi_S = \dfrac{1}{\rho_{ma} - \rho_t}\left(1 - \dfrac{\Delta t_{ma}}{\Delta t}\right)$

$\phi_N = \phi_N$ for appropriate matrix

$$\phi_e = \frac{\phi_D \times \phi_{N_{sh}} - \phi_N \times \phi_{D_{sh}}}{\phi_{N_{sh}} - \phi_{D_{sh}}}$$

$$\phi_e = \frac{\phi_D \times \phi_{N_{sh}} - \phi_N \times \phi_{D_{sh}}}{(\phi_{N_{sh}} - \phi_{D_{sh}}) - (\phi_N - \phi_D)}$$

$$\phi_g = \left(\frac{\phi_N^2 + \phi_D^2}{2}\right)^{1/2}$$

$$\phi_X = \frac{\phi_{N(lime)} + \phi_{D(lime)}}{2}$$

V-SHALE

$$(V_{sh})_{GR} = \frac{GR - GR\ clean}{GR\ shale - GR\ clean}$$

$$(V_{sh})_{SP} = \frac{SP - SP\ clean}{SP\ shale - SP\ clean}$$

$$(V_{sh})_{ND} = \frac{\phi_N - \phi_D}{\phi_{N_{sh}} - \phi_{D_{sh}}}$$

LITHOLOGY

$$\rho_{maa} = \frac{\rho_b - \phi_X \times \rho_f}{1 - \phi_X}$$

$$\Delta t_{maa} = \frac{\Delta t - \phi_X \times \Delta t_f}{1 - \phi_X}$$

$$U_{ma} = \frac{P_e \times \rho_e - \phi_X U_f}{1 - \phi_X}$$

$$M = \frac{\Delta t_f - \Delta t}{(\rho_b - \rho_f) \times 100}$$

$$N = \frac{\phi_{Nf} - \phi_N}{\rho_b - \rho_f}$$

RESISTIVITY

$$R_2 = R_1 \times \frac{T_1 + 7}{T_2 + 7} \qquad \text{Ionic solutions}$$

$$R_{we} = R_{mfe}(10)^{SP/k}$$

$$R_w = R_{mf} \times \frac{R_t}{R_{xo}} \qquad \text{Water-bearing}$$

PERMEABILITY

$$k = \left(\frac{250 \times \phi^3}{(S_w)_i} \right)^2 \qquad \text{Oils}$$

$$k = \left(\frac{79 \times \phi^3}{(S_w)_i} \right)^2 \qquad \text{Dry gas}$$

VOLUMETRICS

$$\text{HCPV} = \phi(1 - S_w)h \qquad \text{Hydrocarbon feet}$$

$$\text{OIP} = \text{HCPV} \times 7758 \qquad \text{bbl/acre}$$

$$\text{GIP} = \text{HCPV} \times 43560 \qquad \text{scf/acre}$$

$$N = \frac{\text{OIP} \times \text{area} \times \text{recovery factor}}{\beta_o} \qquad \text{STB}$$

$$N = \frac{\text{GIP} \times \text{area} \times \text{recovery factor}}{\beta_g} \qquad \text{scf}$$

MOVABLE OIL

$$\% \text{ OIP moved} = \frac{S_{xo} - S_w}{1 - S_w}$$

$$(\text{HCPV})_{\text{moved}} = \phi(S_{xo} - S_w)$$

$$S_{xo} = S_w^{1/5} \text{ (approx.)}$$

HYDROCARBON DENSITY

$$\rho_{hy} = 1 - \left[\frac{\phi_D - \phi_T}{0.7(1 - S_{xo})\phi_T} \right]$$

LOGARITHMIC SCALE INTERPRETATION

TO READ Sw:
PLACE 100 ON THIS SCALE ON TOP OF
THE NORMALIZED F CURVE.
READ Sw WHERE THE Rt CURVE INTER-
SECTS THE SCALE.

TO READ φ FROM F CURVE:
PLACE 10 ON THIS SCALE ON TOP OF THE
F=100 LINE ON THE F LOG.
READ φ WHERE THE F CURVE INTER-
SECTS THIS SCALE.

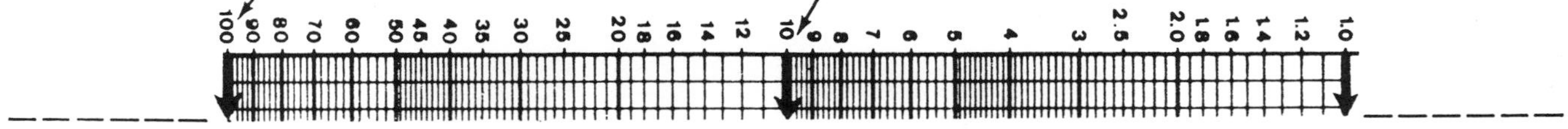

FIGURE 1A *Tear-out Logarithmic Scaler for use with F Overlays as Discussed in Text.*

INDEX